Sustainable Textiles: Production, Processing, Manufacturing & Chemistry

Series Editor

Subramanian Senthilkannan Muthu, Head of Sustainability, SgT and API, Kowloon, Hong Kong

This series aims to address all issues related to sustainability through the lifecycles of textiles from manufacturing to consumer behavior through sustainable disposal. Potential topics include but are not limited to: Environmental Footprints of Textile manufacturing; Environmental Life Cycle Assessment of Textile production; Environmental impact models of Textiles and Clothing Supply Chain; Clothing Supply Chain Sustainability; Carbon, energy and water footprints of textile products and in the clothing manufacturing chain; Functional life and reusability of textile products; Biodegradable textile products and the assessment of biodegradability; Waste management in textile industry; Pollution abatement in textile sector; Recycled textile materials and the evaluation of recycling; Consumer behavior in Sustainable Textiles; Eco-design in Clothing & Apparels; Sustainable polymers & fibers in Textiles; Sustainable waste water treatments in Textile manufacturing; Sustainable Textile Chemicals in Textile manufacturing. Innovative fibres, processes, methods and technologies for Sustainable textiles; Development of sustainable, eco-friendly textile products and processes; Environmental standards for textile industry; Modelling of environmental impacts of textile products; Green Chemistry, clean technology and their applications to textiles and clothing sector; Eco-production of Apparels, Energy and Water Efficient textiles. Sustainable Smart textiles & polymers, Sustainable Nano fibers and Textiles; Sustainable Innovations in Textile Chemistry & Manufacturing; Circular Economy, Advances in Sustainable Textiles Manufacturing; Sustainable Luxury & Craftsmanship; Zero Waste Textiles.

Subramanian Senthilkannan Muthu · Ali Khadir
Editors

Textile Wastewater Treatment

Sustainable Bio-nano Materials and Macromolecules, Volume 2

Springer

Editors
Subramanian Senthilkannan Muthu
SgT Group and API
Hong Kong, Kowloon, Hong Kong

Ali Khadir
Western University
London, ON, Canada

ISSN 2662-7108 ISSN 2662-7116 (electronic)
Sustainable Textiles: Production, Processing, Manufacturing & Chemistry
ISBN 978-981-19-2854-3 ISBN 978-981-19-2852-9 (eBook)
https://doi.org/10.1007/978-981-19-2852-9

This Springer imprint is published by the registered company Springer Nature Singapore Pte Ltd.
The registered company address is: 152 Beach Road, #21-01/04 Gateway East, Singapore 189721, Singapore

Contents

About the Editors

Dr. Subramanian Senthilkannan Muthu currently works for SgT Group as Head of Sustainability and is based out of Hong Kong. He earned his Ph.D. from The Hong Kong Polytechnic University and is a renowned expert in the areas of Environmental Sustainability in Textiles & Clothing Supply Chain, Product Life Cycle Assessment (LCA) and Product Carbon Footprint Assessment (PCF) in various industrial sectors. He has five years of industrial experience in textile manufacturing, research and development and textile testing and over a decade's of experience in life cycle assessment (LCA), carbon and ecological footprints assessment of various consumer products. He has published more than 100 research publications, written numerous book chapters and authored/edited over 100 books in the areas of Carbon Footprint, Recycling, Environmental Assessment and Environmental Sustainability.

Ali Khadir is an environmental engineer and a member of the Young Researcher and Elite Club, Islamic Azad University of Shahre Rey Branch, Tehran, Iran. He has published several articles and book chapters in reputed international publishers. He also has been the reviewer of journals and international conferences. His research interests center on emerging pollutants, dyes and pharmaceuticals in aquatic media, advanced water and wastewater remediation techniques and technology.

Treatment of Textile Wastewater by Nanoparticles

P. Vithalani, P. Mahla, and N. Bhatt

Abstract Rapid globalisation and industrialisation are leading towards huge production of anthropogenic synthetic products. Synthetic products contain many hazardous compound, exposure of these compounds result into changes in environment and living being. After the production, 10–15% are going as waste which are creating major issues for the environment. So, removal of compounds is major task for all industrialists. Nanotechnology plays major role in removal of hazardous compounds from wastewater. Nanoparticles are in small size playing enormous role in catalysis and removal of toxic compound. Synthesis of nanoparticles is carried out by physical, chemical and biological method. Biological approach for synthesis of nanoparticles plays enormous role in treatment. Biological approach is highly efficient, stable, economic and less detrimental for the environment. Still, metal and metal oxide-based nanoparticles are having toxic properties and are transformed into environment and disturb the whole ecosystem. Modification in treatment strategies is mandatory to increase efficiency. Numerous research work is carried out and still going on.

Keywords Nanotechnology · Metal and metal oxide-based nanoparticles · Treatment strategies · Synthesis · Toxic effect

Abbreviations

°C	Degree Celsius
Kg	Kilogram
%	Percentage
mg/L	Milligram per litre
nm	Nanometer
Fe_2O_3	Iron (III) oxide

P. Vithalani · P. Mahla · N. Bhatt (✉)
Department of Biogas Research and Microbiology, Gujarat Vidyapith, Sadra, Gandhinagar, Gujarat, India
e-mail: bhatt@gujaratvidyapith.org; bhattnikhil2114@gmail.com

S. S. Muthu and A. Khadir (eds.), *Textile Wastewater Treatment*, Sustainable Textiles: Production, Processing, Manufacturing & Chemistry,
https://doi.org/10.1007/978-981-19-2852-9_1

MgO	Magnesium Oxide
ppm	Parts per million
COD	Chemical Oxygen Demand
BOD	Biological Oxygen Demand
mW cm^{-2}	Milliwatt per square centimetre

1 Introduction

Rapid investigation in research leads to modification in technology and incorporation of new methods for treatment of textile effluent. As industrialisation increased after World War II, production of synthetic compound increased, emphasising pollution of ecosystem. Coloured compound having detrimental effect on ecosystem and accumulated in water bodies lead to disturbance in food chain [31]. The exposure of these compounds led to 14 lakh complicated cases of respiratory infection and around 4000 cases of death were reported in India in year of 2018 [60]. Treatment of wastewater embedded with synthetic compound must be treated before discharge. In past decades, many physical, chemical and biological methods were utilised for the treatment of textile effluent [15]. Recently, the researchers are focusing on nanotechnology for the treatment of textile effluent.

The term nanotechnology was firstly used by Norio Taniguchi in 1974. Nanotechnology is combination of material science and biology and deals with removal of pollutants from wastewater. Nanomaterials are building block of nanotechnology, having size of 10–100 nm which are having higher surface area for reaction and having attractive shapes [57]. The classification of nanoparticle is carried out in different ways described in Fig. 1 [57]. Nanostructures have their wide application in different areas such as electronics, magnetic, biomedical and material science. Nanostructures are utilised for improving, purifying and reserving quality of soil, water and air in environment [8]. Nanoparticles increase membrane permeability, resistance, strength and innovative mechanism for treatment. Nanoparticles are synthesised by chemical and biological methods which have wide application in remediation process. In this chapter, synthesis of nanoparticles and how it is useful for removal of pollutants from wastewater and toxicity of nanoparticles are described in detail.

2 Synthesis of Nanoparticles

2.1 Physical–Chemical Mediated Synthesis

The synthesis of nanoparticles is carried out by two methods; one is bottom-up and second one is top-down. In top-down method, large particles are converted into fine particles by various physical methods and in bottom-up method condensation

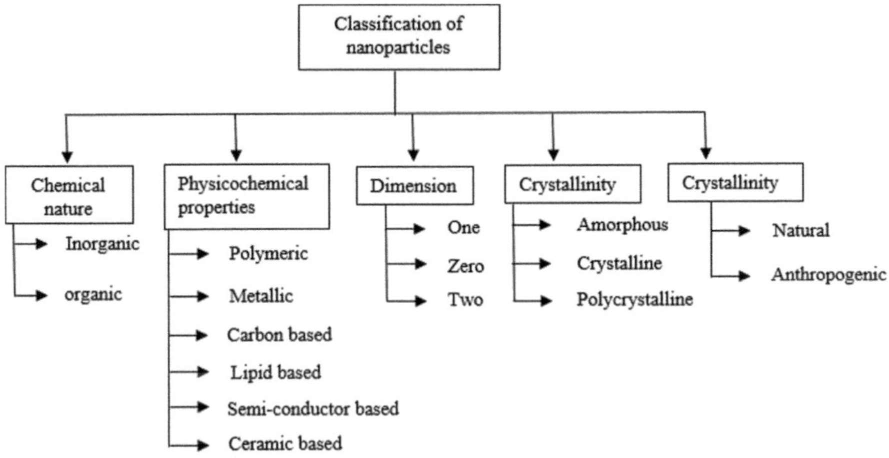

Fig. 1 Classification of nanoparticles

of more than two particles is carried out. Mainly chemical mediated synthesis and biological mediated synthesis of nanoparticles is following bottom-up method [57]. The detailed approach for synthesis is given in Fig. 2 [31].

In top–bottom approach, more instruments and high cost is needed for disruption of structure. So, the shape and size of nanoparticles are not uniform which may affect its applicability. Bottom approach is based on chemicals developed to overcome this issue. But the synthesis of nanoparticles is carried out in controlled manner for desirable size, shape and effectivity. Chemical mediated synthesis contains disadvantages such as increase in volume size, problem with mixing, change in mass, heat transfer

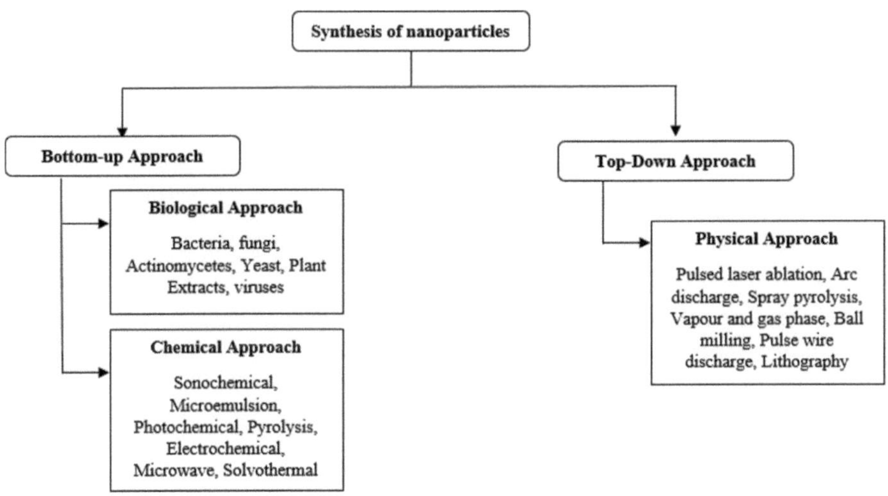

Fig. 2 Approach for synthesis of nanoparticles

and growth kinetics of particle formation etc. Chemical mediated synthesis of metal oxide creates toxic effect [42]. To overcome such problem currently researchers are focusing on biological mediated synthesis of nanoparticles.

2.2 *Green Synthesis of Nanoparticles*

Assembly of nanoparticles is carried out by using biological approach. Biological approach for synthesising nanoparticles is eco-friendly, cost-effective and less harmful to environment. In recent research, biological mediated synthesis of nanoparticles is carried out and applied in treatment of textile wastewater for removal of hazardous compound from wastewater [62].

2.2.1 Mechanism of Synthesis

Synthesis of nanoparticles is carried out by three different mechanisms: proteins (reductase enzymes), polysaccharides and electron shuttle quinones. Proteins, polysaccharides and enzymes are natural reducing agent and capable to reduce metal ions which are used in synthesis mechanism. Polysaccharide have different compounds which can reduced different metal ions and convert them into metal nanoparticles. Enzymes and proteins are utilised to make intracellular nanoparticles. Electrons are transferred through low molecular weight mediators which are helpful for synthesis of extracellular nanoparticles [20]. Mechanism of green nanoparticle synthesised by biological system is described in Fig. 3.

3 Role of Nanotechnology in Textile Wastewater Treatment

Clean and freshwater are requirement of all living organism. Earth is known as blue planet because 70% portion is covered under water. Out of 70%; 2.5% consists of freshwater from which only 0.3% are assessable. Rest remaining water is in the form of snow. Urbanisation and industrialisation play a major role in deteriorating water resources [26]. Increase in industrial, technical progress and diverse products lead to pollution of environment. The wastewater of industries is containing many heavy metal, chlorinated compounds, softeners, dyestuff which are causing air, water and soil pollution. The wastewater of textile industry having this compounds discharge without any precaution will affect the whole ecosystem [60]. The compounds are causing many health issues as it is having toxic nature. Accumulation of compounds takes place after entering in living body which increases in concentration after entering in food chain.

Textile wastewater having many colourant compounds, softeners, chlorinated compounds, detergents, heavy metals and surface-active agents upon discharging

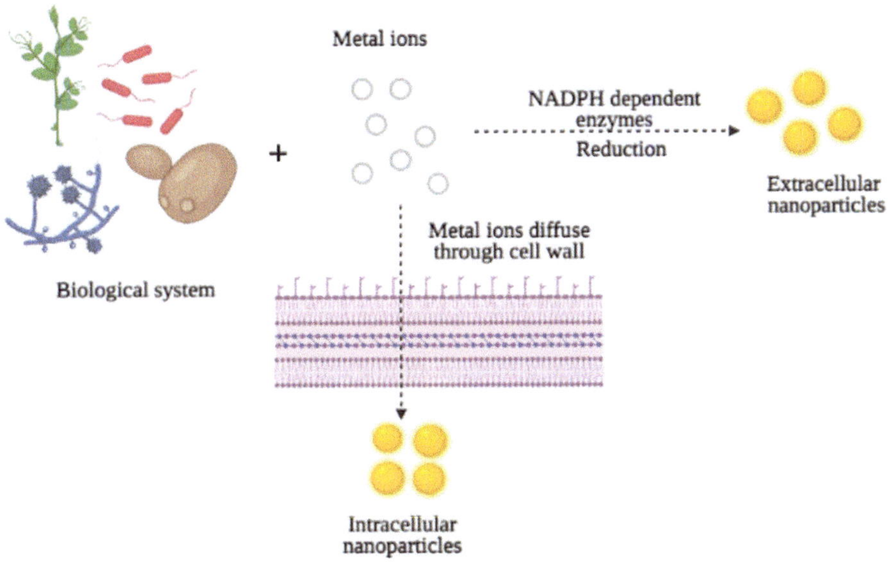

Fig. 3 Intracellular and extracellular synthesis of nanoparticles by biological system

affects ecosystem. So, treatment of this wastewater is enforced for all industrialists. Many physicochemical treatments are performed such as coagulation, precipitation, distillation, ultrafiltration, chemical transformation and ultraviolet treatment [32].

In recent time, nanotechnology is emerging technology for wastewater treatment. Nanomaterial with increased surface area to volume ratio is effectively used in treatment. Nanostructures are newly functional structures which are thermally and mechanically stable and resistance for the treatment used in removal of hazardous compound from wastewater [56]. Through improvement in conventional technologies, different nano-based structures are utilised for the treatment of textile wastewater. The details of nano-based structures and their role in treatment are described.

3.1 Adsorption Based Techniques

3.1.1 Carbon-Based Nanoadsorbants

Adsorption-based removal of contaminant from waste is applied since many years. Traditional adsorbent, such as activated carbon and polymers, is simple and cost effective for treatment, but it is not always efficient for removal of all contaminants. Carbon-based nanomaterials such as graphene, carbon-based nanotubes (CNT) are utilised for removal of contaminants [65]. Adsorption mechanism for removal of contaminant is described in Fig. 4. CNT was utilised for adsorption of zinc ion.

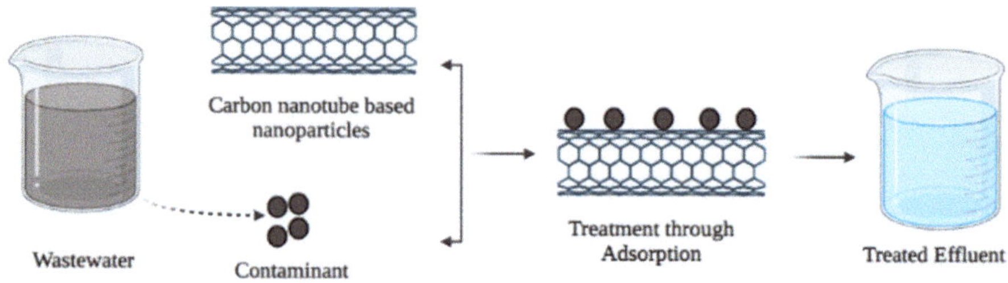

Fig. 4 Adsorption phenomenon

The results showed that electrostatic force between CNT and zinc ion controls the adsorption. This study presented that pH, temperature, CNT mass and modification in surface area of CNT can affect adsorption phenomenon [3]. Reversibility of nickel ion was carried out with single wall carbon nanotubes (SWCNT) and multiwall carbon nanotube (MWCNT) and GAC. Results displayed that better reversibility was found in SWCNT and MWCNT in comparison with GAC [37]. Adsorption of heavy metal lead was studied using MWCNT. The result was 73% of lead was adsorbed on surface of MWCNT [71].

3.1.2 Metal-Based Nanoadsorbants

Metal-based nanoparticles are used as nanoadsorbant. Metals can efficiently remove pollutant compounds from wastewater. Fe_2O_3-based nanoparticles were synthesised and applied for adsorption of Methylene blue dye. The result indicated that 7.25 mg/g was maximum adsorption capacity of Methylene blue [63]. Fe_2O_3/MgO-based nanoparticles were synthesised by different methods and used for removal of Remazol red textile dye. The result directed that 50 ppm dye was removed within 19 min based on Langmuir and Freundlich adsorption model [38]. Silica-based nanoparticles were developed for adsorption of Rhodamine B dye. The result was 234.61 mg/g dye was adsorbed within 5 min of reaction based on Langmuir, Freundlich and Temkin isotherm models [12].

3.1.3 Polymer-Based Nanoadsorbants

Natural copolymer is utilised for treatment of wastewater. Natural polymer is eco-friendly, degradable and cost effective. Cross-linking mechanism was exploited for the synthesis of polyhedral oligomeric silsesquioxanes (POSS) and chitosan-based nanocomposite materials (PSCS). The result found was 98% adsorption of cadmium heavy metal based on Langmuir model. Natural polymer has smooth surface, high solubility and dispersibility treats metal containing wastewater efficiently [34]. Saponin was extracted from *Bacopa monniera* leaf with graphene oxide, synthesised sliver nanoparticles which were applied for adsorption of Methylene blue and

Congo red [22]. Hybrid polymer β cyclodextrin multi-decorated halloysite nanotubes revealed high efficiency for Methylene blue dye adsorption [9].

3.1.4 Zeolites

Coal fly ash was utilised for synthesis of nanocrystalline zeolite for adsorption of crystal violet from aqueous solution. The result found was, at optimised condition of pH 9, temperature 50 °C, adsorption was 99% of initial concentration 100 mg/L crystal violet at contact time 60 min. [64]. The application of nano-zeolite was performed on caesium containing wastewater. Caesium was efficiently absorbed by nano-zeolite [35]. Sodalite zeolite was functional for adsorption of Crystal violet. The results showed that sodalite zeolite was efficient at very low concentration of dye [33].

3.2 Nano-based Membrane Techniques

Membrane filtration is a traditional way to remove pollutant from wastewater. Many polymeric membranes were utilised for filtration purpose, major drawback of this technique is membrane fouling. Nanostructure mediated filtration is quite efficient [32]. Mechanism of nanofiltration is described in Fig. 5. Polyether sulfonate membrane modified with $CaCl_2$ was used for experiment on retardation of selected dye. Results showed, way of retardation of dye was Methylene blue > Orange II

Fig. 5 Nanomembrane-based filtration

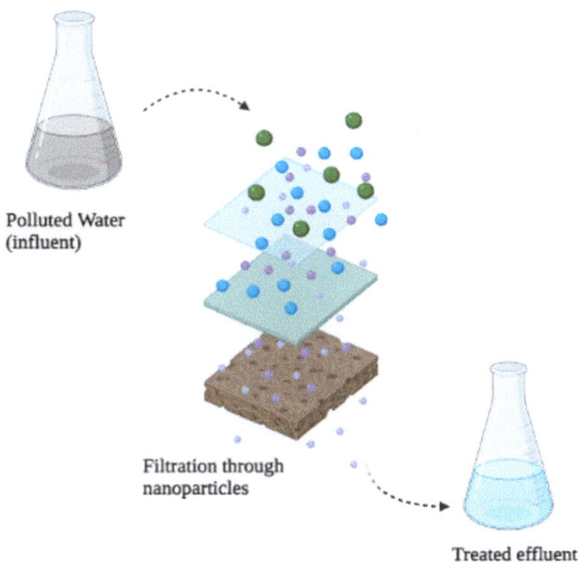

> Crystal violet > Congo red and fouling analysis presented that membrane incorporated with $CaCl_2$ has better antifouling effect [52]. Microporous membrane poly (vinylidene fluoride) fabricated with nonwoven support was used for separating dyes. Resulted in high rejection of Naphthol blue black (89%), Erichrome black T (83%), Methylene blue (45%) and Rhodamine B (1%) [68]. Polyamide-based nanomembrane was useful for filtration of wastewater embedded with five different dyes, namely Reactive blue 15, Reactive red 194, Reactive yellow 145, Reactive black 5 and Reactive orange 16. Results showed more than 90% removal of dye is found in 60 min. filtration process with low concentration of dyes. Nanofiltration was functional for removal of low concentration of dye [55].

3.3 *Microorganisms Mediated Nanotechnology*

Microorganisms such as bacteria, fungi, yeast, algae, actinomycetes are utilised for green nanotechnology. Microorganisms are potential source for synthesis of nanostructure and have application in removal of dyes. The details of synthesis of nanoparticles and removal of pollutant are described in Table 1.

3.4 *Plant Mediated Nanotechnology*

Plants and extracts of leaves, fruits, peel, seed and bark are utilised for green synthesis of nanoparticles. Plants and their extracts have natural reducing and capping agent which are useful for synthesis and condensing multistep processes. Plants are utilised for large-scale production of nanoparticles as it is cost-effective, easy, simple and fast. Different plants are utilised for synthesis of nanoparticles and applied for various dye degradation as described in Table 2.

3.5 *Enzymes Mediated Nanotechnology*

The application of two different areas, enzymes and nanoparticles, plays important role in treatment of textile effluent. As enzymes are utilised as catalysts, lowers down activation energy and make rate of reaction faster. The applicability of enzymes with nanoparticles reduces steric hinderance and minimise the cell interaction. Enzymes are eco-friendly approach, utilised with nanoparticles for effective, adaptable treatment of effluent. Immobilisation of enzymes on nanoparticles provides high resistance towards unfolding, increases efficiency and is reusable [62].

Immobilisation of horseradish peroxidase is done on two-dimension nanosheet of copper oxide supported with poly methyl methacrylate by using encapsulation technique. Result observed was decolourisation of dye increased in immobilised enzyme

Table 1 Application of microorganism mediated nanoparticles in dye degradation

Microorganism	Pollutant	Name of nanoparticles	References
Chlorella vulgaris	Methylene blue	Sliver nanoparticles	[51]
Penicillium corylophilum As-1	Methylene blue	CuO/ZnO nanocomposite	[19]
Escherichia sp. SINT7	Congo red, malachite green, direct blue 1, reactive black 5	Copper nanoparticles	[43]
Deinococcus radiodurans	Malachite green	Sliver nanoparticles	[72]
Sargassum myriocystum	Methylene blue	Sliver nanoparticles	[4]
Marinospirillum alkaliphilum	Disperse blue 183	Sliver nanoparticles	[41]
Cordyceps militaris	Methylene blue	Zinc oxide	[36]
Bacillus licheniformis M09	Methylene blue	Silver nanoparticles	[39]
Aspergillus niger	Bismarck brown	Zinc oxide	[30]
Aspergillus niger L3 and Trichoderma longibrachiatum L2	Methylene blue, Malachite green	Sliver nanoparticles	[18]
Clostridium pasteurianum. C. pasteurianum	Methyl orange	Molybdenum Nanoparticles	[46]
Leuconostoc lactis	Congo red, methyl orange	Sliver nanoparticles	[58]
Bacillus marisflavi	Congo red, methylene blue	Gold nanoparticles	[40]
Erwinia herbicola	Methylene blue, methyl orange, and erichrome black T	SnO_2 nanoparticles	[66]

than free enzyme and effectively at higher concentration of dye [2]. Rod-shaped Fe_2O_3 nanostructures were synthesised with one pot microwave method. Metal salt precursors were utilised for shape directing agents. On polyethylene glycol with high temperature graphite layer was deposited which showed mimicry of peroxidase enzyme and was applied for decolourisation of different dyes [27]. Horse peroxidase enzymes were entrapped on polyacrylamide gel and checked for the potential of degradation. Entrapped horse peroxidase enzyme was applied for degradation of methyl orange. Results showed 90% decolourisation of methyl orange with immobilised enzyme in packed bed reactor [7]. NH_2-MIL-88B iron-based catalysed made by rapid microwave heating method which possesses peroxidase activity were utilised for degradation of Methylene blue. Complete elimination of Methylene blue was observed in 45 min and 80% removal after five cycle which shows high reusability of catalyst [25]. Laccase enzyme was immobilised on nano-silica to increase reusability and stability. With immobilised laccase results obtained were, 96% degradation of reactive violet 1, 79% activity after 20 cycle and 91% initial activity, was maintained at lower temperature for 60 days [21].

Table 2 Application of plant mediated nanoparticles in dye degradation

Microorganism	Pollutant	Name of nanoparticles	References
Cannabis sativa Leaf extract	Congo red, methyl orange	Zinc oxide nanoparticles	[10]
Syzygium Cumini	Rhodamine B	Zinc oxide nanoparticles	[49]
Cynometra Ramiflora leaf extract	Rhodamine B	Zinc oxide nanoparticles	[69]
Calliandra haematocephala	Methylene blue	Zinc oxide nanoparticles	[70]
Taraxacum officinale	Direct yellow-142 and methyl orange	Cobalt oxide nanoparticles	[54]
Alpinia nigra	Methyl orange, Rhodamine B and orange G	Sliver nanoparticles	[5]
Spondias dulcis	Methylene blue	Sliver nanoparticles	[73]
Albizia procera leaf extract	Methylene blue	Sliver nanoparticles	[48]
Camellia sinensis	Bromophenol blue	Copper nanoparticles	[1]
Averrhoa bilimbi fruit *and Pandanus amaryllifolius* leaf extracts	Methylene blue	Gold and titanium oxide composite	[74]
Pomegranate peel extract	Methylene blue	Sliver nanoparticles	[29]
Tabernaemontana divaricata	Methylene blue	Zinc oxide	[50]
Ferulago angulata (schlecht)	Rhodamine B	Zinc oxide and Copper oxide	[61]

4 Recent Advancement in Treatment of Textile Wastewater

Textile industrial development is generating large amount of liquid waste, which is released into environment, leads to water pollution. In all processes of textile industry, water is utilised in large amount, about 200 L of water is utilised in production of 1 kg of textile [23]. The liquid waste of textile industry has many organic-inorganic constituents, dyes and dyestuff and heavy metals. Before releasing it is necessary to remove all constituents. The constituents present in wastewater may cause skin related issues. Accumulation of organic pollutants in environment leads to disturbance in food chain. Treatment of textile wastewater should be efficient at large scale. Recently, many different reactors, aerobic, anaerobic, microaerophilic and photocatalytic reactors, are constructed or fabricated for the treatment of textile effluent.

4.1 Photocatalytic Reactor

Photocatalysis is a remarkable method for removal of pollutants from wastewater. Photocatalysis is defined as catalytic reaction involving production of catalyst by light absorption [47]. Mechanism of photocatalyst involves photon as activated catalysts to generate hydroxyl ions and radicals are utilised for oxidation of organic compounds and modify them into less harmful product. Efficient performance of reactor depends upon mass transfer and photon transfer. Different types of photocatalytic reactor are developed such as immobilised, suspended, visible light based, LED based, batch and continuous reactor [67].

Artificially synthesised composite made up from titanium oxide, graphene oxide and graphitic carbon nitride immobilised on polystyrene film evaluated for degradation of Remazol turquoise blue using airlift photocatalytic reactor. Result found was, 60% decolourisation and 51% degradation of dye in 90 min. of reaction time, measured by total organic carbon analysis. Updated study was carried out by same researcher with airlift photocatalytic reactor and analysed the performance. The result presented was 92% total organic carbon reduction and 94% Remazol turquoise blue dye decolourisation in 140 min in immobilised catalyst [13, 14]. Photobiodegradation of Basic red 46 was carried out by pilot-scale solar photoreactor. The solar UV/immobilised with TiO_2 was evaluated with artificial intelligence. Results in optimised condition with initial concentration of dye 10.65 mg/L, pH 10.82, rate of fluid flow of 852 L/h, 99% of Basic red 46 dye removal were observed. [6]. Zinc oxide capped polyethylene glycol (ZnO-PEG) and polypipeazine amide tight ultrafiltration membrane was utilised in photocatalytic reactor for treatment of textile wastewater. The optimum operating condition of reactor was pH 11, ZnO-PEG 0.1 g/L, more than 95% degradation of synthetic wastewater and COD removal were observed with 75% dilution of industrial wastewater was obtained [16]. Photocatalytic degradation of Acid Orange 7 was carried out in baffled reactor with titanium oxide as nano-catalyst. Complete mineralisation of dye was achieved after 30 h of reaction [53].

4.2 Nanotechnology with Electricity Generation

As real textile effluent contains many hazardous pollutants, hence, any one technology cannot be applied for efficient treatment. Therefore, combination of two or more is required for efficient treatment of wastewater. Different types of nanocatalysts were structured with modified z scheme system to increase efficiency for treatment [17]. Currently, hybrid technologies are in demand due to higher efficiency in treatment. Nanotechnology plays a very crucial role in degradation of pollutants from wastewater. Combination of nanotechnology with fuel cell has major impact on treatment of textile effluent [28]. Hybrid photocatalytic and peroxicoagulation were utilised for removal of organic pollutant p-nitrosodimethylanilline and generation of electricity. The 11% rate of nitrosodimethylanilline degradation was observed with maximum power density 3.14 mW cm^{-2} and iron plate size 30 cm^2 [44]. Integration of oxidation of photocatalyst and biodegradation approach were applied for treatment of CETP effluent. Reactor was setup with synthesised TiO_2 combined with activated carbon (AC) and biofilm. Results gained removal of COD, BOD, TSS and ammonical nitrogen within 6 h as 61, 92, 88 and 50%, respectively, and decolourisation was 78% in presence of solar light, TiO_2/AC and biofilm [11]. Photocatalytic reactor and peroxicoagulation were applied for degradation of Amaranth dye. The performance of photocatalytic reactor, maximum power density of 9.53 mW/m^2 and current density 0.0178 mA/m^2 was attained at degradation of 10 mg/L of Amaranth [45].

5 Toxicity of Nanoparticles and Impact on Environment

Increase in production of nanoparticles and practice in different areas is leading towards difficulties for humans and environment. Increased usage of nanostructure led to increment in exposure to humans and environment. Nanoparticles can enter through various path and cross the barriers of human body, resulting into damage in lung, liver, kidney, mitochondria and genetical modifications, leading eventually to apoptosis and death [59]. The details of nanoparticles and their toxic effect are described in Table 3 [24]. Recently nanoparticles are applied in large area, directly or indirectly the presence of nanoparticles lead to environmental transformation. In air, water and soil the exposure of nanoparticles converts them into aggregate form and changes their characteristics. Different environmental conditions such as pH, temperature, organisms, water content, light and organisms will lead to physical, chemical and biological transformation [75]. Details of types of toxicity and transformation are described in Fig. 6.

Table 3 Nanoparticles and their toxic effect

Nanoparticles	Daily usage	Toxicity effect
ZnO	Sunscreen, food additives, pigments and biosensors	Decrease cell viability, reduce enzyme activity in liver
TiO_2	Pigment, thickener, cosmetic care products	Cytotoxicity and genotoxicity in various cell lines
Silica	Food additive	Cytotoxicity in gastrointestinal cells
Sliver	Antimicrobial agent	Can damage DNA and caused *invitro* toxicity, effect in human microvascular endothelial cells
Gold	Drug delivery and diagnosis	Genotoxicity in human HepG2 cell, cellular toxicity and *invivo* toxicity
Polymeric	Drug delivery system	Nonspecific cytotoxicity

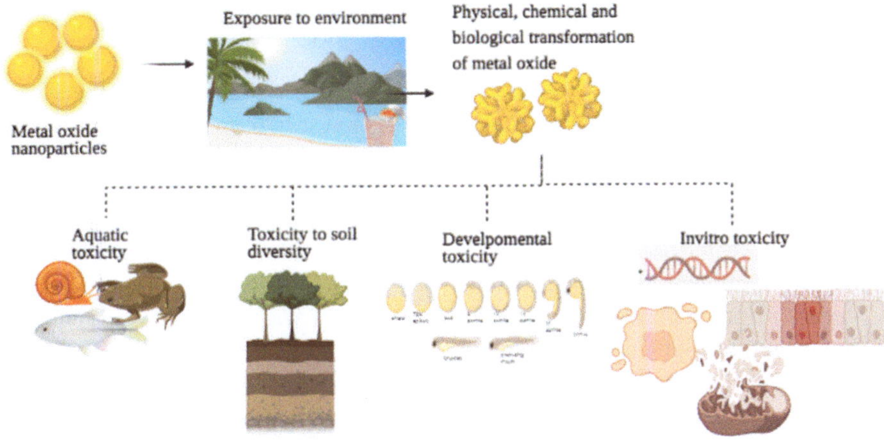

Fig. 6 Environmental transformation and types of toxicity

6 Future Aspects

Nanotechnology is a crucial technology emerging recently with numerous innovations in field of environmental applications. Enormous usage of toxic compounds in various industries led towards water pollution. The scarcity of pure water is major task faced by world recently which required new technologies of water conservation and wastewater treatment. Massive amount of wastewater generated from industries is mandatory to undergo treatment process before discharging in ecosystem. Cost-effective technologies, such as filtration, precipitation, adsorption, microbial degradation and chemical oxidation, are adopted for removal of contaminants from wastewater, but the conventional methods are not that much effective. Nowadays, nanostructures are more efficient for removal of contaminates from wastewater as it has high stability and thermal resistance. Usage of nanoparticles for treatment is applied for small-scale studies. The nanoparticles mediated removal of hazardous compounds is effective for small scale only. Major challenge is facing in nanostructure mediated treatment is required supplementary treatment for removal of nanoparticles because metals containing nanostructure have toxic nature and not utilised for large-scale studies. So, treatment of enormous amount of wastewater needs modification and improvement in current strategies. To overcome treatment difficulties many researchers are working in this field since long. For large-scale treatment, reactors are constructed but are not that much effective for treatment of produced wastewater per day. Therefore, desires of some advancement in treatment strategies; operative in large scale as well as eco-friendly to the environment should be adopted.

7 Conclusion

Water is playing a major part in the whole ecosystem. Increment in industrialisation and production of synthetic compounds led towards water, air and soil pollution. Elimination of toxic compounds before discharge into ecosystem is obligatory. Therefore, many physical, chemical and biological methods are utilised for treatment among them, nanotechnology plays an important role in remediation in which nanostructures are synthesised and applied for removal of various synthetic compounds. Treatment of effluent by nanoparticles is more efficient than other technologies but it is applied for small scale. To treat huge amount of wastewater any one method cannot be applied, hence, combination of two or more than two methods is applied for efficient treatment of effluent. Lots of research work is going on to find out best technology for treatment of wastewater. Many researchers and academic institutes are working in this area.

References

1. Ahmed A, Usman M, Liu QY, Shen YQ, Yu B, Cong HL (2019) Plant mediated synthesis of copper nanoparticles by using *Camelia sinensis* leaves extract and their applications in dye degradation. Ferroelectrics 549(1):61–69. https://doi.org/10.1080/00150193.2019.1592544
2. Aldhahri M, Almulaiky YQ, El-Shishtawy RM, Al-Shawafi WM, Salah N, Alshahrie A, Alzahrani HAH (2021) Ultra-thin 2D CuO nanosheet for HRP immobilization supported by encapsulation in a polymer matrix: characterization and dye degradation. Catal Lett 151(1):232–246. https://doi.org/10.1007/s10562-020-03289-7
3. Ansari A, Mehrabian MA, Hashemipour H (2012) Zinc ion adsorption on carbon nanotubes in an aqueous solution. Pol J Chem Technol 14(3):29–37. https://doi.org/10.2478/v10026-012-0081-6
4. Balaraman P, Balasubramanian B, Kaliannan D, Durai M, Kamyab H, Park S, Chelliapan S, Lee CT, Maluventhen V, Maruthupandia A (2020) Phyco-synthesis of silver nanoparticles mediated from marine algae *Sargassum myriocystum* and Its potential biological and environmental applications. Waste Biomass Valorization 11(10):5255–5271. https://doi.org/10.1007/s12649-020-01083-5
5. Baruah D, Yadav RNS, Yadav A, Das AM (2019) *Alpinia nigra* fruits mediated synthesis of silver nanoparticles and their antimicrobial and photocatalytic activities. J Photochem Photobiol B Biol 201(October):111649. https://doi.org/10.1016/j.jphotobiol.2019.111649
6. Berkani M, Bouchareb MK, Bouhelassa M, Kadmi Y (2020) Photocatalytic degradation of industrial dye in semi-pilot scale prototype solar photoreactor: optimization and modeling using ANN and RSM based on box-wilson approach. Top Catal 63(11–14):964–975. https://doi.org/10.1007/s11244-020-01320-0
7. Bilal M, Rasheed T, Iqbal HMN, Hu H, Wang W, Zhang X (2018) Horseradish peroxidase immobilization by copolymerization into cross-linked polyacrylamide gel and its dye degradation and detoxification potential. Int J Biol Macromol 113(2017):983–990. https://doi.org/10.1016/j.ijbiomac.2018.02.062
8. Biswas P, Wu CY (2005) Nanoparticles and the environment. J Air Waste Manag Assoc 55(6):708–746
9. Cao XT, Showkat AM, Kim DW, Jeong YT, Kim JS, Lim KT (2015) Preparation of β-cyclodextrin multi-decorated halloysite nanotubes as a catalyst and nanoadsorbent for dye removal. J Nanosci Nanotechnol 15(11):8617–8621. https://doi.org/10.1166/jnn.2015.11482

10. Chauhan A, Verma R, Kumari S, Sharma A, Shandilya P, Li X, Batoo KM, Imran A, Kulshrestha S, Kumar R (2020) Photocatalytic dye degradation and antimicrobial activities of Pure and Ag-doped ZnO using *Cannabis sativa* leaf extract. Sci Rep 10(1):1–16. https://doi.org/10.1038/s41598-020-64419-0
11. Chavan A, Fulekar MH (2018) Integration of photocatalytic oxidation and biodegradation treatment processes to enhance degradation efficiency of CETP wastewater contaminants. Bionanoscience 8(3):761–768. https://doi.org/10.1007/s12668-018-0534-3
12. Chen J, Sheng Y, Song Y, Chang M, Zhang X, Cui L, Meng D, Zhu H, Shi Z, Zou H (2018) Multimorphology mesoporous silica nanoparticles for dye adsorption and multicolor luminescence applications. ACS Sustain Chem Eng 6(3):3533–3545. https://doi.org/10.1021/acssuschemeng.7b03849
13. Das S, Mahalingam H (2019) Dye degradation studies using immobilized pristine and waste polystyrene-TiO_2/rGO/g-C_3N_4 nanocomposite photocatalytic film in a novel airlift reactor under solar light. J Environ Chem Eng 7(5):103289. https://doi.org/10.1016/j.jece.2019.103289
14. Das S, Mahalingam H (2020) Novel immobilized ternary photocatalytic polymer film based airlift reactor for efficient degradation of complex phthalocyanine dye wastewater. J Hazard Mater 383(September 2019):121219. https://doi.org/10.1016/j.jhazmat.2019.121219
15. Deng D, Lamssali M, Aryal N, Ofori-Boadu A, Jha MK, Samuel RE (2020) Textiles wastewater treatment technology: a review. Water Environ Res 92(10):1805–1810. https://doi.org/10.1002/wer.1437
16. Desa AL, Hairom NHH, Ng L, Ng CY, Ahmad MK, Mohammad AW (2019) Industrial textile wastewater treatment via membrane photocatalytic reactor (MPR) in the presence of ZnO-PEG nanoparticles and tight ultrafiltration. J Water Process Eng 31(January):100872. https://doi.org/10.1016/j.jwpe.2019.100872
17. Du M, Cao S, Ye X, Ye J (2020) Recent advances in the fabrication of all-solid-state nanostructured TiO_2 based Z-scheme heterojunctions for environmental remediation. J Nanosci Nanotechnol 20(9):5861–5873. https://doi.org/10.1166/jnn.2020.18719
18. Elegbede JA, Lateef A, Azeez MA, Asafa TB, Yekee TA, Oladipo IC, Adebayo EA, Beukes LS, Gueguim-Kana EB (2018) Fungal xylanases-mediated synthesis of silver nanoparticles for catalytic and biomedical applications. IET Nanobiotechnol 12(6):857–863. https://doi.org/10.1049/iet-nbt.2017.0299
19. Fouda A, Salem SS, Wassel AR, Hamza MF, Shaheen TI (2020) Optimization of green biosynthesized visible light active CuO/ZnO nano-photocatalysts for the degradation of organic methylene blue dye. Heliyon 6(9):e04896. https://doi.org/10.1016/j.heliyon.2020.e04896
20. Gahlawat G, Choudhury AR (2019) A review on the biosynthesis of metal and metal salt nanoparticles by microbes. RSC Adv 9:12944–12967. https://doi.org/10.1039/c8ra10483b
21. Gahlout M, Rudakiya DM, Gupte S, Gupte A (2017) Laccase-conjugated amino-functionalized nanosilica for efficient degradation of Reactive Violet 1 dye. Int Nano Lett 7(3):195–208. https://doi.org/10.1007/s40089-017-0215-1
22. Ganguly S, Mondal S, Das P, Bhawal P, Das TK, Bose M, Choudhary S, Gangopadhyay S, Das AK, Das NC (2018) Natural saponin stabilized nano-catalyst as efficient dye-degradation catalyst. Nano-Struct Nano-Obj 16:86–95. https://doi.org/10.1016/j.nanoso.2018.05.002
23. Ghaly AE, Ananthashankar R, Alhattab VVR, Ramakrishnan VV (2014) Production, characterization and treatment of textile effluents: a critical review. J Chem Eng Proc Technol 05(01):1–18. https://doi.org/10.4172/2157-7048.1000182
24. Gupta R, Xie H (2018) Nanoparticles in daily life: applications, toxicity and regulations. J Environ Pathol Toxicol Onco 37(3):209–230. https://doi.org/10.1615/JEnvironPatholToxicolOncol.2018026009
25. He J, Zhang Y, Zhang X, Huang Y (2018) Highly efficient Fenton and enzyme-mimetic activities of NH_2-MIL-88B(Fe) metal organic framework for methylene blue degradation. Sci Rep 8(1):1–8. https://doi.org/10.1038/s41598-018-23557-2
26. Jain K, Patel AS, Pardhi VP, Jeet S, Flora S (2021) Nanotechnology in wastewater management: a new paradigm towards wastewater treatment. Molecules 26:1–26. https://doi.org/10.3390/molecules26061797

27. Jain S, Panigrahi A, Sarma TK (2019) Counter anion-directed growth of iron oxide nanorods in a polyol medium with efficient peroxidase-mimicking activity for degradation of dyes in contaminated water. ACS Omega 4(8):13153–13164. https://doi.org/10.1021/acsomega.9b01201
28. Jeon TH, Koo MS, Kim H, Choi W (2018) Dual-functional photocatalytic and photoelectron catalytic systems for energy- and resource-recovering water treatment. ACS Catal 8(12):11542–11563. https://doi.org/10.1021/acscatal.8b03521
29. Joshi SJ, Geetha SJ, Al-Mamari S, Al-Azkawi A (2018) Green synthesis of silver nanoparticles using pomegranate peel extracts and its application in photocatalytic degradation of methylene blue. Jundishapur. J. Nat. Pharm. Prod 13(3):4. https://doi.org/10.5812/jjnpp.67846
30. Kalpana VN, Kataru BAS, Sravani N, Vigneshwari T, Panneerselvam A, Devi Rajeswari V (2018) Biosynthesis of zinc oxide nanoparticles using culture filtrates of *Aspergillus niger*: Antimicrobial textiles and dye degradation studies. OpenNano 3:48–55. https://doi.org/10.1016/j.onano.2018.06.001
31. Kalra A, Gupta A (2020) Materials today: proceedings recent advances in decolourization of dyes using iron nanoparticles: a mini review. Mater Today. https://doi.org/10.1016/j.matpr.2020.04.677
32. Khan NA, Khan SU, Ahmed S, Farooqi IH, Dhingra A, Hussain A, Changani F (2019) Applications of nanotechnology in water and wastewater treatment: a review. Asian J Water Environ Pollut 16(4):81–86. https://doi.org/10.3233/AJW190051
33. Klunk M, Dasgupta S, Das M, Cunha MG, Wander PR (2019) Synthesis of sodalite zeolite and adsorption study of crystal violet dye. ECS J Solid State Sci Technol 8(10):N144–N150. https://doi.org/10.1149/2.0131910jss
34. Kucuk AC, Urucu OA (2019) Silsesquioxane-modified chitosan nanocomposite as a nanoadsorbent for the wastewater treatment. React Funct Polym 140(April):22–30. https://doi.org/10.1016/j.reactfunctpolym.2019.04.011
35. Lee KY, Kim KW, Park M, Kim J, Oh M, Lee EH, Chung DY, Moon JK (2016) Novel application of nanozeolite for radioactive cesium removal from high-salt wastewater. Water Res 95:134–141. https://doi.org/10.1016/j.watres.2016.02.052
36. Li JF, Rupa EJ, Hurh J, Huo Y, Chen L, Han Y, Ahn JC, Park JK, Lee HA, Mathiyalagan R, Yang DC (2019) *Cordyceps militaris* fungus mediated Zinc Oxide nanoparticles for the photocatalytic degradation of methylene blue dye. Optik 183:691–697. https://doi.org/10.1016/j.ijleo.2019.02.081
37. Lu C, Liu C, Rao GP (2008) Comparisons of sorbent cost for the removal of Ni^{2+} from aqueous solution by carbon nanotubes and granular activated carbon. J Hazard Mater 151(1):239–246. https://doi.org/10.1016/j.jhazmat.2007.05.078
38. Mahmoud HR, El-Molla SA, Saif M (2013) Improvement of physicochemical properties of Fe_2O_3/MgO nanomaterials by hydrothermal treatment for dye removal from industrial wastewater. Powder Technol 249:225–233. https://doi.org/10.1016/j.powtec.2013.08.021
39. Momin B, Rahman S, Jha N, Annapure US (2019) Valorization of mutant *Bacillus licheniformis M09* supernatant for green synthesis of silver nanoparticles: photocatalytic dye degradation, antibacterial activity, and cytotoxicity. Bioprocess Biosyst Eng 42(4):541–553. https://doi.org/10.1007/s00449-018-2057-2
40. Nadaf NY, Kanase SS (2016) Biosynthesis of gold nanoparticles by *Bacillus marisflavi* and its potential in catalytic dye degradation. Arab J Chem 12(8):4806–4814. https://doi.org/10.1016/j.arabjc.2016.09.020
41. Nazari N, Jookar Kashi F (2020) A novel microbial synthesis of silver nanoparticles: Its bioactivity, Ag/Ca-Alg beads as an effective catalyst for decolorization Disperse Blue 183 from textile industry effluent. Sep Purif Technol 259:118117. https://doi.org/10.1016/j.seppur.2020.118117
42. Nikam AV, Prasad BLV, Kulkarni AA (2018) Wet chemical synthesis of metal oxide nanoparticles: a review. CrystEngComm 20(35):5091–5107. https://doi.org/10.1039/C8CE00487K

43. Noman M, Shahid M, Ahmed T, Niazi MBK, Hussain S, Song F, Manzoor I. (2020) Use of biogenic copper nanoparticles synthesized from a native *Escherichia* sp. as photocatalysts for azo dye degradation and treatment of textile effluents. Environ Pollut 257. https://doi.org/10.1016/j.envpol.2019.113514
44. Nordin N, Ho LN, Ong SA, Ibrahim AH, Abdul Rani AL, Lee SL, Ong YP (2020) Hydroxyl radical formation in the hybrid system of photocatalytic fuel cell and peroxi-coagulation process affected by iron plate and UV light. Chemosphere 244:125459. https://doi.org/10.1016/j.chemosphere.2019.125459
45. Nordin N, Ho LN, Ong SA, Ibrahim AH, Wong YS, Lee SL, Oon YS, Oon YL (2017) Influence of Amaranth dye concentration on the efficiency of hybrid system of photocatalytic fuel cell and Fenton process. Environ Sci Pollut Res 24(29):23331–23340. https://doi.org/10.1007/s11356-017-9964-7
46. Nordmeier A, Merwin A, Roeper DF, Chidambaram D (2018) Microbial synthesis of metallic molybdenum nanoparticles. Chemosphere 203:521–525. https://doi.org/10.1016/j.chemosphere.2018.02.079
47. Pandey S, Mandari KK, Kim J, Kang M, Fosso-Kankeu E (2020) Recent advancement in visible-light-responsive photocatalysts in heterogeneous photocatalytic water treatment technology. In: Photocatalysts in advanced oxidation processes for wastewater treatment, pp 167–196. https://doi.org/10.1002/9781119631422.ch6
48. Rafique M, Sadaf I, Tahir MB, Rafique MS, Nabi G, Iqbal T, Sughra K (2019) Novel and facile synthesis of silver nanoparticles using *Albizia procera* leaf extract for dye degradation and antibacterial applications. Mater Sci Eng C 99:1313–1324. https://doi.org/10.1016/j.msec.2019.02.059
49. Rafique M, Tahir R, Gillani SSA, Tahir MB, Shakil M, Iqbal T, Abdellahi MO (2020) Plant-mediated green synthesis of zinc oxide nanoparticles from *Syzygium Cumini* for seed germination and wastewater purification. Int J Environ Anal Chem 1–16:23–38. https://doi.org/10.1080/03067319.2020.1715379
50. Raja A, Ashokkumar S, Pavithra Marthandam R, Jayachandiran J, Khatiwada C, Kaviyarasu K, Ganapathi Raman R, Swaminathan M (2018) Eco-friendly preparation of zinc oxide nanoparticles using *Tabernaemontana divaricata* and its photocatalytic and antimicrobial activity. J Photochem Photobiol B Biol 181(2017):53–58. https://doi.org/10.1016/j.jphotobiol.2018.02.011
51. Rajkumar R, Ezhumalai G, Gnanadesigan M (2021) A green approach for the synthesis of silver nanoparticles by *Chlorella vulgaris* and its application in photocatalytic dye degradation activity. Environ Technol Innov 21:101282. https://doi.org/10.1016/j.eti.2020.101282
52. Rambabu K, Bharath G, Monash P, Velu BF, Naushad M, Arthanareeswaran G, Loke Show P (2019) Effective treatment of dye polluted wastewater using nanoporous $CaCl_2$ modified polyethersulfone membrane. Process Saf Environ Prot 124:266–278. https://doi.org/10.1016/j.psep.2019.02.015
53. Ranjbar PZ, Ayati B, Ganjidoust H (2019) Kinetic study on photocatalytic degradation of Acid Orange 52 in a baffled reactor using TiO_2 nanoparticles. J Environ Sci 79:213–224. https://doi.org/10.1016/j.jes.2018.06.012
54. Rasheed T, Nabeel F, Bilal M, Iqbal HMN (2019) Biogenic synthesis and characterization of cobalt oxide nanoparticles for catalytic reduction of direct yellow-142 and methyl orange dyes. Biocatal Agric Biotechnol 19:101154. https://doi.org/10.1016/j.bcab.2019.101154
55. Rashidi HR, Sulaiman NMN, Hashim NA, Hassan CRC, Ramli MR (2015) Synthetic reactive dye wastewater treatment by using nano-membrane filtration. Desalination Water Treat 55(1):86–95. https://doi.org/10.1080/19443994.2014.912964
56. Roy A, Bhattacharya J (2019) Nanotechnology in industrial wastewater treatment. IWA Publishing
57. Salem S, Fouda A (2020) Green synthesis of metallic nanoparticles and their prospective biotechnological applications: an overview. Biol Trace Elem Res 199(1):344–370
58. Saravanan C, Rajesh R, Kaviarasan T, Muthukumar K, Kavitake D, Shetty PH (2017) Synthesis of silver nanoparticles using bacterial exopolysaccharide and its application for degradation of azo-dyes. Biotechnol Rep 15:33–40. https://doi.org/10.1016/j.btre.2017.02.006

59. Sengul AB, Asmatulu E (2020) Toxicity of metal and metal oxide nanoparticles: a review. Environ Chem Lett 18(5):1659–1683. https://doi.org/10.1007/s10311-020-01033-6
60. Shahi MP, Kumari P, Mahobiya D, Shahi SK (2021) Chapter 4—nano-bioremediation of environmental contaminants: applications, challenges, and future prospects. In: Bioremediation for environmental sustainability. Elsevier B.V. https://doi.org/10.1016/B978-0-12-820318-7/00004-6
61. Shayegan Mehr E, Sorbiun M, Ramazani A, Taghavi Fardood S (2018) Plant-mediated synthesis of zinc oxide and copper oxide nanoparticles by using *ferulago angulata* (schlecht) boimass extract and comparison of their photocatalytic degradation of Rhodamine B (RhB) under visible light irradiation. J Mater Sci Mater Electron 29(2):1333–1340. https://doi.org/10.1007/s10854-017-8039-3
62. Shukla P (2020) Microbial nanotechnology for bioremediation of industrial wastewater. Front Microbiol 11(November). https://doi.org/10.3389/fmicb.2020.590631
63. Singh KK, Senapati KK, Sarma KC (2017) Synthesis of superparamagnetic Fe_3O_4 nanoparticles coated with green tea polyphenols and their use for removal of dye pollutant from aqueous solution. J Environ Chem Eng 5(3):2214–2221. https://doi.org/10.1016/j.jece.2017.04.022
64. Sivalingam S, Sen S (2019) Efficient removal of textile dye using nanosized fly ash derived zeolite-x: kinetics and process optimization study. J Taiwan Inst Chem Eng 96:305–314. https://doi.org/10.1016/j.jtice.2018.10.032
65. Smith SC, Rodrigues DF (2015) Carbon-based nanomaterials for removal of chemical and biological contaminants from water: a review of mechanisms and applications. Carbon 91:122–143. https://doi.org/10.1016/j.carbon.2015.04.043
66. Srivastava N, Mukhopadhyay M (2014) Biosynthesis of SnO_2 nanoparticles using bacterium *erwinia herbicola* and their photocatalytic activity for degradation of dyes. Ind Eng Chem Res 53(36):13971–13979. https://doi.org/10.1021/ie5020052
67. Sundar KP, Kanmani S (2020) Progression of photocatalytic reactors and it's comparison: a review. Chem Eng Res Des 154:135–150. https://doi.org/10.1016/j.cherd.2019.11.035
68. Tran TT, Van Kumar SR, Lue SJ (2019) Separation mechanisms of binary dye mixtures using a PVDF ultrafiltration membrane: donnan effect and intermolecular interaction. J Membr Sci 575:38–49. https://doi.org/10.1016/j.memsci.2018.12.070
69. Varadavenkatesan T, Lyubchik E, Pai S, Pugazhendhi A, Vinayagam R, Selvaraj R (2019) Photocatalytic degradation of Rhodamine B by zinc oxide nanoparticles synthesized using the leaf extract of *Cyanometra ramiflora*. J Photochem Photobiol B Biol 199:111621. https://doi.org/10.1016/j.jphotobiol.2019.111621
70. Vinayagam R, Selvaraj R, Arivalagan P, Varadavenkatesan T (2020) Synthesis, characterization and photocatalytic dye degradation capability of Calliandra haematocephala-mediated zinc oxide nanoflowers. J Photochem Photobiol B Biol 203:111760. https://doi.org/10.1016/j.jphotobiol.2019.111760
71. Wang H, Zhou A, Peng F, Yu H, Yang J (2007) Mechanism study on adsorption of acidified multiwalled carbon nanotubes to Pb(II). J Colloid Interface Sci 316(2):277–283. https://doi.org/10.1016/j.jcis.2007.07.075
72. Weng Y, Li J, Ding X, Wang B, Dai S, Zhou Y, Pang R, Zhao Y, Xu H, Tian B, Hua Y (2020) Functionalized gold and Silver bimetallic nanoparticles using *Deinococcus radiodurans* protein extract mediate degradation of toxic dye malachite green. Int J Nanomed 15:1823–1835. https://doi.org/10.2147/IJN.S236683
73. Yadav P, Manjunath H, Selvara R (2019) Antibacterial and dye degradation potential of zero-valent silver nanoparticles synthesised using the leaf extract of *Spondias dulcis*. IET Nanobiotechnol 13(1):84–89. https://doi.org/10.1049/iet-nbt.2018.5058
74. Yulizar Y, Sudirman Apriandanu DOB, Wibowo AP (2019) Plant extract mediated synthesis of Au/TiO_2 nanocomposite and its photocatalytic activity under sodium light irradiation. Compos Commun 16(May):50–56. https://doi.org/10.1016/j.coco.2019.08.006
75. Zhang J, Guo W, Li Q, Wang Z, Liu S (2018) The effects and the potential mechanism of environmental transformation of metal nanoparticles on their toxicity in organisms. Environ Sci Nano 5(11):2482–2499. https://doi.org/10.1039/c8en00688a

Magnetic-Based Biocomposites in Dye Adsorption

Adewale Adewuyi

Abstract Adsorption is an important method used in the removal of dyes from the water system. The application of magnetic-based biocomposite as an adsorbent in the adsorption process to remove dyes from water is a possible technique in wastewater purification. This chapter focuses on the production and application of magnetic-based biocomposite in the removal of dyes from water. It was evident that magnetic-based biocomposites are potential adsorbent for the removal of dyes from water. However, there is scanty information available on the use of magnetic-based biocomposites on a large scale, as most studies reported were on a laboratory scale. Despite the promise of high efficiency exhibited by magnetic-based biocomposites towards dyes in water, there is a need to evaluate their economic viability and prospects on a large-scale application.

Keywords Adsorption · Biocomposites · Desorption · Dye · Magnetite · Wastewater

1 Introduction

Pollution of water resources by organic molecules such as dyes is a serious global challenge. Many dyes are produced yearly, which end up in the environment during or after use. Dyes have found application in several industries such as paper, leather, textile, cosmetic, food and plastics [25]. Most dyes are toxic and carcinogenic [39], which can cause havoc to health and, when they get into the environment, may cause long-term adverse effect on plants, animals and human beings. About two-thirds of over 100,000 commercially known dyes are consumed by textile industries, which generate a large volume of coloured wastewater [74]. The colour produced by the dyes in water is highly discernible and inadmissible due to its ability to prevent reoxygenation in water. Furthermore, it disrupts microorganism activities in aquatic

---

A. Adewuyi (✉)
Department of Chemical Sciences, Faculty of Natural Sciences, Redeemer's University, Osun State, Ede, Nigeria
e-mail: walexy62@yahoo.com

S. S. Muthu and A. Khadir (eds.), *Textile Wastewater Treatment*, Sustainable Textiles: Production, Processing, Manufacturing & Chemistry,
https://doi.org/10.1007/978-981-19-2852-9_2

system because it inhibits the penetration of sunlight [79]. Most dyes have complex chemical structures, soluble in aqueous systems and persistent when they get into the environment. When in the environment, they are often exposed to environmental factors that may degrade them to form intermediates that are more toxic than their original forms. Therefore, it is crucial to understand the fate of these dyes when they get into the environment. Moreover, the need to develop efficient techniques for their complete removal in water cannot be overemphasized. Although some of the dyes have been banned in many countries, they are still widely used in other countries [49]. Among the numerous dyes known, the azo dye remains the largest synthetic dyes class with a wide application [49]. It is aromatic, and because of the possibility of reductive cleavage taking place at the azo linkages that may lead to the formation of amines, azo dyes are considered toxic. The diazo dye groups are also common and are toxic. An example of this is congo red dye which may be prepared by coupling tetrazotized benzidine with two molecules of naphthionic acid. Previous studies have reported dyes in surface and underground water system, as presented in Table 1. When the dyes get into the water bodies, they cause an increase in the levels of biochemical oxygen demand (BOD) and chemical oxygen demand (COD); the main concern is in the non-biodegradability of some of the dyes [47].

Although dyes may be grouped as natural or synthetic, their classification may also be based on colour, chemical structure, particle charge in solution and application. Natural dyes are sourced from plant and animals; however, as scientific knowledge advanced, the properties of natural dyes were improved by way of chemical synthesis, which led to synthetic dyes. Over time, synthetic dyes have gradually replaced their natural counterparts due to improved performance and aesthetics. Synthetic dyes may be ionic or non-ionic, with examples such as direct, acid, vat, basic, dispersed and reactive dyes. The reactive, direct, acid and basic dyes are

Table 1 Reported presence of dye in environmental water system

Source	Dye type	Country	References
Cristais river	Disperse azo, disperse Blue 373, disperse orange 37 and disperse violet 93	Brazil	[10]
Ribeirão das Cruzes river	Basic red 51	Brazil	[15]
River water	Disperse dyes	China	[82]
Pearl river delta	*N*-nitrosamines	China	[12]
Raw river water and drinking water treatment plant	*p*-aminophenol	Brazil	[19]
Cristais river	Disperse blue 373, disperse violet 93 and disperse orange 37	Brazil	[18]
Ripoll river	Textile dye	Spain	[14]
Archaeological sites	Natural dyes	China	[34]
River water and sediment	Azo dyes	Japan	[27]
Khniss and Hamdoun rivers	Disperse dyes	Tunisia	[40]

mostly water-soluble [36], making it difficult to remove them from water [26]. They persist in the environment; a recent investigation has shown that when azo dyes are used, about 15–50% remain in water without binding to fabric, making azo dyes a significant contributor of dye to the environment [31]. The presence of these dyes in the environment is undesirable; therefore, it is important to remove them from water systems before they get into the environment. Approaches reported as potential methods of removal included oxidation, membrane separation, advanced oxidation, electrochemical process, flocculation, ion exchange and adsorption.

Varieties of methods reported for the removal of dyes from water are known to rely on chemical, physical or biological techniques [8] in their operations. The chemical technique has the advantage of being effective at both high and low concentration of dye, however, it leads to the production of unwanted chemical deposition of sludge as secondary pollutants, which require additional cost for removal and sometimes may be far more problematic than the dye to be removed. Apart from this, the process requires a large number of chemical reagents and energy consumption [65, 84]. The biological technique is simple and requires the use of microorganisms to decolourize the dye contaminated water system; however, the process needs to be under specific conditions such as controlled process pH, temperature and nutrition, which may make it complex and expensive to start up or maintain [46]. Adsorption and membrane separation are typical examples of the physical techniques used [56]. Since the use of a membrane is limited with poor reusability, adsorption is a better process in terms of physical technique. Most dyes are resistant to degradation due to their complex chemical structure. Therefore, adsorption remains outstanding among the methods developed over time for removing dyes in the water system. Apart from being efficient, adsorption is cheap, reliable and can be easily set up. The use of adsorption is sustainable, and environmentally friendly. Since the adsorbents used can be regenerated and reused as appropriate, adsorption is considered a viable option for removing dyes from water.

Generally, an adsorbent is considered suitable for removing dyes from water system when its effective surface area for the sorption process is large with high adsorption capacity, excellent regeneration capacity and outstanding performance at a wide pH range. Several adsorbents have been reported in time past to remove dyes in the water system, as shown in Table 2. Despite the high adsorption capacities exhibited by some of these adsorbents, they still suffer from certain limitations that border on effective regeneration for long-term use and separation from the water system after the sorption process. Recent studies have shown that the use of magnetic-based biocomposite can help circumvent these limitations. In line with this, this chapter focuses on the use of magnetic-based biocomposites to remove dyes in solution.

2 Adsorption as a Process for Dye Removal

Adsorption as a process in dye removal from solution may be described as the decolourization of water contaminated with dye by attachment of dye molecules

Table 2 Selected adsorbents previously reported for the removal of named dyes in solution

Adsorbent	Dye type	Adsorption capacity (mg g^{-1})	References
Mesoporous zeolite	Crystal violet	1217	[9]
Montmorillonite/graphene oxide composite	Crystal violet	746	[48]
Montmorillonite/graphene oxide composite	Methylene blue	641	[76]
Kaolin-based mesoporous silica	Methylene blue	653	[33]
Mesoporous zeolite	Methylene blue	548	[9]
Cellulose/clay composite hydrogel	Methylene blue	277	[63]
Smectite rich natural clays	Basic yellow 28	77	[11]
Chitosan sponge	Rose Bengal	602	[61]
Chitosan/surfactant composite	Orange G	1452	[80]
Chitosan/polyvinyl alcohol/zeolite	Methyl orange	153	[24]
Polyurethane/chitosan foam	Food red 17	267	[16]

to a solid surface by chemical or physical interaction [44]. The solid surface is known as adsorbent, while the dye adsorbed from solution is known as adsorbate. If the interaction between the adsorbate and the adsorbent is physical, the adsorption process is referred to as physisorption. In contrast, when the interaction forces are due to chemical bonding, it is referred to as chemisorption [73]. The removal of adsorbate from the surface of the adsorbent is known as desorption which is the reverse of adsorption. In the case of physisorption, the interaction forces are weak, such as Van der-Waals forces. The process is easily reversed in physisorption resulting in desorption due to the weak forces of interaction, but desorption is difficult in the case of chemisorption.

It is crucial to determine the amount of adsorbate an adsorbent can accumulate, which helps understand its adsorption capacity, which may be calculated from adsorption isotherm. Since dyes are not easily broken down, they can be removed from the solution via adsorption. The adsorbent should have a short removal time, high uptake kinetic and good surface character for removal [6]. The adsorption capacity is the mass of adsorbate (dye) adsorbed per unit mass of adsorbent (magnetic-based biocomposite). Lots of isotherm models have been developed to describe the adsorption process; such isotherm models include Langmuir, Freundlich, Temkin, Hill, etc.

The Langmuir isotherm model is a model that assumes that the molecules adsorbed on the surface of the adsorbent form a monolayer, and each adsorbate on the surface of the adsorbent has the same adsorption activation energy; the model can be expressed as [59].

$$\frac{C_e}{q_e} = \frac{1}{Q_o}C_e + \frac{1}{Q_o K_L} \tag{1}$$

where C_e (mg L^{-1}) is the equilibrium amount of adsorbate, q_e (mg g^{-1}) is the amount of adsorbate removed at equilibrium, Q_o (mg g^{-1}) is the maximum monolayer coverage capacity and K_L (L mg^{-1}) represents the Langmuir isotherm constant. To better understand the Langmuir isotherm model, it is important to determine the value of R_L, which is an essential feature of the Langmuir isotherm that can be expressed as:

$$R_L = \frac{1}{1 + K_L C_o} \tag{2}$$

It has been reported that when $R_L > 1$, the sorption process is unfavourable, when $R_L = 1$, the sorption is considered linear when $0 < R_L < 1$, the process is favourable, and when $R_L = 0$, the process is taken to be irreversible [2, 3].

The Hill model is different from Langmuir; unlike the Langmuir model, which assumes that each adsorption site on the adsorbent can only accept one mole, the Hill model was developed on the fact that each adsorption site can accept n molecules of the adsorbate, which can be expressed as [6]

$$Q = \frac{nM_m}{1 + \left(\frac{C_{1/2}}{C}\right)^n} \tag{3}$$

Q is the equilibrium adsorption capacity (mg g^{-1}), n represents the number of molecules connected at each adsorption site, $C_{1/2}$ is the adsorbate concentration at half-saturation (mg g^{-1}) and where N_m represents the number of occupied adsorption sites.

Adsorption studies have also been subjected to the Freundlich isotherm model, which may be expressed as:

$$q_e = K_f C_e^n \tag{4}$$

where the equilibrium concentration of the adsorbate is represented as C_e (mg L^{-1}), K_f (mg g^{-1}) represents the Freundlich isotherm constant, q_e (mg g^{-1}) is the amount of the adsorbate adsorbed at equilibrium on the adsorbent and the adsorption intensity is represented as n. The Freundlich model is a multilayer sorption which describes the sorption process on a heterogeneous surface as an exponential distribution of active sites and their energies. It is also important to check the effect of certain parameters on the sorption capacity of an absorbent; such parameters include the effect of temperature, pH, adsorbate concentration, adsorbent particles and adsorbent dose. Despite the several adsorbents known, the use of magnetic-based biocomposites in removing dye from solution remains outstanding because of the ease of separation after the adsorption of dye by the simple application of magnet [71].

Different types of adsorption kinetic and diffusion models have been proposed, among which pseudo-first-order, pseudo-second-order, intra-particle diffusion and Elovich models are common. The pseudo-first-order model may be represented as:

$$\log(q_e - q_t) = \log q_e - \frac{K_1}{2.303}t \tag{5}$$

where q_e (mg g^{-1}) is the equilibrium amount of dye ions, q_t (mg g^{-1}) represents the dye ions adsorbed at a specific time, k_1 (min^{-1}) stands for the rate constant for pseudo-first-order adsorption and t is the time (min). From this expression, k_1 and q_e can be determined from the plot of log (q_e–q_t) against t.

The pseudo-second-order model may be described as:

$$\frac{t}{q_t} = \frac{1}{k_2 q_e^2} + \frac{1}{q_e}t \tag{6}$$

where q_e (mg g^{-1}) is the equilibrium amount of dye ion, q_t (mg g^{-1}) is the amount of dye ion adsorbed at a specific time, and k_2 (g mg^{-1} min^{-1}) represents the rate constant for the pseudo-second-order model. The value of q_e and k_2 may be determined from the various plots as described in Table 3.

The sorption may be further described using the intra-particle diffusion model, as expressed:

$$q_t = k_{\text{id}} t^{0.5} + C \tag{7}$$

where the k_{id} is the diffusion constant (mg g^{-1} min$^{1/2}$) while C (mg g^{-1}) represents a constant that shows the layer thickness. The parameters can be determined from the plot of q_t against $t^{1/2}$. C is the value of the intercept, which is related to the boundary layer thickness. The larger the intercept value from the plot, the greater the boundary effect will be [28]. The data generated from the sorption process can also be subjected to the Elovich model, which can be linearized as:

$$q_t = \frac{1}{\beta}\text{In}(\alpha\beta) + \frac{1}{\beta}\text{Int} \tag{8}$$

Table 3 Forms of pseudo-second-order expression from which q_e and k_2 may be determined

Type	Form	Graph plot	Parameters
1	$\frac{t}{q} = \frac{1}{kq_e^2} + \frac{1}{q_e}t$	$\frac{t}{q_t}$VS.t	$q_e = \frac{1}{\text{Slope}}$; $K_2 = \frac{\text{Slope}^2}{\text{Intercept}}$; $h = \frac{1}{\text{Intercept}}$
2	$\frac{1}{q} = \left(\frac{1}{kq_e^2}\right)\frac{1}{t} + \frac{1}{q_e}$	$\frac{1}{q_t}$VS.$\frac{1}{t}$	$q_e = \frac{1}{\text{Intercept}}$; $K_2 = \frac{\text{Intercept}^2}{\text{Slope}}$; $R = \frac{1}{\text{Slope}}$
3	$\frac{1}{t} = \frac{K_2 q_e^2}{q} - \frac{K_2 q_e^2}{q_e}$	$\frac{1}{t}$VS.$\frac{1}{q_t}$	$q_e = \frac{-\text{Slope}}{\text{Intercept}}$; $K_2 = \frac{\text{Intercept}^2}{\text{Slope}}$; $R = \text{Slope}$
4	$\frac{q}{t} = K_2 q_e^2 - \frac{K_2 q_e^2 \cdot q}{q_e}$	$\frac{q_t}{t}$VS.q_t	$q_e = \frac{-\text{Intercept}}{\text{Slope}}$; $K_2 = \frac{\text{Slope}^2}{\text{Intercept}}$; $R = \text{Intercept}$

where α is the rate (mg g^{-1} min^{-1}), and β is the magnitude of coverage (g mg^{-1}) that are determined from the intercept and slope of the plot of q_t vs Int.

3 Preparation of Magnetic-Based Biocomposite

Preparation of magnetic-based biocomposite may be achieved using several methods which cover chemical vapour deposition, co-precipitation, solvothermal, microwave-assisted, hydrothermal, photolysis, sonochemical, sol–gel, laser pyrolysis, annealing, microemulsion, electrodeposition and thermal deposition [17, 32, 43, 45, 50, 67, 68, 70]. Several starting materials as feedstock have been reported in the synthesis of magnetic-based biocomposites; some of this includes biomass such as agricultural products and wastes containing lignin, cellulose and hemicellulose with polyphenolic functional groups which bind with dye via different mechanisms. Recently, the biomasses are converted to carbon nanoparticles with the inclusion of magnetic metal oxide to impose magnetic properties on the composite. Biocomposites produced this way have improved particle size and shape distribution, crystallinity, structure, pore size, surface area and stability. A large number of magnetic nanomaterials have been prepared and used in the removal of dyes from wastewater [54]; some of these include the use of metallic nanomaterials (nano-Fe_2O_4, nano-$MnFe_2O_4$, nano-ZnO, nano-TiO_2, etc.), carbonaceous nanomaterials (graphene, carbon nanotubes, etc.) and bionanomaterials (nanochitosan, nanocellulose, etc.).

A study by Wang et al. [62] reported the preparation of a magnetic composite adsorbent using graphene multi-walled carbon nanotube for the removal of methylene blue from solution. The composite exhibited an adsorption capacity of 65.79 mg g^{-1} with good regeneration capacity. Another study [13] made use of β-Cyclodextrin and carbon in the preparation of a magnetic composite to remove methylene blue from the solution. The magnetic properties exhibited by the biocomposites may be determined by applying an external magnetic field to them. The orientation of magnetic moment displayed by the biocomposites helps identify the magnetism they possess, which may be ferromagnetism, diamagnetism, ferrimagnetism, ferromagnetism or anti-ferromagnetism [5]. The magnetism exhibited by magnetic biocomposites can be attributed to their high surface-area-to-volume ratio [57]. The property varies with the size and shape of the composite. The specific magnetic property displayed by a biocomposite is mainly determined by its composition, which depends on the presence or absence of unpaired valence electrons located on the metal atoms or ions found in the composite [83]. Magnetic metal oxide biocomposites have shown high adsorption capacity towards dye contaminated water. They were earlier prepared by co-precipitation, and have been found to be environmentally friendly and cost-effective [4]. Their particle size, porous structure and outstanding magnetic properties make their recovery by magnetic separation technology after adsorption or regeneration easy. This ease of recovery gives the use of magnetic-based biocomposite an advantage over powder adsorbents in the removal of dyes from wastewater.

4 Magnetic-Based Biocomposite as Dye Adsorbent

One of the major advantages of using magnetic-based biocomposites is the short treatment time. A large volume of dye contaminated water can be cleaned within a short time, and process recovery is superb. Since synthetic magnetic materials may be expensive, attention has shifted towards the use of bio-based materials in order to obtain low-cost magnetic material, which has led to the preparation of magnetic-based biocomposite. A previous study [42] revealed xanthan gum grafted poly (acrylic acid)-based Fe_3O_4 hydrogel nanocomposite as a suitable material for the sorption of methyl violet with an adsorption capacity of 642 mg g^{-1}. A reported study [41] further showed the synthesis and application of gum Ghatti–based iron oxide hydrogel nanocomposite as remediation means for methylene blue contaminated water; the biocomposite exhibited a maximum adsorption capacity of 671.14 mg g^{-1}. Recently, the preparation of magnetic hydrogel biocomposite via free radical polymerization method for the sorption of methylene blue was reported [38]. The hydrogel biocomposite exhibited a high maximum adsorption capacity (1109.55 mg g^{-1}) in an adsorption process that may be described spontaneous and endothermic in nature. A study [37] revealed the synthesis of magnetic composite based on saponite clays with magnetite (2–7 wt%) content; characterization of the composite showed a more developed mesoporosity and microporosity when compared to saponite clay with a capacity to remove malachite green, congo red and indigo carmine from solution.

In order to reduce the production cost of magnetic-based biocomposite, it is preferable to use waste materials as feedstock. Industrial and agricultural waste materials could be used. They can be used as a low-cost adsorbent; however, the introduction of magnetic properties helps in improving the performance of the resulting biocomposite. The agricultural waste may include leaves, fruits peels, seeds and fibres. This may also include waste materials like sawdust, stem bark and wood shavings which may serve as cheap sources of lignin, cellulose and hemicellulose required for the biocomposite preparation. An example of this is the preparation of magnetically modified spent coffee ground for the sorption of safranin O, amido black 10B, crystal violet, bismarck brown Y, congo red, acridine orange and malachite green [50]. Organic dyes have also been removed from the solution using ferrofluid-modified peanut husks with encouraging adsorption capacity in a process that may be described by Langmuir isotherm [51]. Biodegradable magnetic-based adsorbent has also been prepared mainly from glutamic acid and chitosan [75]; the resulting glutamic acid-modified chitosan magnetic biocomposite microspheres were used for the removal of methylene blue, light yellow and crystal violet from solution with an adsorption capacity which was as high as 375.4 mg g^{-1}. The impressive adsorption capacity exhibited by the glutamic acid-modified chitosan magnetic biocomposite was attributed to its high surface area. The dyes were further desorbed from the composite, which was regenerated using simple regenerating agents (HCl and NaOH). Novel magnetic carbon-iron oxide nanocomposite was reported [58] for the removal of crystal violet dye from solution; the sorption capacity was determined to be 113.31 mg $g^{-}1$ at an optimal condition in a process that obeys Langmuir isotherm. Recently, a safe and cost-effective method was developed for the treatment of dye contaminated water system [53]. It involved the synthesis of magnetic nanocomposite-based polyacrylic acid cross-linked with magnetic 3D cross-linkers and carboxylated cellulose, as described in Fig. 1. During the synthesis, acrylic functionalized Fe_3O_4 particles were covalently linked to the polymer chains to introduce the magnetic properties; the inclusion of highly dispersed carboxylated cellulose nanocrystals reduced the gel-like properties of the nanocomposite. The composite was found efficient towards the removal of cationic dyes from water with a maximum adsorption capacity of 332 mg g^{-1}. Different magnetic-based composites previously reported in the past and their adsorption capacities are presented in Table 4.

The use of gum xanthan-grafted-polyacrylic acid and Fe_3O_4 magnetic nanoparticles based nanocomposite hydrogel have also been documented for the removal of methyl violet from aqueous solution [41]. The proposed mechanism for the synthesis is as shown in Fig. 2. The synthesized Fe_3O_4 magnetic nanoparticles based nanocomposite hydrogel exhibited a 99% removal of methyl violet from solution with an adsorption capacity of 642 mg g^{-1}, which may be described by a pseudo-second-order kinetics model. Similarly, a study [60] recently reported a new hydrogel nanocomposite based on chitosan-grafted-polyacrylic acid/oxidized electrospun carbon nanofibres; an illustration of the preparation steps is shown in Fig. 3. The hydrogel composite exhibited a high adsorption capacity of 1095 mg g^{-1} towards methylene blue and good regeneration capacity with promising potentials as a prospective low-cost adsorbent for treating wastewater.

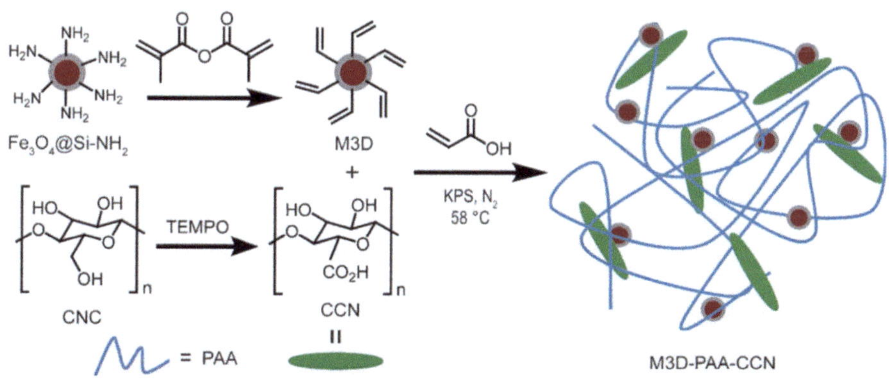

Fig. 1 Schematic illustration of the synthesis of M3D–PAA–CCN nanocomposite [53]. PAA = Polyacrylic acid, CNC = Carboxylated cellulose nanocrystals, CCN = carboxylated cellulose nanocrystals, M3D–PAA–CCN = Magnetic 3D-polyacrylic acid-carboxylated cellulose nanocrystals

Table 4 Previously reported magnetic-based composites for dye removal, their adsorption isotherm, kinetic models and capacities

Adsorbent	Dye	Isotherm model	Kinetic model	Adsorption capacity (mg g^{-1})	References
Magnetic Fe_3O_4 @ GPTMS and Lys	Orange I	Langmuir	Pseudo 2nd order	83	[81]
Magnetic Fe_3O_4 @ GPTMS and Lys	Acid red 18	Langmuir	Pseudo 2nd order	71	[81]
Magnetic Fe_3O_4 @ GPTMS and Lys	Azure I	Langmuir	Pseudo 2nd order	190	[81]
Magnetic Fe_3O_4 @ GPTMS and Lys	Methyl blue	Langmuir	Pseudo 2nd order	185	[81]
Graphene-Fe_3O_4 nanocomposite	Pararosaniline	Langmuir	Pseudo 2nd order	198.23	[66]
Magnetic Fe_3O_4 @ graphene	Congo red	Langmuir	Pseudo 2nd order	33.06	[77]
Magnetic graphene oxide	orange G	Langmuir	Pseudo 2nd order	20.85	[20]
Magnetic Zn Fe_2O4 nanoparticles	Acid red 88	Langmuir	Pseudo 2nd order	111.1	[30]
Chitosan supported CNT-MNPs	Acid red 18	Langmuir	Redlich–Peterson	809.9	[64]
EDTAD modified magnetic chitosan	Methylene blue	Sipss	Pseudo 2nd order	113.26	[69]
Fe_3O_4/CeO_2 (Fe/Ce) nanocomposite	Acid black 210	Langmuir	Pseudo 2nd order	90.50	[23]
M3D–PAA–CCN	Methylene blue	Langmuir	Pseudo 2nd order	332	[53]
Magnetic hydrogel nanocomposite	Methylene blue	Langmuir	Pseudo 2nd order	1081.60	[38]
Magnetic graphene oxide composite	Acid blue 113	Langmuir	Pseudo 2nd order	32.2	[78]
Fe_3O_4/AC/CD/Alg	Methylene blue	Langmuir	Pseudo 2nd order	10.63	[72]
S-doped Fe_2O_3/C nanocomposite	Congo red	Langmuir	Pseudo 1st order	270.20	[29]
Rice bran-based magnetic composite	Reactive blue 4	Langmuir	Pseudo 2nd order	218.82	[35]
Rice bran-based magnetic composite	Crystal violet	Langmuir	Pseudo 2nd order	159.24	[35]
Strontium ferrite-bentonite-composite	Eriochrome black T	Freundlich	Pseudo 2nd order	329.61	[21]

(continued)

Table 4 (continued)

Adsorbent	Dye	Isotherm model	Kinetic model	Adsorption capacity (mg g^{-1})	References
Strontium ferrite-bentonite-composite	Methyl orange	Freundlich	Pseudo 2nd order	219.56	[21]

Fe3O4/AC/CD/Alg = magnetic iron oxide (Fe3O4)/activated charcoal (AC)/β-cyclodextrin (CD)/sodium alginate (Alg) polymer nanocomposite, M3D–PAA–CCN = Magnetic 3D-polyacrylic acid-carboxylated cellulose nanocrystals, GPTMS = toxic 3-glycidoxypropyltrimethoxysilane, Lys = Lysine, MNPs = magnetic nanoparticles

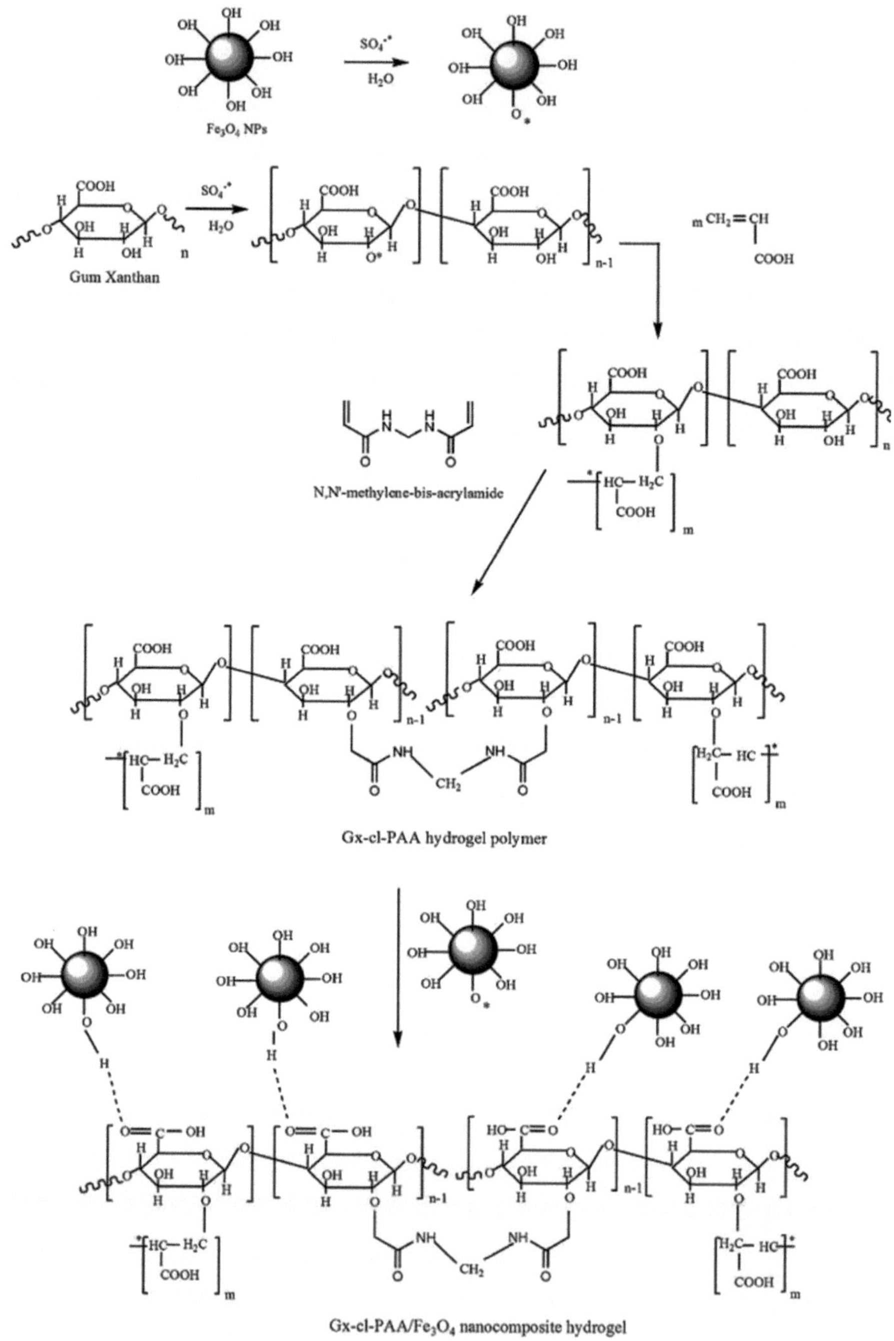

Fig. 2 Proposed mechanism for the synthesis of Fe_3O_4 magnetic nanoparticles based nanocomposite hydrogel [41]

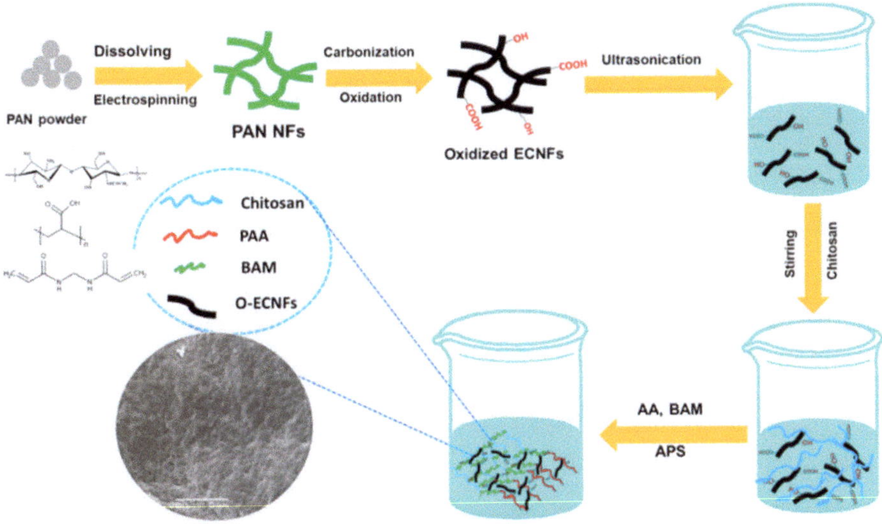

Fig. 3 Illustration for the preparation of hydrogel nanocomposite [60]

5 Factors Affecting the Efficiency of Magnetic-Based Biocomposites for the Adsorption of Dye

5.1 Physicochemical Properties of Magnetic-Based Biocomposite as Dye Adsorbent

The physicochemical properties exhibited by magnetic-based biocomposite plays a key role in their efficiency. In this regard, the types of functional groups on the surface of the biocomposite determines the kind of interaction that will be involved in the adsorption process. The type of functional group expressed by the biocomposite is also determined by the composition of the composite; some of this may have groups such as amine, carboxyl, ester, hydroxyl and so on. The presence of heteroatoms in the functional groups contributes to nonbonding electrons that may promote the interaction required to remove the dyes from the solution. However, these functional groups may be modified in order to improve the efficiency of magnetic-based biocomposite towards the dyes. The surface and porosity of magnetic-based biocomposite play an important role in sorption capacity expressed towards dyes. The larger the active surface area available for adsorption, the better the efficiency expressed towards the dyes. The distribution of pore in the structure of magnetic-based biocomposite is also important as they can trap dye molecules during the sorption process. The more and better distributed the pores are, the better the sorption process for dye removal.

The hydrophilicity of biocomposite is essential. The better the hydrophilic property, the better they interact with dye in solution for removal. Magnetic-based biocomposite with polar functional groups exhibits better hydrophilicity, thus better wettability which reduces the repulsive forces between the surface of the biocomposite and the dye molecules in solution. The nature of dye in solution has a vital role to play in the adsorption process. The molecular size, functional groups and molecular structure of the dyes play important role as well during such interaction. The charge on the surface of the dyes when ionized in solution is important as this determines the kind of interaction that will promote its removal from solution; while some dyes are neutral, some are cationic or anionic.

6 Effect of Operating Parameters on the Sorption of Dyes by Magnetic-Based Biocomposite

Several operation parameters are known to affect the performance of magnetic-based biocomposite towards removing dyes in solution. These parameters include initial dye concentration, temperature, contact time, solution pH and adsorbent dosage.

Efficiency exhibited by magnetic-based biocomposite depends on the interaction between the initial concentrations of dye in solution and the available sorption site on the surface of magnetic-based biocomposite. Previous studies have shown that the adsorption capacity of magnetic-based biocomposite increases as the initial concentration of dye in the solution increase [54]. It has become apparent that the high initial dye concentration promotes high mass transfer driving force for the sorption process [22].

The rate of a chemical reaction can be greatly influenced by temperature; therefore, temperature plays a significant role in understanding whether the sorption process is exothermic or endothermic. The adsorption process is considered to be endothermic if the adsorption capacity increases with an increase in temperature [55]. For the endothermic process, the interaction between the surface of magnetic-based biocomposite and the dyes becomes stronger with an increase in mobility of the dyes as temperature increases. However, if the adsorption capacity increases with a decrease in temperature, the process is considered to be exothermic. Therefore, temperature is a vital contributing factor in understanding the sorption process.

The effect of contact time helps in ascertaining the optimum time required for the dyes to completely occupy the active site for adsorption at the surface of the magnetic-based biocomposite. It may also relate to the time it takes to remove the dye in solution completely. Studies have shown that adsorption capacity and percentage removal improves with time [1–3]. The process is known to increase until equilibrium is attained; at equilibrium, it becomes steady; however, if allowed for long, desorption of dye from the surface of the biocomposite may set in.

Due to the vital impact of pH value on the ionization degree of functional groups, pH is considered to play an important role in understanding the mechanism and efficiency of adsorption. In most cases, a change in pH value brings about a change in the charges on the surface of magnetic-based biocomposite. The change in charges helps to develop the most suitable conditions for the sorption of dye from solution. A study has shown that the removal efficiency of anionic dyes increase at low pH value while the reverse is the case for cationic dyes [52]. The percentage of removal increases as the adsorbent dose increase while the adsorption capacity reduces. The observation has been related to the fact that as adsorbent dose increases, more active sites become available on the surface of the adsorbent for sorption with the dyes in solution; however, the reduction in adsorption capacity may be attributed to a decrease in the mass transfer of the dye ions from solution to the surface of the adsorbent [7].

7 Desorption

Apart from outstanding performance exhibited by magnetic-based biocomposites, it is much more important that it should have a high regeneration capacity. Therefore, the ability to reuse a biocomposite for a long time before being discarded is vital. The desorption process facilitates the recovery of the dyes in a form that can be re-concentrated and used again. The desorption process further helps to minimize waste generated as the dye adsorbed is recovered without being thrown into the environment. The fact that the dyes may be recovered in a form that can be reused minimizes the process cost. The desorption mode may also aid the understanding of the adsorption process. The desorption process may be achieved using methods that include an acid, organic solvent, base, thermal and mechanical treatments. Some of the acids used include H_2SO_4, HCl, HNO_3 and H_3PO_4, while base may be NaOH and liquid NH_3. The typical organic solvent used may include methanol, acetonitrile, hexane and ethanol.

In most cases, a weak solution of the acid or base is used or a solution of a mixture of acid and organic solvent, depending on the polarity of the adsorbed dye. This may involve agitating the dye-saturated magnetic-based biocomposite with the appropriate solvent to desorb the dye under specified operating conditions. Most desorption or regeneration methods may have shortcomings that can be circumvented; moreover, most studies were conducted at laboratory levels which suggest the need for more studies to be conducted on desorption at a large-scale level.

8 Challenges and Economic Perspective

The economic viability of an adsorbent is essential in its consideration for practical application. However, there is scanty information on the cost evaluation of most reported magnetic-based biocomposites in the literature. Most studies reported laboratory-based experimental studies and not cost implications or large-scale studies. One major challenge with the practical industrial application of magnetic-based biocomposite is that information on large-scale application is limited, making it difficult to make a detailed comparison with conventional or known adsorbents currently found in the market or in use on a large scale for the removal of dyes in solution. In most cases where low-cost is used to refer to magnetic-based biocomposite, it only meant the initial cost of production at a small scale. However, there is a need to consider and include local availability, transportation, desorption, pretreatment, regeneration cycle and running cost in order to better estimate the actual production cost of magnetic-based biocomposites.

9 Conclusion

This chapter considers the role magnetic-based biocomposite plays in the removal of dyes from solution. It evaluated the preparation, characteristics and application of magnetic-based biocomposite as adsorbents for dye removal in solution. It became evident that magnetic-based biocomposites are potential adsorbents for the removal of dyes in water. However, little information is available on the evaluation and application on a large scale, as most reports have been on laboratory-scale studies. Therefore, there is a need to conduct more studies on the large-scale application of magnetic-based biocomposite as an adsorbent for the removal of dyes in water as well as its economic viability.

References

1. Adewuyi A, Gopfert A, Adewuyi OA, Wolff T (2016) Adsorption of 2-chlorophenol onto the surface of underutilized seed of Adenopus breviflorus: A potential means of treating waste water. J Envron Chem Eng 4:664–672. https://doi.org/10.1016/j.jece.2015.12.012
2. Adewuyi A, Oderinde RA (2019) Chemically modified vermiculite clay: A means to removing emerging contaminant from polluted water system in developing nation. Polym Bull 76:4967–4989. https://doi.org/10.1007/s00289-018-2643-0
3. Adewuyi A, Oderinde RA (2019) Chemically modified vermiculite clay: a means to remove emerging contaminant from polluted water system in developing nation. J Polym Bull 76:4967–4989. https://doi.org/10.1007/s00289-018-2643-0
4. Adewuyi A, Yusuf A, Lau WJ, Hojamberdiev M, Oderinde RA (2020) Synthesis of amine imprinted manganese ferrite and its application in the removal of free fatty acid from waste vegetable oil. Surf Interf 21:100715. https://doi.org/10.1016/j.surfin.2020.100715
5. Akbarzadeh A, Samiei M, Davaran S (2012) Magnetic nanoparticles: preparation, physical properties, and applications in biomedicine. Nanoscale Res Lett 7:144. https://doi.org/10.1186/1556-276X-7-144
6. Azha SF, Sellaoui L, Engku EHE, Yee CJ, Bonilla- A, Lamine AB, Ismail S (2019) Iron-modified composite adsorbent coating for azo dye removal and its regeneration by photo-Fenton process: synthesis, characterization and adsorption mechanism interpretation. Chem Eng J 361:31–40. https://doi.org/10.1016/j.cej.2018.12.050

7. Banerjee S, Chattopadhyaya MC (2017) Adsorption characteristics for the removal of a toxic dye tartazine from aqueous solutions by a low cost agricultural by-product. Arab J Chem 10:S1629–S1638. https://doi.org/10.1016/j.arabjc.2013.06.005
8. Bharathi KS, Ramesh ST (2013) Removal of dyes using agricultural waste as low-cost adsorbents: a review. Appl Water Sci 3:773–790. https://doi.org/10.1007/s13201-013-0117-y
9. Briao GV, Jahn SL, Foletto EL, Dotto GL (2018) Highly efficient and reusable mesoporous zeolite synthetized from a biopolymer for cationic dyes adsorption. Colloids Surf A: Physicochem Eng Aspects 556:43–50. https://doi.org/10.1016/j.colsurfa.2018.08.019
10. Carneiro PA, Umbuzeiro GA, Oliveira DP, Zanoni MVB (2010) Assessment of water contamination caused by a mutagenic textile effluent/dyehouse effluent bearing disperse dyes. J Hazard Mat 174:694–699. https://doi.org/10.1016/j.jhazmat.2009.09.106
11. Chaari I, Fakhfakh E, Medhioub M, Jamoussi F (2019) Comparative study on adsorption of cationic and anionic dyes by smectite rich natural clays. J Mol Struct 1179:672–677. https://doi.org/10.1016/j.molstruc.2018.11.039
12. Chen W, Chen Y, Huang H, Mahdi Y, Khorram S, Zhao W, Wang D, Qi S, Zhang BJG (2019) Occurrence of *N*-Nitrosamines in the Pearl River delta of China: characterization and evaluation of different sources. Water Res 164:114896. https://doi.org/10.1016/j.watres.2019.114896
13. Cheng J, Chang PR, Zheng PW, Ma XF (2014) Characterization of magnetic carbon nanotube-cyclodextrin composite and its adsorption of dye. Ind Eng Chem Res 53:1415–1421. https://doi.org/10.1021/ie402658x
14. Colin N, Maceda-Veiga A, Flor- N, Mora J, Fortuño P, Vieira C, Prat N, Cambra J, de Sostoa A (2016) Ecological impact and recovery of a Mediterranean river after receiving the effluent from a textile dyeing industry. Ecotoxicol Environ Saf 132:295–303. https://doi.org/10.1016/j.ecoenv.2016.06.017
15. Corrêa GT, de Souza JC, Silva JP, Pividori MI, Zanoni MVB (2020) Determination of temporary dye Basic Red 51 in commercial hair dye, river water and wastewater from hairdressing salon using graphite-epoxy composite electrode modifed with magnetic nanoparticles. Microchemical J 159:105485. https://doi.org/10.1016/j.microc.2020.105485
16. da Rosa SR, da Rosa BC, Goncalves JO, Pinto LAA, Mallmann ES, Dotto GL (2019) Synthesis of a bio-based polyurethane/chitosan composite foam using ricinoleic acid for the adsorption of Food Red 17 dye. Int J Biol Macromol 121:373–380. https://doi.org/10.1016/j.ijbiomac.2018.09.186
17. Daengsakul S, Mongkolkachit C, Thomas C, Siri S, Thomas I, Amornkitbamrung V, Maensiri S (2009) A simple thermal decomposition synthesis, magnetic properties, and cytotoxicity of $La_{0.7}Sr_{0.3}MnO_3$ nanoparticles. Appl Phys A 96:691–699. https://doi.org/10.1007/s00339-009-5151-0
18. de Aragao UG, Freeman HS, Warren SH, Oliveira DP, Terao Y, Watanabe T, Claxton LD (2005) The contribution of azo dyes to the mutagenic activity of the Cristais river. Chemosphere 60:55–64. https://doi.org/10.1016/j.chemosphere.2004.11.100
19. de Souza JC, da Silva BF, Morales DA, Umbuzeiro GA,Zanoni MVB (2020) Assessment of p-aminophenol oxidation by simulating the process of hair dyeing and occurrence in hair salon wastewater and drinking water from treatment plant. 387:122000. https://doi.org/10.1016/j.jhazmat.2019.122000
20. Deng JH, Zhang XR, Zeng GM, Gong JL, Niu QY, Liang J (2013) Simultaneous removal of Cd (II) and ionic dyes from aqueous solution using magnetic graphene oxide nanocomposite as an adsorbent. Chem Eng J 226:189–200. https://doi.org/10.1016/j.cej.2013.04.045
21. Elkhider KHA, Ihsanullah I, Zubair M, Manzar MS, Mu'azu ND, Al-Harthi MA (2020) Synthesis characterization and dye adsorption performance of strontium ferrite decorated bentonite-CoNiAl magnetic composite. Arab J Sci Eng 45:7397–7408. https://doi.org/10.1007/s13369-020-04544-0
22. Eren Z, Acar FN (2006) Adsorption of reactive black 5 from an aqueous solution: equilibrium and kinetic studies. Desalination 194:1–10. https://doi.org/10.1016/j.desal.2005.10.022
23. Gao S, Zhang W, An Z, Kong S, Chen D (2019) Adsorption of anionic dye onto magnetic Fe_3O_4/CeO_2 nanocomposite: equilibrium, kinetics, and thermodynamics 37:185–204. https://doi.org/10.1177/0263617418819164

24. Habiba U, Siddique TA, Li Lee JJ, Joo TC, Ang BC, Afifi AM (2018) Adsorption study of methyl orange by chitosan/polyvinyl alcohol/zeolite electrospun composite nanofibrous membrane. Carbohydr Polym 191:79–85. https://doi.org/10.1016/j.carbpol.2018.02.081
25. Hashemian S, Dehghanpor A, Moghahed M (2015) $Cu_{0.5}Mn_{0.5}Fe_2O_4$ nano spinels as potential sorbent for adsorption of brilliant green. J Ind Eng Chem 24:308–314. https://doi.org/10.1016/j.jiec.2014.10.001
26. Hassan MM, Carr CM (2018) A critical review on recent advancements of the removal of reactive dyes from dyehouse effluent by ion-exchange adsorbents. Chemosphere 209:201–219. https://doi.org/10.1016/j.chemosphere.2018.06.043
27. Ito T, Adachi Y, Yamanashi Y, Shimada Y (2016) Long–term natural remediation process in textile dye-polluted river sediment driven by bacterial community changes. Water Res 100:458–465. https://doi.org/10.1016/j.watres.2016.05.050
28. Kannan N, Rengasamy G (2005) Comparison of cadmium ion adsorption on various activated carbons. Water Air Soil Pollut 163:185–201. https://doi.org/10.1007/s11270-005-0277-y
29. Khoshsang H, Ghaffarinejad A, Kazemi H, Wang Y, Arandiyan H (2018) One-pot synthesis of S-doped Fe_2O_3/C magnetic nanocomposite as an adsorbent for anionic dye removal: equilibrium and kinetic studies. J Nanostruct Chem 8:23–32. https://doi.org/10.1007/s40097-017-0251-4
30. Konicki W, Sibera D, Mijowska E, Lendzion-Bielun Z, Narkiewicz U (2013) Equilibrium and kinetic studies on acid dye acid red 88 adsorption by magnetic $ZnFe_2O_4$ spinel ferrite nanoparticles. J Colloids Interface Sci 398:152–160. https://doi.org/10.1016/j.jcis.2013.02.021
31. Lellis B, Fávaro-Polonio CZ, Pamphile JA, Polonio JC (2019) Effects of textile dyes on health and the environment and bioremediation potential of living organisms. Biotechnol Res Innov 3:275–290. https://doi.org/10.1016/j.biori.2019.09.001
32. Li C, Wei Y, Liivat A, Zhub Y, Zhu J (2013) Microwave-solvothermal synthesis of Fe_3O_4 magnetic nanoparticles. Mater Lett 107:23–26. https://doi.org/10.1016/j.matlet.2013.05.117
33. Li TT, Shu Z, Zhou J, Chen Y, Yu DX, Yuan XM, Wang YX (2015) Template free synthesis of kaolin-based mesoporous silica with improved specific surface area by a novel approach. Appl Clay Sci 107:182–187. https://doi.org/10.1016/j.clay.2015.01.022
34. Liu J, Li W, Kang X, Zhao F, He M, She Y, Zhou Y (2021) Profiling by HPLC-DAD-MSD reveals a 2500-year history of the use of natural dyes in Northwest China. Dyes Pigm 187:109143. https://doi.org/10.1016/j.dyepig.2021.109143
35. Ma CM, Hong GB, Wang YK (2020) Performance evaluation and optimization of dyes removal using rice bran-based magnetic composite adsorbent. Materials 13:2764. https://doi.org/10.3390/ma13122764
36. Mahapatra NN (2016) Textile dyes. Woodhead Publishing India Pvt, Boca Raton, CRC Press, New Delhi
37. Makarchuk OV, Dontsova TA, Astrelin IM (2016) Magnetic nanocomposites as efficient sorption materials for removing dyes from aqueous solutions. Nanoscale Res Lett 11:161. https://doi.org/10.1186/s11671-016-1364-2
38. Malatji N, Makhado E, Ramohlola KE, Modibane KD, Maponya TC, Monama GR, Hato MJ (2020) Synthesis and characterization of magnetic clay-based carboxymethyl cellulose-acrylic acid hydrogel nanocomposite for methylene blue dye removal from aqueous solution. Environ Sci Pollut Res 27:44089–44105. https://doi.org/10.1007/s11356-020-10166-8
39. Mekkawy HA, Ali MO, El- AM (1998) Toxic effect of synthetic and natural food dyes on renal and hepatic functions in rats. Toxicol Lett 95:155–161. https://doi.org/10.1016/S0378-4274(98)80621-8
40. Methneni N, Gonzalez JAM, Loco JV, Anthonissen R, Maele JV, Verschaeve L, Fernandez-Serrano M, Mansour HB (2021) Ecotoxicity profile of heavily contaminated surface water of two rivers in Tunisia. Environ Toxicol Pharmacol 82:103550. https://doi.org/10.1016/j.etap.2020.103550
41. Mittal H, Kumar V, Saruchi RSS (2016) Adsorption of methyl violet from aqueous solution using gum xanthan/Fe_3O_4 based nanocomposite hydrogel. Int J Biol Macromol 89:1–11. https://doi.org/10.1016/j.ijbiomac.2016.04.050

42. Mittal H, Ray S (2016) A study on the adsorption of methylene blue onto gum ghatti/TiO_2 nanoparticles-based hydrogel nanocomposite. Inter j biol macromol 88:66–80. https://doi.org/10.1016/j.ijbiomac.2016.03.032
43. Mojica Pisciotti ML (2015) Desarrollo de nanopartículas magnéticas para su utilización en el tratamiento médico: Hipertermia. Universidad Nacional de Cuyo
44. Munir M, Nazar MF, Zafar MN, Zubair M, Ashfaq M, Hosseini- A, Khan SUD, Ahmad A (2020) Effective adsorptive removal of methylene blue from water by didodecyldimethylammonium bromide-modified brown clay. ACS Omega 5:16711–16721. https://doi.org/10.1021/acsomega.0c01613
45. Nejati K, Zabihi R (2012) Preparation and magnetic properties of nano size nickel ferrite particles using hydrothermal method. Chem Cent J 6:23
46. Ngulube T, Gumbo JR, Masindi V, Maity A (2017) An update on synthetic dyes adsorption onto clay based minerals: a state-of-art review. J Environ Manag 191:35–57. https://doi.org/10.1016/j.jenvman.2016.12.031
47. Orts F, del Río AI, Molina J, Bonastre J, Cases F (2018) Electrochemical treatment of real textile wastewater: Trichromy Procion HEXL®. J Electroanaly Chem 808:387–394. https://doi.org/10.1016/j.jelechem.2017.06.051
48. Puri C, Sumana G (2018) Highly effective adsorption of crystal violet dye from contaminated water using graphene oxide intercalated montmorillonite nanocomposite. Appl Clay Sci 166:102–112. https://doi.org/10.1016/j.clay.2018.09.012
49. Raval NP, Shah PU, Shah NK (2016) Adsorptive amputation of hazardous azo dye congo red from wastewater: a critical review. Environ Sci Pollut Res 23:14810–14853. https://doi.org/10.1007/s11356-016-6970-0
50. Safarik I, Horska K, Svobodova B, Safarikova M (2012) Magnetically modified spent coffee grounds for dyes removal. Eur Food Res Technol 234:345–350. https://doi.org/10.1007/s00217-011-1641-3
51. Safarik I, Safarikova M (2010) Magnetic fluid modified peanut husks as an adsorbent for organic dyes removal. Phys Proc 9:274–278. https://doi.org/10.1016/j.phpro.2010.11.061
52. Salleh MAM, Mahmoud DK, Karim WAWA, Idris A (2011) Cationic and anionic dye adsorption by agricultural solid wastes: a comprehensive review. Desalination 280:1–13. https://doi.org/10.1016/j.desal.2011.07.019
53. Samadder R, Akter N, Roy AC, Uddin MM, Hossen MJ, Azam MS (2020) Magnetic nanocomposite based on polyacrylic acid and carboxylated cellulose nanocrystal for the removal of cationic dye. RSC Adv 10:11945–11956. https://doi.org/10.1039/D0RA00604A
54. Santhosh C, Velmurugan V, Jacob G, Jeong SK, Grace AN, Bhatnagar A (2016) Role of nanomaterials in water treatment applications: a review. Chem Eng J 306:1116–1137. https://doi.org/10.1016/j.cej.2016.08.053
55. Senthilkumaar S, Kalaamani P, Subburaam CV (2006) Liquid phase adsorption of crystal violet onto activated carbons derived from male flowers of coconut tree. J Hazard Mater 136:800–808. https://doi.org/10.1016/j.jhazmat.2006.01.045
56. Shamraiz U, Hussain RA, Badshah A, Raza B, Saba S (2016) Functional metal sulfides and selenides for the removal of hazardous dyes from Water. J Photochem Photobiol B 159:33–41. https://doi.org/10.1016/j.jphotobiol.2016.03.013
57. Singamaneni S, Bliznyuk VN, Binek C, Tsymbal EY (2011) Magnetic nanoparticles: recent advances in synthesis, self-assembly and applications. J Mater Chem 21:16819–16845. https://doi.org/10.1039/C1JM11845E
58. Singh KP, Gupta S, Singh AK, Sinha S (2011) Optimizing adsorption of crystal violet dye from water by magnetic nanocomposite using response surface modeling approach. J Hazard Mater 186:1462–1473. https://doi.org/10.1016/j.jhazmat.2010.12.032
59. Terdputtakun A, Arqueropanyo O, Sooksamiti P, Janhom S, Naksata W (2017) Adsorption isotherm models and error analysis for single and binary adsorption of Cd(II) and Zn(II) using leonardite as adsorbent. Environ Earth Sci 76:777. https://doi.org/10.1007/s12665-017-7110-y
60. Thamer BM, Aldalbahi A, Meera MA, El- MH (2020) In situ preparation of novel porous nanocomposite hydrogel as effective adsorbent for the removal of cationic dyes from polluted water. Polymers 12:3002. https://doi.org/10.3390/polym12123002

61. Wang M, Ma YF, Sun Y, Hong SY, Lee SK, Yoon B, Chen L, Ci LJ, Nam JD, Chen XY, Suhr J (2017) Hierarchical porous chitosan sponges as robust and recyclable adsorbents for anionic dye adsorption. Sci Rep 7:18054. https://doi.org/10.1038/s41598-017-18302-0
62. Wang PF, Cao MH, Wang C, Ao YH, Hou J, Qian J (2014) Kinetics and thermodynamics of adsorption of methylene blue by a magnetic graphene-carbon nanotube composite. Appl Surf Sci 290:116–124. https://doi.org/10.1016/j.apsusc.2013.11.010
63. Wang Q, Wang Y, Chen L (2019) A green composite hydrogel based on cellulose and clay as efficient absorbent of colored organic effluent. Carbohydr Polym 210:314–321. https://doi.org/10.1016/j.carbpol.2019.01.080
64. Wang S, Zhai YY, Gao Q, Luo WJ, Xia H, Zhou CG (2014) Highly efficient removal of acid red 18 from aqueous solution by magnetically retrievable chitosan/carbon nanotube: batch study, isotherms, kinetics, and thermodynamics. Chem Eng Data 59:39–51. https://doi.org/10.1021/je400700c
65. Wang T, Zhou Y, Cao S, Lu J, Zhou Y (2019) Degradation of sulfanilamide by Fenton-like reaction and optimization using response surface methodology. Ecotoxicol Environ Saf 172:334–340. https://doi.org/10.1016/j.ecoenv.2019.01.106
66. Wu Q, Feng C, Wang C, Wang Z (2013) A facile one-pot solvothermal method to produce superparamagnetic graphene–Fe_3O_4 nanocomposite and its application in the removal of dye from aqueous solution. Colloids Surf B: Biointerf 101:210–214. https://doi.org/10.1016/j.colsurfb.2012.05.036
67. Wu S, Sun A, Zhai F, Wang J, Xu W, Zhang Q, Volinsky AA (2011) Fe3O4 magnetic nanoparticles synthesis from tailings by ultrasonic chemical co-precipitation. Mat Lett 65:1882–1884. https://doi.org/10.1016/j.matlet.2011.03.065
68. Wu W, He Q, Chen H, Tang J, Nie L (2007) Sonochemical synthesis, structure and magnetic properties of air-stable Fe_3O_4/Au nanoparticles. Nanotechnol 18:145609. https://doi.org/10.1088/0957-4484/18/14/145609
69. Xia Y, Dai X, Huang S, Tian X, Yang H, Li Y, Liu Y, Zhao M (2013) Fast and highly efficient removal of methylene blue by a novel EDTAD-modified magnetic chitosan material. Desalin Water Treat 51:7586–7595. https://doi.org/10.1080/19443994.2013.777368
70. Xu J, Yang H, Fu W, Du K, Sui Y, Chen J, Zeng Y, Li M, Zou G (2007) Preparation and magnetic properties of magnetite nanoparticles by sol–gel method. J Magn Magn Mater 309:307–311. https://doi.org/10.1016/j.jmmm.2006.07.037
71. Xu J, Zhang F, Sun J, Sheng J, Wang F, Sun M (2014) Bio and nanomaterials based on Fe_3O_4. Molecules 19:21506–21528. https://doi.org/10.3390/molecules191221506
72. Yadav S, Asthana A, Chakraborty R, Jain B, Singh AK, Carabineiro SAC, Susan MABH (2020) Cationic dye removal using novel magnetic/activated charcoal/β-cyclodextrin/alginate polymer nanocomposite. Nanomaterials 10:170. https://doi.org/10.3390/nano10010170
73. Yadla SV, Sridevi V, Lakshmi MVVC (2012) A review on adsorption of heavy metals from aqueous solution. J Chem Biol Phys Sci 2:1585–1593
74. Yagub MT, Sen TK, Ang HM (2012) Equilibrium, kinetics, and thermodynamics of methylene blue adsorption by pine tree leaves. Water Air Soil Pollut 223:5267–5282. https://doi.org/10.1007/s11270-012-1277-3
75. Yan H, Li H, Yang H, Li A, Cheng R (2013) Removal of various cationic dyes from aqueous solutions using a kind of fully biodegradable magnetic composite microsphere. Chem Eng J 223:402–411. https://doi.org/10.1016/j.cej.2013.02.113
76. Yang Y, Yu WY, He SJ, Yu SX, Chen Y, Lu LH, Shu Z, Cui HD, Zhang Y, Jin HY (2019) Rapid adsorption of cationic dye-methylene blue on the modified montmorillonite/graphene oxide composites. Appl Clay Sci 168:304–311. https://doi.org/10.1016/j.clay.2018.11.013
77. Yao Y, Miao S, Liu S, Ma LP, Sun H, Wang S (2012) Synthesis, characterization, and adsorption properties of magnetic Fe_3O_4@graphene nanocomposite. Chem Eng J 184:326–332. https://doi.org/10.1016/j.cej.2011.12.017
78. Ying TY, Raman AAA, Bello MM, Buthiyappan A (2020) Magnetic graphene oxide-biomass activated carbon composite for dye removal. Korean J Chem Eng 37:2179–2191. https://doi.org/10.1007/s11814-020-0628-9

79. Zaini MAA, Cher TY, Zakaria M, Kamaruddin MJ, Setapar SHM, Yunus MAC (2014) Palm oil mill effluent sludge ash as adsorbent for methylene blue dye removal. Desalination Water Treat 52:3654–3662. https://doi.org/10.1080/19443994.2013.854041
80. Zhang L, Cheng ZJ, Guo X, Jiang XH, Liu R (2014) Process optimization, kinetics and equilibrium of orange G and acid orange 7 adsorptions onto chitosan/surfactant. J Mol Liq 197:353–367. https://doi.org/10.1016/j.molliq.2014.06.007
81. Zhang YR, Shen SL, Wang SQ, Huang J, Su P, Wang QR, Zhao BX (2014) A dual function magnetic nanomaterial modified with lysine for removal of organic dyes from water solution. Chem Eng J 239:250–256. https://doi.org/10.1016/j.cej.2013.11.022
82. Zhao YG, Li XP, Yao SS, Zhan PP, Liu JC, Xu CP, Lu YY, Chen XH, Ji MC (2016) Fast throughput determination of 21 allergenic disperse dyes from river water using reusable three-dimensional interconnected magnetic chemically modified graphene oxide followed by liquid chromatography–tandem quadrupole mass spectrometry. J Chromatogr A 1431:36–46. https://doi.org/10.1016/j.chroma.2015.12.089
83. Zhen G, Muir BW, Moffat BA, Harbour P, Murray KS, Moubaraki B, Suzuki K, Madsen I, Agron-Olshina N, Waddington L, Mulvaney P, Hartley PG (2011) Comparative study of magnetic behavior of spherical and cubic superparamagnetic iron oxide nanoparticles. J Phys Chem C 115:327–334. https://doi.org/10.1021/jp104953z
84. Zhou Y, Fang X, Wang T, Hu Y, Lu J (2017) Chelating agents enhanced CaO_2 oxidation of bisphenol A catalyzed by Fe^{3+} and reuse of ferric sludge as a source of catalyst. Chem Eng J 313:638–645. https://doi.org/10.1016/j.cej.2016.09.111

Application of Wheat Bran/Straw-Based Biomaterials in Textile Wastewater

Chandra Wahyu Purnomo, Lalak Tarbiyatun N. Maleiva, Oky Amelia Sandra, and Alexander H. Soeriyadi

Abstract Abundantly available biomass waste especially from main crops such as wheat can be utilized as a part of low-cost wastewater treatment technique. Wheat harvesting and post processes will produce at least two major solid wastes which are wheat straw and bran or husk. Wheat is the most important grain source for human food consumption with straw and bran as unused part of the wheat industry. On the other hand, we have a large amount of dye wastewater that is difficult to be treated from textile industries around the globe. In this chapter, these two potentials will be linked to provide a greener approach to address the challenges. Wheat straw (WS) and wheat bran (WB) possess unique natural characteristic that can be used directly as bio-sorbent of various dyes. The direct use of these material can be coupled with the utilization of saturated WS or WB as substrate for fungus or bacteria to be completely degraded the dye molecules to become nontoxic ones. Advanced treatment of WS and WB can also be done to prepare super bio-sorbent to improve the sorption performance as a perfect dye removal. Several modifications techniques are surfactant modification, acid of alkali treatment, steam and milling treatment, and esterification. However, the cost of modification should be always in the consideration and should be compared with the enhancement that can be achieved. Indeed, almost all the processes are still in a lab scale and there is no sign for large-scale implementation yet. This chapter will investigate the challenges, feasibility, and scalability to utilize WS and WB as a low-cost bio-sorbent in the near future.

C. W. Purnomo (✉) · O. A. Sandra
Chemical Engineering Department, Universitas Gadjah Mada, Jl. Grafika no 2, Bulaksumur, Mlati, Sleman, Yogyakarta, Indonesia 55281
e-mail: chandra.purnomo@ugm.ac.id

L. T. N. Maleiva
Department of Chemical Engineering, Faculty of Engineering, Tanjungpura University, Prof. Dr.H. Hadari Nawawi street, Bansir Laut, Pontianak Tenggara, Pontianak City, West Kalimantan, Indonesia 78124
e-mail: lalaktnm@teknik.untan.ac.id

A. H. Soeriyadi
School of Chemistry, University of New South Wales, Sydney, NSW 2052, Australia
e-mail: alexander.soeriyadi@unsw.edu.au

S. S. Muthu and A. Khadir (eds.), *Textile Wastewater Treatment*, Sustainable Textiles: Production, Processing, Manufacturing & Chemistry,
https://doi.org/10.1007/978-981-19-2852-9_3

Keywords Wheat bran · Wheat straw · Bio-sorbent · Dye removal

1 Introduction

Decolorization of textile wastewater is necessary prior to discharge it into the nearby local waterways. It is essential to degrade textile wastewater's toxic chemicals and hinder the harmful environmental impacts. More than 50% of cotton manufacturing is colored with reactive dyes delivering effluents heavily colored, high concentrations of salts, and high biological oxygen demand/chemical oxygen demand (BOD/COD) [5].

From around 700,000 tons of synthetic dyes being produced in the world, about 200,000 tons of these dyes are wasted as effluents because of the ineffective dyeing and finishing processes [31]. Moreover, dyes applied in textile fabric manufacturing are mixed with various additives such as solvents, antifoaming, whitening agents, and pH conditioners [42]. The various reaction may occur among the components; hence the complexity of wastewater in textile industry.

Azo dyes constitute the largest and the most prominent commercial dyes employed in the textile, printing, tannery, paper manufacture, and photography industries. Azo dye effluents bring a severe environmental impact because their precursors and decomposition products (such as aromatic amines) are carcinogenic [40]. Many studies shows that the azo dyes are resistant to degrade under aerobic conditions and induce acute toxicity to the receiving water bodies [6, 12, 29]. The physical and chemical parameters of textile effluent referred to Bureau of Indian Standard (BIS) Permissible Limits are performed in Table 1.

Considering the large quantity and complexity of textile wastewater as described above, it is necessary to always seek any alternative method to deliver efficient and economical treatment methods. In this chapter, wheat bran and wheat straw are proposed to be used as bio-sorbents for textile wastewater.

2 Wheat Brand and Straw Characteristics

Wheat (*Triticum aestivum*) is one of the four cereal products (wheat, maize, rice, barley) which is widely consumed in the world [12]. The high protein content and the number of calories needed by humans make wheat farming the first rank of cereal agriculture in the world [1]. The agricultural industry sector will never be separated from the problem of agricultural waste. The agricultural sector constantly runs into problems in the form of the production of very large amounts of solid biomass waste. Examples of solid biomass waste are straw, bran, corn cobs, husks, and so on. The biomass waste can be used as raw material or made into a bio-based product for certain utilization.

Table 1 Physical and chemical parameters of textile effluent and BIS permissible limits [42]

Parameters	Effluent	Limits
Color	Dark violet	–
Odor	Bad odor	–
Electrical conductivity (EC)(μmho/cm)	360	600
Temperature (°C)	40	> 40
Total suspended solids (mg/L)	2000	100
Total dissolved solids (mg/L)	8000	2100
Total solids (mg/L)	10,000	–
pH	9.5	5.5–9.0
Alkalinity	430	200–600
Total hardness	460	600
Dissolved oxygen	3.5	–
BOD	90	30
COD	571	250
Chloride	1298	1000
Sulfate	1118	1000
Phosphate	2.18	5
Nitrate nitrogen	80	50
Lead	1.3	0.1
Nickel	2.50	3
Zinc	5.46	5
Chromium	1.83	2
Copper	3.21	3
Oil and gases	14.7	10

2.1 *Wheat Straw Characteristic*

Commercial-scale wheat farming has problems with the storage and treatment of hay waste. The presence of wheat straw is plentiful but has deficient nutrients. Most of the efforts in wheat straw waste have focused on the management of straw waste dealing with its abundant availability [23, 48]. Wheat straw waste is a renewable, abundant, and economical plant-based material. Complex carbohydrates in wheat straw are a source of cellulose which can be processed into bioethanol. Bioethanol derived from wheat straw is also known as 2nd generation bioethanol. Wheat straw can also be used as an adsorbent in wastewater treatment because it has a large pore surface area, economical, cost-effective, and environmentally friendly. Various studies on wastewater treatment have shown the promise of wheat straw as an impactful cheaper alternative of bio-sorbent with great performance in removal of dye waste than what is currently available in the market [2].

Wheat straw has 3 main organic components, namely cellulose (38.6%), hemicellulose (32.6%), and lignin (14.1%) or better known as the lignocellulose component. This component serves as an adsorbent in the dye removing process in the textile industry [16]. Cellulose is one of the main components of lignocellulose which consists of D-glucose monomer units that are bound to 1, 4-glycosidic bonds and have a soluble (hydrophilic) structure. Hemicellulose is a collection of several sugar units or called heteropolysaccharides and is more soluble (hydrophilic) than cellulose. Lignin is a major part of plant cell walls and is insoluble in water (hydrophobic). The adsorption process in the dye removing process using wheat straw was slightly slower than the use of apple pomace adsorbent due to the lignin complex structure which resulted in inhibition of dye absorption by wheat straw cellulose [35]. Therefore, it is necessary to have a wheat straw delignification process to optimize the adsorption process. Three different agricultural wastes, namely barley husks, corn cobs, and wheat straw have been given a delignification procedure, the results of the delignification process are effectual in increasing the surface area of the dye molecule adsorption process [34].

2.2 *Wheat Bran Characteristic*

Wheat bran is a by-product of the process of making wheat flour. Bran is famed as a food ingredient with high nutritional content and fiber, so it is great for a high-fiber food source. Wheat flour is used as a raw material for making assorted kinds of food ingredients such as bread and noodles. High market demand causes the wheat milling industry to generate large quantities of wheat bran [9]. Therefore, apart from being used as a raw material for high-fiber diet foods, wheat bran is also considered as a wastewater treatment bio-sorbent which is inexpensive, effective, and abundant. The treatment of textile waste conceiving heavy metal compounds usually uses an adsorbent of activated charcoal or uses the ion-exchange method which are costly [13]. Various studies on cheap bio-sorbents have been carried out, one of which is using wheat bran. The results of this study can be concluded that modified wheat bran is efficacious in absorbing heavy metals in liquid waste [9, 12, 15, 37].

Bran comes from the thin shell or sheath that covers the grain of wheat and has the highest nutritional content of all parts. Bran contains complex carbohydrates, fiber, iron, B vitamins, and other nutrients that are good for diet and prevent cancer [9]. Wheat bran contain carbohydrates (58%), starch (19%), protein (18%), fat (0.5%), and minerals (2%) [12]. Wheat bran does not contain lignin, so it is more cost-effective than wheat straw, which requires a delignification process. The use of wheat bran as a bio-sorbent has been proven effective in removing heavy metal content in wastewater such as cadmium [37], lead [9], and chromium (VI) [27]. Heavy metal content is found in all textile waste so that wheat bran can be used as an alternative as an adsorbent in wastewater treatment of textile waste. Wheat bran needs to be given a pretreatment in order to expand its surface area and porosity [9]. In addition, pretreatment is also required to create wheat bran at certain pH and temperature conditions before the adsorption process depending on the element or metal ion to be absorbed [15, 37].

3 Textile Wastewater Treatment Using Wheat Bran and Wheat Straw

Due to the low biodegradability of most of the dyes and chemicals utilized in the textile industry, they unbelievably resist to aerobic biological treatment of conventional activated sludge processes [30]. Thus, the adsorption process offered several merits over other treatments such as high efficiency, low-cost, selectivity, high level of purification, and retrieval of adsorbate and adsorbent [8, 38].

Adsorption process has been employed to eliminate dyes from the waste effluents. The process of adsorption is an applicable method rather than others due to its sludge-less, clean procedure and entirely eliminated dyes even from diluted solution [3]. Recently the applications of affordable adsorbents for the dyes and metal ions elimination have been observed [7].

3.1 Wheat Bran Application to Dye Remediation

The most economical adsorbents are from agricultural bio-product, for example, wheat bran (WB) that can be collected from flour mills. Structurally, wheat comprises of cellulose, hemicellulose, and lignin with large amount of hydroxyl (OH) groups [47]. These fibers also have large surface area which makes them favorable for their binding with other compounds [47]. The findings of WB utilization to remove dyes are summarized in Table 2.

WB sorption to certain dye is relatively vast, it has been reported that the equilibrium time for malachite green (MG) dye is only 40 min with removal efficiency of almost 90% [32]. This fast uptake and high capacity of dye is an important parameter for real scale application. After being saturated, WB packed with dyes can be degraded as a substrate for the growth certain fungi.

The extensive adsorption of WB is due to the carboxylic acid, alcohol, phenolic groups present in pectin and cellulose, it reveals the occurrence of hydroxyl groups on exterior of wheat bran employed for the adsorption Crystal Violet (CV) dye. CV is an eminent cancer-causing poison. CV stimulates the eyes sore, 2 queasiness, vomiting, oral herpes [43], cancer, jaundice, tissue necrosis, renal failure, everlasting injury to the cornea and oculus, dermal irritation, and alimentary tract irritation [19]. When ingested it might cause peritonitis, weight loss, birth defects, reproductive defects, and mutations in genetic material [20].

Previous work by Haq et al. [20], investigate various parameters of the WB characteristics on its efficacy, for instance, granular size of adsorbent (105, 210, and

Table 2 Wheat bran (WB) application in dye removal

Dye (conc.)	WB modification (dosage)	% Removal (contact time)	References
Malachite green (MG)	Unmodified	90	[32]
Astrazon yellow 7GL (50 mg/L)	Unmodified (4 g/L)	71,83 (30 min)	[39]
Coomassie brilliant blue (CBB) (100 mg/L)	Unmodified (2 g/L)	95.70 (30 min)	[4]
Reactive blue 19 (RB 19), reactive red 195 (RR 195), and reactive yellow 145 (RY 145) (50 mg/L)	Unmodified (1 g/L)	RB 19: 97,5 RR 195: 94,2 RY 145: 98.9 (300 min)	[13]
Acid red 18 (50 mg/L)	Surfactant CTAB, TTAB and STAB modified	80	[46]
Acid red 18, acid orange 7, and acid black 1 (50 mg/L)	Gemini surfactant modified	95	[45]
Methylene blue (100 mg/L)	Sulfuric acid dehydration (2 g/L)	99,5 (300 min)	[28]
Methylene blue (100 mg/L)	Tartaric acids modified (5 g/L)	85.76 (1 h)	[44]
Cristal violet (100 mg/L)	Acid modified and alkali modified (10 g/L)	88 (180 min)	[14]
Reactive blue RB 19 and reactive red RR 195 (10 mg/L)	Nitric acid treated (1 g/L)	RB19: 68 RR 195: 70 (90 min)	[41]

500 mesh sizes), pH (1–11), duration of contact (5–120 min), amount of wheat bran (0.1–1.0 g), initial dye concentration (300–550 ppm), and temperature (0–60°C). The computations indicate that data fits efficiently with the Langmuir isotherm model. Moreover, the adsorption data i.e., the regression value ($R^2 = 1$) clarified the good correlation with pseudo second order kinetic equation rather than others. To figure out nature of adsorption, free energy (ΔG) of adsorption of CV on WB, enthalpy (ΔH) and entropy (ΔS) exhibited that the adsorption of CV on WB was feasible, spontaneous, and endothermic in nature. Maximum removal efficiency of WB toward CV i.e., 86.57% is investigated at optimum conditions i.e., pH 2, 0.6 g of WB, 30 min of contact time and 400 ppm of initial dye concentration Optimum uptake capacity was calculated, i.e., 116.6 mg/g. Thus, the yields reveal adsorption of CV at the surface of WB is directly proportional to exposure time, amount of adsorbent, and temperature, however, inversely to the initial dye concentration [20].

Further proof removal performance of various dyes, Çiçek et al. [13] applied WB to treat various dyes such as Reactive Blue 19 (RB 19), Reactive Red 195 (RR 195), and Reactive Yellow 145 (RY 145). It is observed that electrostatic attractions between negatively charged dye anion and positively charged wheat bran surface are responsible for high uptake capacity at low pH. From kinetic adsorption results that fitted with pseudo second order kinetic, it can be a hint that the surface adsorption and intraparticle diffusion contributed to dye adsorption.

3.1.1 Modified WB

Surfactant modified WB is synthesized to improve the WB uptake performance to a specific dye. Anionic dye is difficult to be adsorbed by raw WB since the surface of the materials is full of hydroxyl groups. To change the surface to be positive charged, some surface modification of WB using cationic surfactant such as Sodium dodecyl sulfate (SDS), decyl trimethyl ammonium bromide (DeTAB), dodecyl trimethyl ammonium bromide (DTAB), tetradecyl trimethyl ammonium bromide (TTAB), hexadecyl trimethyl ammonium bromide (CTAB), and stearyl trimethyl ammonium bromide (STAB) has been performed [46]. Furthermore, the novel developed surfactant of cationic Gemini surfactant (N1-dodecyl-N1, N1, N2, N2-tetramethyl-N2-octylethane-1, 2-diaminium bromide, 12-2-12) has also been used to modify WB and able to achieve maximum removal capacity of several anionic dye (AR-18. AO-7. AB-1) of 95% with 12 h contact time and initial concentration of 50 mg/L [45].

Other surface modification to produce positive WB residue surface can be achieved using amine molecules such as dimethylamine [47] and ammonium salt [46]. The most efficacious method to date of modified WB was obtained by incorporation of multiple quaternary ammonium salts (MQAS) for Acid Red (AR-18) dye removal that can achieve almost 100% removal from 50 g/L dye concentration with 1 g/L dose of modified WB for 12 h contact time [46].

Simple pretreatment of WB has been intensively conducted as well to improve the dye uptake performance such as dehydration using strong acids. Özer and Dursun [28] have demonstrated that the dehydrated WB using concentrated sulfuric acids leads to complete removal of methylene blue form 1.0 L solution with initial concentration of 100 mg/L by only 1 gr of hydrated WB addition after 25 h treatment time. The high removal of the hydrated WB is claimed due to sulfonate functional group attachment during acid treatment. Meanwhile, other researchers were using different acids to modify WB i.e., tartaric acids which can improve the removal performance from 55.32 to 85.76% of Methylene Blue [44].

Further novel modification of WB for composite preparation has been reported as well. Pooladi et al. [33] have prepared wheat brand/Fe_3O_4 composite to remove Methylene blue (MB) and methyl violet (MV) form solution with impressive result of more than 98% removal by only 50 min of treatment time. The magnetite incorporation to WB will create an ease of separation between used adsorbent and treated solution at the end of process.

3.1.2 WB as Bacterial Growth Media

Indirect use of WB for color removal is by using WB as a growth media for specific bacteria that can decolorize certain dyes. Wheat bran is an ideal growth substrate because it is available at low cost and contains a compound of carbohydrates, proteins, amino acids, and trace elements that is essential for microorganism [17].

Recently, Lade et al. [22] have isolated decolorizing bacteria of *bacterium Providencia rettgeri strain* HSL1 in WB medium. This bacterial strain with fungus *Aspergillus ochraceus* will create a unique fungal-bacterial consortium that can decolorize 92% of textile effluent. This biodegradation is catalyzed by several enzymes i.e., laccase, lignin peroxidase, azoreductase, and NADH-DCIP reductase. This finding revealed that the fungal-bacterial consortium's decolorization performance is higher than individual cultures due to the synergistic reaction among the individual culture [11].

Previously, Lade et al. [21] also performed enzymatic decolorization of dye method to treat the carcinogenic benzidine-based dye Trypan Blue. This utilization of wheat bran provides further evidence that agricultural waste can be used as novel low-cost growth medium for biodegradation of dyes, and indicate the potential of the microbial consortium in bioremediation.

3.2 Wheat Straw (WS) Application to Dye Remediation

The current utilization of WS for dye removal is compiled in Table 3. Robinson et al. [35] have used mixture of WS with other biomass waste as bio-sorbents (apple pomace) for treating the synthetic dye wastewater. The experiments revealed that the higher the initial concentration (C_o) of dye of from 50, 100, 150, 200 mg/L, the higher the equilibrium capacity of WS i.e., 48, 64, 75, and 80%, respectively. In terms of particle size, both apple pomace and wheat straw had the same results which 600 μm was the appropriate sorbent size. In an optimum condition, the bio-sorbent could achieve 91% dye removal [35].

Further work using natural wheat straw was also conducted for reactive blue textile dye remediation [25]. The concentration of dye effluent was varied into 5, 20, 30, 50, 70, and 90 mg/L. The adsorption kinetic and isotherm were studied at room temperature. The yields indicated that WS is an economically effective material with the capacity of 68% dye removal or equal to 1.63 mg/g adsorption capacity.

3.2.1 Treated Wheat Straw (WS)

Robinson et al. [34] investigated biological residue materials, i.e., wheat straw, corncob, and barley husk as the bio-sorbent for various dyes (Cibacron Yellow C-2R, Cibacron Red C-2G, Cibacron Blue C-R, Remazol Black B, and Remazol Red RB). The objective of this study was to observe the impact of delignification process

Table 3 Wheat straw (WS) application in dye removal

Dye (conc.)	WS pretreatment (dosage)	% removal (contact time)	References
Dye mixture (cibacron yellow, cibacron red, cibacron blue, remazol black, remazol red) (200 mg/L)	Untreated	80 (72 h)	[35]
Remazol red (200 mg/L), remazol black (500 mg/L)	Untreated	70–75 (45 h)	[26]
Solar orange 7 GLL, Solar jade green FFB (50 mg/L)	Untreated	74.9 (7 h)	[10]
Cibacron yellow C-2R, cibacron red C-2G, cibacron blue C-R, remazol black B, and remazol red RB	Milled, steamed NaOH treated, NH_4OH treated (10 g/L)	56 (102 h)	[34]
Cationic (methylene blue, MB) and anionic (orange II, ORII and methyl orange, MO) (25 mg/L)	Acid–base treated and cationic DMAEMA grafted (200 ml/L)	98.38% (30 min)	[23]
Reactive blue (1000 mg/L)	Surfactant modified, carbonized (4 g/L)	92.17 (>300 min)	[25]
Methylene blue (MB) crystal violet (CV) (250 mg/L)	Citric acid esterified (2 g/L)	98.58 (6 h)	[18]
Eriochrome black T (100 mg/L)	Pyrolyzed and aminated (40/100 ml)	54.5% (60 min)	[36]

through chemical (NaOH and NH_4OH) and physical (steeping and milling) pretreatments in order to upgrade those agricultural waste performances as dye adsorbents. The experiments found that the milled, steamed, and NaOH treated wheat straw had the identical ability in decolorization (54, 56, and 53%, respectively), compared to untreated WS which merely achieved 26% of dye removal. Although steamed WS effectively enhance the surface area by high pressure heating, it required 150 h to reach the equilibrium state, while the milled WS exhibited a faster period after 102 h contact time. Over all of the research, the milling process was considered as the most feasible treatment compared to other chemical and physical pretreatments.

The performance of NWS (Natural Wheat Straw) was subsequently compared to surfactant modified wheat straw (MWS) and activated carbon (AC) in terms of reactive blue decolorization [25]. In comparison with NWS, MWS and AC had a greater removal of 68, 92.17 and 90.5%, respectively. Additionally, the adsorption capacity of MWS and AC was also a much higher than NWS, at 11.79 and 10.53 mg/g, consecutively. These results indicated that the more the adsorbent amount, the less the adsorption capacity at the constant volume and concentration of the dye. This phenomenon may be rendered by the saturation condition of adsorbent active sites because of the particulate aggregation which led to the decrease in adsorbent surface. Furthermore, the effect of pH was observed that NWS and AC exhibited better performance at lower pH, while MWS achieved higher remediation at higher level of pH. The different optimum pH of each material may open a wider application for various type of dyes that can be treated.

3.2.2 WS as Substrate Media

Various agricultural wastes, i.e., wood chips, wheat straw, and corn-cob shreds also had been employed by Nigam et al. [26] to observe the adsorption ability of these agricultural wastes toward both single and multiple dyes. Out of 500 ppm the dye mixture, wheat straw, and corn-cob shreds could remove 70–75% of dye concentration at room temperature. After saturated with dye molecules, the biomass was used as fermentation substrate. The two strains of *Phanerochaete chrysosporium* and *Coriolus versicolor* used for solid-state fermentation (SSF) exhibited the higher level decolorization. Dye degradation products in aerobic circumstance could hinder the carcinogenic amines. These preliminary experiments performed that dye adsorption onto wheat straw or corn-cob shreds may exhibit a low-cost and cleaner treatment for the large volumes of textile dye-containing effluents, on the contrary, most physicochemical treatments have several drawbacks which render excessive amounts of chemical consumption or sludge production with conspicuous discharge issues, pricey plant necessary or operating expenses.

Meanwhile, a recent utilization of WS as substrate for dye removal was presented by Majul et al. [24], by treating various dyes using immobilized ligninolytic white-rot fungus *Peniophora laxitexta* in lignocellulosic support (WS and soybean hulls). The inoculated biocarrier can be used repeatedly without significant reduction of the dye removal efficiency. This unique feature of bio-sorbent is due to the combined effect of adsorption to the substrate and enzymatic decolorization. This novel application of WS as biocarrier could be applied for industrial scale dye wastewater treatment.

4 Conclusion

Wheat straw and brand have a great potential and flexibility for dye removal. The potential is in terms of availability in nature as side products of agriculture, while the flexibility is due to the large possibility in application whether directly used (unmodified) for bio-sorbent or pretreated (acid/base treated, impregnated, grafted, reacted, carbonized, and so on) to become high capacity adsorbents for various dye contaminated wastewater treatment. After being utilized for absorbing dyes, saturated straw or brand still can be used for microbial or fungal substrate ensuring a complete removal of hazardous dye molecules from the environment.

References

1. Ahmad N, Fazal H, Abbasi BH, Farooq S, Ali M, Khan MA (2012) Biological role of Piper nigrum L. (Black pepper): a review 3(7):1945–1953. https://doi.org/10.1016/S2221
2. Ahmed MJ, Hameed BH, Hummadi EH (2021) Insight into the chemically modified crop straw adsorbents for the enhanced removal of water contaminants: a review. J Mol Liq 330:115616. https://doi.org/10.1016/j.molliq.2021.115616
3. Anastopoulos I, Pashalidis I, Orfanos AG, Manariotis ID, Tatarchuk T, Sellaoui L, Núñez-Delgado A (2020) Removal of caffeine, nicotine and amoxicillin from (waste)waters by various adsorbents. a review. J Environ Manage 261(February). https://doi.org/10.1016/j.jenvman.2020.110236
4. Ata S, Imran Din M, Rasool A, Qasim I, Ul Mohsin I (2012) Equilibrium, thermodynamics, and kinetic sorption studies for the removal of coomassie brilliant blue on wheat bran as a low-cost adsorbent. J Anal Meth Chem 1(1). https://doi.org/10.1155/2012/405980
5. Babu BR, Parande AK, Raghu S, Prem Kumar T (2007) Cotton textile processing: waste generation and effluent treatment. J Cotton Sci 11(3):141–153
6. Basibuyuk M, Forster CF (1997) The use of sequential anaerobic/aerobic processes for the biotreatment of a simulated dyeing wastewater. Environ Technol (United Kingdom) 18(8):843–848. https://doi.org/10.1080/09593330.1997.9618562
7. Bhatnagar A, Jain AK (2005) A comparative adsorption study with different industrial wastes as adsorbents for the removal of cationic dyes from water. J Colloid Interface Sci 281(1):49–55. https://doi.org/10.1016/j.jcis.2004.08.076
8. Bhatti HN, Safa Y, Yakout SM, Shair OH, Iqbal M, Nazir A (2020) Efficient removal of dyes using carboxymethyl cellulose/alginate/polyvinyl alcohol/rice husk composite: adsorption/desorption, kinetics and recycling studies. Int J Biol Macromol 150:861–870. https://doi.org/10.1016/j.ijbiomac.2020.02.093
9. Bulut Y, Baysal Z (2006) Removal of Pb(II) from wastewater using wheat bran. J Environ Manage 78(2):107–113. https://doi.org/10.1016/j.jenvman.2005.03.010
10. Chandran CB, Singh D, Nigam P (2002) Remediation of textile effluent using agricultural residues. Appl Biochem Biotechnol - Part A Enzym Eng Biotechnol 102–103:207–212. https://doi.org/10.1385/ABAB:102-103:1-6:207
11. Chen BY, Chang JS (2007) Assessment upon species evolution of mixed consortia for azo dye decolorization. J Chin Inst Chem Eng, 38(3–4):259–266. https://doi.org/10.1016/j.jcice.2007.04.002
12. Chowdhury S, Saha PD (2013) Citric acid modified wheat bran as a potential adsorbent for removal of Cu(II) and malachite green from aqueous solutions. Desalin Water Treat 51(31–33):6038–6048. https://doi.org/10.1080/19443994.2013.764352
13. Çiçek F, Özer D, Özer A, Özer A (2007) Low cost removal of reactive dyes using wheat bran. J Hazard Mater 146(1–2):408–416. https://doi.org/10.1016/j.jhazmat.2006.12.037
14. Das S, Singh S, Garg S (2019) Evaluation of wheat bran as a biosorbent for potential mitigation of dye pollution in industrial waste waters. Orient J Chem 35(5):1565–1573. https://doi.org/10.13005/ojc/350514
15. Farajzadeh MA, Monji AB (2004) Adsorption characteristics of wheat bran towards heavy metal cations. Sep Purif Technol 38(3):197–207. https://doi.org/10.1016/j.seppur.2003.11.005
16. Ghaffar SH, Fan M (2013) Structural analysis for lignin characteristics in biomass straw. Biomass Bioenerg 57:264–279. https://doi.org/10.1016/j.biombioe.2013.07.015

17. Gomathi D, Muthulakshmi C, Kumar DG, Ravikumar G, Kalaiselvi M, Uma C (2012) Submerged fermentation of wheat bran by Aspergillus flavus for production and characterization of carboxy methyl cellulase. Asian Pac J Trop Biomed 2:S67–S73. https://doi.org/10.1016/S2221-1691(12)60132-4
18. Gong R, Zhu S, Zhang D, Chen J, Ni S, Guan R (2008) Adsorption behavior of cationic dyes on citric acid esterifying wheat straw: kinetic and thermodynamic profile. Desalination 230(1–3):220–228. https://doi.org/10.1016/j.desal.2007.12.002
19. Gupta VK, Khamparia S, Tyagi I, Jaspal D, Malviya A (2015) Decolorization of mixture of dyes: a critical review. Glob J Environ Sci Manage 1(1):71–94. https://doi.org/10.7508/gjesm.2015.01.007
20. Haq HA, Javed T, Abid MA, Zafar S, Din MI (2021) Adsorption of crystal violet dye from synthetic textile effluents by utilizing wheat bran (Triticum aestivum). Desalin Water Treat 224:395–406. https://doi.org/10.5004/dwt.2021.27178
21. Lade H, Kadam A, Paul D, Govindwar S (2015) A low-cost wheat bran medium for biodegradation of the benzidine-based carcinogenic dye trypan blue using a microbial consortium. Int J Environ Res Public Health 12(4):3480–3505. https://doi.org/10.3390/ijerph120403480
22. Lade H, Kadam A, Paul D, Govindwar S (2016) Exploring the potential of fungal-bacterial consortium for low-cost biodegradation and detoxification of textile effluent. Arch Environ Prot 42(4):12–21. https://doi.org/10.1515/aep-2016-0042
23. Lin Q, Wang K, Gao M, Bai Y, Chen L, Ma H (2017) Effectively removal of cationic and anionic dyes by pH-sensitive amphoteric adsorbent derived from agricultural waste-wheat straw. J Taiwan Inst Chem Eng 76:65–72. https://doi.org/10.1016/j.jtice.2017.04.010
24. Majul L, Wirth S, Levin L (2020) High dye removal capacity of Peniophora laxitexta immobilized in a combined support based on polyurethane foam and lignocellulosic substrates. Environ Technol (United Kingdom). https://doi.org/10.1080/09593330.2020.1801851
25. Mousa KM, Taha AH (2016) Adsorption of reactive blue dye onto natural and modified wheat straw. Am J Chem Eng 4(1):9. https://doi.org/10.11648/j.ajche.20160401.12
26. Nigam P, Armour G, Banat IM, Singh D, Marchant R (2000) Physical removal of textile dyes from effluents and solid-state fermentation of dye-adsorbed agricultural residues. Biores Technol 72(3):219–226. https://doi.org/10.1016/S0960-8524(99)00123-6
27. Ogata F, Nagai N, Itami R, Nakamura T, Kawasaki N (2020) Potential of virgin and calcined wheat bran biomass for the removal of chromium(VI) ion from a synthetic aqueous solution. J Environ Chem Eng 8(2):103710. https://doi.org/10.1016/j.jece.2020.103710
28. Özer A, Dursun G (2007) Removal of methylene blue from aqueous solution by dehydrated wheat bran carbon. J Hazard Mater 146(1–2):262–269. https://doi.org/10.1016/j.jhazmat.2006.12.016
29. O'Neill C, Hawkes FR, Hawkes DL, Esteves S, Wilcox SJ (2000) Anaerobic-aerobic biotreatment of simulated textile effluent containing varied ratios of starch and azo dye. Water Res 34(8):2355–2361. https://doi.org/10.1016/S0043-1354(99)00395-4
30. Pala A, Tokat E (2002) Color removal from cotton textile industry wastewater in an activated sludge system with various additives. Water Res 36(11):2920–2925. https://doi.org/10.1016/S0043-1354(01)00529-2
31. Palamutcu S (2017) Sustain Text Technol 1–22. https://doi.org/10.1007/978-981-10-2474-0_1
32. Papinutti L, Mouso N, Forchiassin F (2006) Removal and degradation of the fungicide dye malachite green from aqueous solution using the system wheat bran-Fomes sclerodermeus. Enzyme Microb Technol 39(4):848–853. https://doi.org/10.1016/j.enzmictec.2006.01.013
33. Pooladi H, Foroutan R, Esmaeili H (2021) Synthesis of wheat bran sawdust/Fe3O4 composite for the removal of methylene blue and methyl violet. Environ Monitor Assess 193(5). https://doi.org/10.1007/s10661-021-09051-9
34. Robinson T, Chandran B, Nigam P (2002) Effect of pretreatments of three waste residues, wheat straw, corncobs and barley husks on dye adsorption. Biores Technol 85(2):119–124. https://doi.org/10.1016/S0960-8524(02)00099-8
35. Robinson T, Chandran B, Nigam P (2002) Removal of dyes from a synthetic textile dye effluent by biosorption on apple pomace and wheat straw. Water Res 36(11):2824–2830. https://doi.org/10.1016/S0043-1354(01)00521-8

36. Sadeghi S, Zakeri HR, Saghi MH, Ghadiri SK, Talebi SS, Shams M, Dotto GL (2021) Modified wheat straw–derived graphene for the removal of Eriochrome Black T: characterization, isotherm, and kinetic studies. Environ Sci Pollut Res 28(3):3556–3565. https://doi.org/10.1007/s11356-020-10647-w
37. Singh KK, Singh AK, Hasan SH (2006) Low cost bio-sorbent "wheat bran" for the removal of cadmium from wastewater: kinetic and equilibrium studies. Biores Technol 97(8):994–1001. https://doi.org/10.1016/j.biortech.2005.04.043
38. Soni S, Bajpai PK, Mittal J, Arora C (2020) Utilisation of cobalt doped Iron based MOF for enhanced removal and recovery of methylene blue dye from waste water. J Mol Liq 314:113642. https://doi.org/10.1016/j.molliq.2020.113642
39. Sulak MT, Demirbas E, Kobya M (2007) Removal of Astrazon Yellow 7GL from aqueous solutions by adsorption onto wheat bran. Biores Technol 98(13):2590–2598. https://doi.org/10.1016/j.biortech.2006.09.010
40. Szymczyk M, El-Shafei A, Freeman HS (2007) Design, synthesis, and characterization of new iron-complexed azo dyes. Dyes Pigm 72(1):8–15. https://doi.org/10.1016/j.dyepig.2005.07.009
41. Vasu D, Kumar S, Walia YK (2020) Removal of dyes using wheat husk waste as a low-cost adsorbent. Environ Claims J 32(1):67–76. https://doi.org/10.1080/10406026.2019.1669908
42. Vigneshpriya D, Shanthi E (2015) Physicochemical characterization of textile waste water. Int J Innov Res Dev 4(10):48–51
43. Yakout SM, Ali MS (2015) Removal of the Hazardous crystal violet dye by adsorption on corncob-based and phosphoric acid-activated carbon. Part Sci Technol 33(6):621–625. https://doi.org/10.1080/02726351.2015.1016642
44. Yao S, Lai H, Shi Z (2012) Biosorption of methyl blue onto tartaric acid modified wheat bran from aqueous solution. J Environ Health Sci Eng 9(1):1–6
45. Zhang Y, Huang G, An C, Xin X, Liu X, Raman M, Doble M (2017) Transport of anionic azo dyes from aqueous solution to gemini surfactant-modified wheat bran: synchrotron infrared, molecular interaction and adsorption studies. Sci Total Environ 595:723–732. https://doi.org/10.1016/j.scitotenv.2017.04.031
46. Zhang WX, Lai L, Mei P, Li Y, Li YH, Liu Y (2018) Enhanced removal efficiency of acid red 18 from aqueous solution using wheat bran modified by multiple quaternary ammonium salts. Chem Phys Lett 710(September):193–201. https://doi.org/10.1016/j.cplett.2018.09.009
47. Zhong QQ, Yue QY, Li Q, Xu X, Gao BY (2011) Preparation, characterization of modified wheat residue and its utilization for the anionic dye removal. Desalination 267(2–3):193–200. https://doi.org/10.1016/j.desal.2010.09.025
48. Zou S, Li H, Liu L, Wang S, Zhang X, Zhang G (2021) Experimental study on fire resistance improvement of wheat straw composite insulation materials for buildings. J Build Eng 43(March):103172. https://doi.org/10.1016/j.jobe.2021.103172

Application of Peel-Based Biomaterials in Textile Wastewater

Prashant G. Mahajan and Dipika Jaspal

Abstract With the increase in industrialization, there has been a rise in the number of textile industries. Effluent from these textile industries is disposed in various water streams and water bodies. Textile wastewater carries a mixture of several substances, predominantly dyes, and heavy metals. Dyes and heavy metals present in textile wastewater are toxic and hazardous to humans and other living beings. The presence of these substances as contaminants on exposure can lead to skin irritation, enhanced pulse, vomiting, etc. Several materials are being explored as adsorbents for the treatment of dye wastewater and the removal of contaminants. Several peel-based biomaterials are generated as wastes during the course of different domestic material consumption and several activities such as agriculture. Therefore, substantial amount of peel-based biomaterial is generated, which either falls into the soil or is deposited or thrown in streams and lakes, resulting in pollution. Using peel-based biomaterials for textile wastewater treatment, as adsorbents, can minimize emissions from biomaterials as well as treat contaminants in textile wastewater. The capacity of dye removal shown by peel-based materials (banana peel, hydnocarpus pentandra, pineapple peel, and mandarin peel) has been of the order of 85–90% , for a toxic dye Methylene blue. In the case of green pea peels, maximum Methylene blue removal was 96% for particle sizes less than 75 μ. Some other peels, such as rambutan peels, jackfruit peels, and mangosteen peels, have shown a removal capacity of 92–96% for dyes such as Brilliant green. Carbonized orange peels have shown elimination of more than 75% for Direct navy blue 106. As per literature majority of the peels have been converted to activated carbon for the mentioned purpose, the reason being their high porosity and surface area. Some vegetable peels such as potato peel, activated potato

P. G. Mahajan
Research Scholar, Symbiosis Institute of Technology, Symbiosis International (Deemed University), Lavale, Pune, Maharashtra 412115, India

Assistant Professor, Dept. of Engineering Sciences, AISSMS Institute of Information Technology, Savitribai Phule Pune University, Pune 411001, India

D. Jaspal (✉)
Department of Applied Science, Symbiosis Institute of Technology (SIT), Symbiosis International (Deemed University) (SIU), Lavale, Pune, Maharashtra 412115, India
e-mail: dipikaj@sitpune.edu.in

S. S. Muthu and A. Khadir (eds.), *Textile Wastewater Treatment*, Sustainable Textiles: Production, Processing, Manufacturing & Chemistry,
https://doi.org/10.1007/978-981-19-2852-9_4

peel, and calcined potato peel have been successful in eliminating 94%, 76.41%, and 88.6% of Cibacron blue dye from aqueous solution, respectively. In this chapter, studies on the use of fruit peels and vegetable peels utilized for the treatment of textile wastewater have been presented.

Keywords Agricultural waste · Fruit peel · Vegetable peels · Activated carbon · Dyes · Heavy metals · Adsorption · Textile wastewater · Langmuir isotherm model · Freundlich model

1 Introduction

Textile industries are considered to be one of the most polluting industrial sectors, using a variety of materials and chemicals [19]. According to World Bank, textile dyeing and product finishing accounts for 17–20% of industrial wastewater consumption [26]**.** According to statistics, the total dye intake and usage of dyes is over 10 lakh kg per year [21]. Every year, approximately 7 lakh tones of dyes and pigments are consumed globally, with textile industries accounting for two-thirds of that consumption [14, 37]. The textile industry produces the largest amount of dye-containing wastewater (54%) among other industries [28]. Exposure to dye pollutants influences humans, as this leads to skin irritation, increased heart rate, vomiting, shock, eye irritation, causing painful sensitization to light, etc. [4]. The presence of color in wastewater released into the environment influences marine life, as well as causes significant environmental harm [26].

Traditional methods for textile wastewater treatment include desalination, ultrafiltration, coagulation/flocculation, membrane technologies, electrochemical treatment, adsorbent utilization, etc. [13, 18, 35, 38, 39, 49–52]. Adsorption is an eco-friendly, least expensive, and simple technique that does not generate sludge [5]. Adsorption has been identified as a promising method due to its low preliminary cost, ease of operation, and excessive separation performance even at low concentration [10, 58]. In the past few years, researchers have explored different types of adsorbents which include mineral sorbents [20], activated carbons [54], peat [45], chitin [22], fungal biomasses [27], clay adsorbents [46], zeolites [23], waste materials [1], metal [47] and metal oxide-based adsorbents [12] for treating colored wastewater as alternatives to commercially used, expensive activated carbon [15].

Peels are good candidates to substitute activated carbon in effluent treatment because of their typical chemical framework, abundant accessibility, environment-friendly nature, and low cost [53]. Fruit peels, such as rambutan peel, jackfruit peel, mangosteen peel, and spent tea leaves [41], orange peels [29], pineapple peels [33], citrus services fruit peel [32], finger citron, pomelo peel, orange peel and lemon peel [59], banana peels [8], dragon fruit peel [25], pomegranate peel [3], mandarin peel [56], hydnocarpus pentandra [40] and vegetable peels, such as green pea peels [16], water chestnut peels [31], soybean peels, luffa skin peels [11], *Grewia venusta* peel and *Moringa oleifera* seeds [2], cucumber peels [48], potato peels [34], onion peels, and peanut peels [24], potato peel powder [9], tropica peel [57], are examples of low-cost agriculture wastes that have been used to make activated carbon.

Several fruit and vegetable peels have been successfully employed for the treatment of textile wastewater, which have been discussed at length in the subsequent sections of this chapter.

2 Fruit Peels

2.1 *Rambutan Peel, Jackfruit Peel, Mangosteen Peel, and Spent Tea Leaves*

In 2015, Nor et al. used rambutan peel, jackfruit peel, mangosteen peel, and wasted tea leaves to study the adsorption of Brilliant green and Porcion red. Characterization of the adsorbent was carried out by using field emission scanning electron microscope and fourier transform infrared spectroscopy. Rambutan peels removed 96.42% of Brilliant green dye and 41.20% of Porcion red dye. Brilliant green dye was adsorbed from jackfruit peels at the rate of 94.78%, while from Procion red dye at 61.2%. Almost 95.78% of the Brilliant green dye and 48.55% of the Porcion red dye was eliminated using spent tea leaves. Mangosteen peels showed Brilliant green removal efficiency of 92.74% and Porcion red removal efficiency of 45.17% [41].

2.2 *Orange Peel*

Khaled et al. investigated the absorption of Direct navy blue 106 in [30]. Orange peel was utilized to prepare activated carbon. For a concentration of 2000 mg/L adsorbent and 150 mg/L Direct navy blue 106, greatest adsorption observed was 107.53 mg/g. With R^2 values more than 0.97, the Langmuir and Freundlich models both were in line with the adsorption data. According to the kinetic analysis, a pseudo-second-order rate of adsorption was observed, with intraparticle diffusions being one of the rate-determining stages [29].

Arami et al. in 2005 used orange peels to remove Direct red 80 and Direct red 23 from aqueous solutions. Experimental results were analyzed employing Langmuir and Freundlich isotherms. The correlation coefficients for the Langmuir isotherm were better fit for the experimental data than the Freundlich model. For all concentrations of these dyes, the equilibrium period of adsorption was around 15 min at 298 K and pH 2. Based on pseudo-first-order and pseudo-second-order rate mechanisms, kinetic investigations for the two dyes on orange peel were conducted. According to the findings, the adsorption kinetics occurred via. pseudo-second order. Because of the enhanced adsorption kinetics occurring on orange peel, a continuous flow technique to treat colored textile effluent was proposed [6].

Ardejani et al. created a computational finite element model for the direct dyes elimination from textile effluents, in 2007 to specify the adsorption process, taking into accounts both linear and Langmuir isotherms. The model parameter values were derived from an experimental investigation that employed orange peel as an inexpensive adsorbent to eliminate the dyes, Direct red 80 and Direct red 23, from textile wastewaters. The conclusions were relatively consistent with the measured quantities. Pseudo-second-order kinetics was found to be well-suited to the sorption rate. Furthermore, the rate of dye elimination from textile wastewater was influenced by the starting dye concentration in the solution phase. The rate of dye removal using orange peel adsorbent decreased as the starting concentrations of both Direct red 80 and Direct red 23 dyes were raised from 50 to 75 mg/L [7].

Khaled et al. synthesized activated carbon from orange peel in [30] and utilized it to remove textile dyes from synthetic textile dye solution of Direct yellow 12. At 272 °C, the impact of initial dye concentration (25–125 mg/L), pH, contact period, and orange peel carbon concentration (2.5–10.0 g/L) was studied. Several isotherm models, including Langmuir, Redlich-Peterson, Tempkin, Freundlich, Dubinin-Radushkevich, Koble-Corrigan, and generalized isotherm equations, were also analyzed. The results illustrated that an acidic pH (1.5) facilitated Direct yellow 12 adsorption over activated carbon derived from orange peel. The adsorption kinetics of Direct yellow 12 was investigated, and the rate of sorption was confirmed to be pseudo-second-order kinetics, with correlation coefficients equal to unity under all tested conditions. The Langmuir isotherm model showed 75.76 $mg.g^{-1}$ of maximum adsorption [29].

El-Said et al. designed orange peel as a viable low-cost alternative natural waste adsorbent to treat textile effluents. Orange peel was used to remove C.I. Direct yellow 27 and C.I. Direct red 79 from dye solutions. Findings showed that acidic solutions enhanced the adsorption rate of the dyes investigated. The intraparticle diffusion, pseudo-second-order, and pseudo-first-order equations were applied to investigate adsorption kinetic models. It was revealed that the pseudo-second- order equation was reasonably well aligned with the experimental data. The effect of this adsorbent morphology was studied utilizing scanning electron microscopy, which indicated a reasonable agreement with the adsorption data [17].

Disperse dye C.I. removal was also studied by Torres et al. using an orange peel-based adsorbent in 2021. The efficiency of the adsorbent was assessed at various pH values and the adsorption capacity was determined by a comparison between untreated and chemically treated orange peel. Findings indicated a spontaneous coagulation-flocculation phenomenon of the dispersed dye at low pH values, which represented long-term solution for the recovery and a probable, re-use of the dye, collectively with the industrialized valorization of a waste product [55].

2.3 Pineapple Peel

Adsorbent from pineapple waste was synthesized in 2014 by Krishni et al. and was used to adsorb Methylene blue dye from textile effluent. The shape and function of the adsorbent were studied employing scanning electron microscope and fourier transform infrared spectroscopy analyses. The initial dye concentration, solution pH, and contact duration, all influenced the adsorption process. The Langmuir isothermal model was the best fit for the experimental results, with the maximum monolayer adsorption capacity of 97.09 mg/g at 30 °C. Chemisorption was discovered to be the ongoing process based on kinetic studies [33].

2.4 Cucumis Sativus Fruit Peel

In 2017 the adsorption of Acid red 1 dye from textile wastewater onto activated carbon derived from cucumis sativus fruit peel was studied. The porous structures of the adsorbent as well as the chemical groups present on its surface were studied using scanning electron microscope and fourier transform infrared spectroscopy techniques. The findings showed that the solution pH, initial dye concentration, and the quantity of adsorbent used, all had an impact on the process under consideration. As there was a rise in the initial dye concentration, the adsorption capability increased. The adsorption equilibrium data matched the Dubinin-Radushkevich and Langmuir isotherm models better than Freundlich and Temkin isotherms [32].

2.5 Orange Peel, Pomelo Peel, Lemon Peel, and Finger Citron Peel

In 2018, Yu and coworkers utilized adsorbents made from solid waste to study the adsorption of Orange II and Acid chrome blue K dyes from textile effluent. Pseudo-second-order kinetic model was predominant in the process. The equilibrium results were best represented by the Langmuir isotherm model. Orange peel-based nanoporous carbon revealed to have a substantially better adsorption capacity than other forms of biomass nanoporous carbons. Furthermore, orange peel-based nanoporous carbon had a better capacity for absorbing Orange II from binary dye solutions [59].

Pomelo peel was used as the starting material in another investigation, while zinc chloride was used as the activation agent. Pomelo peel biomass-activated carbon was produced using a high-temperature pyrolysis method. Organic dyes, as well as formaldehyde adsorption, were also investigated. When the temperature was 500 °C, the zinc chloride concentration was 20 g/L, the impregnation ratio was 3, the impregnation time was 24 h, and the carbonization time was 45 min. The results showed that biomass-activated carbon had obvious pores and hydrophilic functional groups of C=O, O–H, C–O on the surface. The average pore size of biomass-activated carbon was 2.26 nm, and the total surface area was 983.27 m^2/g, with a total pore volume of 0.56 cm^3/g and 983.27 m^2/g, respectively. The dye adsorption onto biomass-activated carbon kinetics obeyed a pseudo-second-order kinetic model. Methylene blue, Neutral red, and Crystal violet had maximum adsorption capacities of 75.81, 96.19, and 64.06 mg/g, respectively, on biomass-activated carbon. According to this, the dye was absorbed by 1.0 g biomass-activated carbon at a rate of 90% [43].

2.6 *Banana Peel*

A published study in 2018 used an adsorbent made from banana peel to extract Methylene blue from textile effluent at room temperature. With an increase in adsorbent dose from 0.05 to 0.4 g, the percentage removal increased from 34.69 to 86.88% in 100 cm^3 of Methylene blue solution. With a Methylene blue solution clearance rate of 86.22%, the optimum contact time for the adsorption process was 2.5 h. Pseudo-second-order model ($R^2 = 0.9934$) was the best fit for the adsorption kinetic model [8].

Banana and orange peels were made into a combination and tested in 2012 (Mane and Bhusari). When compared to untreated banana and orange peel, acid-treated banana and orange peel considerably enhanced dye adsorption capacity. It revealed color removal efficacy to be highest with a minimum amount of 0.06 g for banana peels, and 0.05 g for orange peel, within a duration of 55 min. The adsorption capacity of orange peel and banana peel, as calculated by Langmuir isotherms, was 0.0647 mg/g and 0.1808 mg/g, respectively [36]. In 2019, Pishgar et al. investigated the elimination of Basic blue 159 from aqueous solution using banana peel. Particle size distribution, field emission scanning electron microscope, and fourier transform infrared spectroscopy analyses were used to explain the adsorbent characteristics. Adsorption was greatly altered by operating parameters such as adsorbent dose, pH of the solution, and mixing speed. Best removal percentage was observed with 0.4 g adsorbent, 200 rpm, and pH 9 [44].

2.7 *Mandarin Peel*

A new methodology for synthetic dye adsorption was developed that utilized mandarin peel for the manufacture of activated carbon. Scanning electron microscopy, x-ray diffraction, thermal gravimetric analysis, fourier transform infrared spectroscopy, and energy-dispersive x-ray spectroscopy were utilized for the characterization of the adsorbent. Carbonized mandarin peel was more successful at eliminating Methylene blue than Methyl orange. An increasing the pH value had a favorable effect on dye adsorption up to a pH of 6.9. An increased adsorption dosage had only a minimal effect on the elimination of Methylene blue. The adsorbent revealed 95.17% of its regeneration activity after being regenerated with the solvent. Pseudo-second-order kinetics was used to investigate the kinetics of Methylene blue adsorption onto carbonized mandarin peel [56].

2.8 *Dragon Fruit Peel*

In 2018, Jawad et al. created a new adsorbent from dragon fruit peel that was utilized to remove Methylene blue. Dragon fruit peel generated from biomass waste showed a tremendous potential as a low-cost adsorbent for removing Methylene blue from an aqueous solution. Dragon fruit peel was characterized using carbon, hydrogen, nitrogen, sulfur, and oxygen analysis, x-ray diffraction, fourier transform infrared spectroscopy, scanning electron microscope, and pHpzc techniques. At pH 5 and dragon fruit peel dosage 0.06 g, better Methylene blue adsorption capacity was attained. The pseudo-second-order model was shown to have a better fit for the kinetic data than the pseudo-first-order model. The equilibrium results resembled the Langmuir isotherm model more closely than the Freundlich model. The standard enthalpy (H°) was 5.40 kJ/mol, the standard entropy (S°) was 37.6 J/mol. K, and the standard free energy (G°) was −6.006, −6.382, and −6.759 kJ/mol, respectively. Thermodynamic adsorption parameters revealed that Methylene blue adsorption by dragon fruit peel was endothermic and spontaneous at the conditions tested [25].

2.9 *Pomegranate Fruit Peel*

In the year 2021, Ahmad et al. used carbon (IV) oxide gasification and potassium hydroxide impregnation procedures to create pomegranate fruit peel-activated carbon for Methylene blue adsorption from an aqueous solution. The ideal radiation power for Methylene blue dye adsorption was 375 W, the impregnation ratio was 0.9, and the activation duration was 4.5 min. As a result, 83.4% Methylene blue dye removal, and 30.8% pomegranate fruit peel-activated carbon removal was achieved. Methylene blue dye absorption was rapid in the early phase (0–3 h) and reached an equilibrium in six hours, at starting doses of 25–300 mg/L, according to kinetic analyses. Greatest monolayer uptake shown using Langmuir model at 30 °C, 45 °C, and 60 °C was 250.00 mg/g, 239.05 mg/g, and 235.58 mg/g, respectively, implying favorable adsorption at low temperatures [3].

Pomegranate peel waste was explored for removing cationic dye from an aqueous solution by Nur Dini et al. in 2019. pH 11, 10 mg/L, and 1.0 g bio-sorbent dose were the best conditions for bio-sorption of 100 mL of Methylene blue onto pomegranate peel powder. The bulk of the Methylene blue onto pomegranate peel powder took two hours to reach equilibrium. The dye removal percentage and adsorption capacity

Table 1 Removal efficiency of various fruit peel-based biomaterials dyes in textile wastewater

Sr. No	Peel used	Dyes removed	Adsorption capacity	References
1	Jackfruit peel	Brilliant green Porcion red	9.47 mg/g 6.12 mg/g	Nor et al. [41]
2	Rambutan peel	Brilliant green Porcion red	9.64 mg/g 4.12 mg/g	Nor et al. [41]
3	Mangsoteen peel	Brilliant green Porcion red	9.27 mg/g 4.51 mg/g	Nor et al. [41]
4	Orange peel	Direct navy blue—106	Greater than 75% (removal efficiency)	Khaled et al.[29]
5	Adsorbent from pineapple peel	Methylene blue	97.09% (removal efficiency)	Krishni et al. [33]
6	Cucumis sativus fruit peel	Acid red 1	18.76 mg.g^{-1}	Khanna and Rattan [32]
7	Orange peel, pomelo peel, Lemon peel and finger citron peel	Orange II Acid chrome blue K	930 mg.g^{-1} 592 mg.g^{-1}	Yu et al. [59]
8	Banana peel waste	Methylene blue	86.22% (removal efficiency)	Bakar et al. [8]
9	Carbonized mandarin peel	Methylene blue	99.77% (removal efficiency)	Unugu land Nigiz [56]
10	Dragon fruit peel	Methylene blue	192.31 mg.g^{-1}	Unugul and Nigiz [25]
11	Pomegranate Fruit peel	Methylene blue	235.58 mg.g^{-1}	Ahmad et al. [3]
12	Hydnocarpus pentandra	Methylene blue	90–95% (removal efficiency)	Nayak et al.[40]

with the maximum dye removal percentage and adsorption capacity was 99.8% and 102.25 mg/g, respectively. Adsorption occurred homogeneously, throughout the adsorption process [42].

2.9.1 Hydnocarpus Pentandra

For the removal of Methylene blue dye, activated charcoal was synthesized from the fruit peel of the hydnocarpus pentandra plant, which demonstrated excellent dye adsorption. At pH 6–7, this adsorption mechanism was the maximum, with highest adsorption at 328 K. When coupled with the surfactant sodium dodecyl sulfate, maximum adsorption occurred. At 2% of the adsorbent concentration, maximum dye adsorption occurred, which was between 90 and 95% [40].

Table 1 summarizes the removal of various dyes using peels such as jackfruit peel, rambutan peel, pomegranate peel, banana peel, mangos teen peel, orange peel, pineapple peel, citrus fruit peel, and dragon fruit peel, for dyes such as Brilliant green, Porcian red, Methylene blue, Methyl orange, Direct navy blue—106 and Acid red 1.

3 Vegetable Peels

3.1 Green Pea Peels

Dod et al. produced activated carbon from green pea peels in 2012, by using a carbonization process in a limited supply of oxygen to drive away volatiles, resulting in a porous carbon structure with enlarged surface. Sulfuric acid was utilized as a chemical reagent throughout the chemical activation procedure to develop a more efficient adsorbent. The equilibrium contact period for highest Methylene blue adsorption for green pea peels adsorbent was found to be substantially shorter (7 h) than for commercial activated carbon (16 h). The quantity of dye taken up by commercial activated carbon, on the other hand, was more than that of green pea peels. Increasing the agitation speed after reaching a specific optimum speed had no discernible effect on the elimination of Methylene blue (200 rpm). The equilibrium time was unaffected by the starting dye (Methylene blue) concentration or agitation speed, implying that adsorption took place primarily on the exterior of the adsorbent particles [16].

3.2 Water Chestnut Peel Waste

Chestnut peel was explored for its adsorption capacity for removal of Rhodamine B dye from aqueous solution. Fourier transform infrared spectroscopy and scanning electron microscopy investigations were used to characterize the samples. Langmuir isotherm plots at 30 °C revealed that water chestnut peel waste had the maximum adsorption capacity (q_m). The Freundlich and Halsey isothermal models were used to examine the equilibrium data. Due to the high correlation coefficients and tight agreement between the experimental and estimated q_e values, the pseudo-second-order kinetic model proved to be the best fit. The thermodynamic characteristics indicated endothermic and spontaneous adsorption (DG_0 and DH_0) [31].

3.3 *Luffa Fruit Peel and Soybean Seed Hulls*

Enzymatic degradation of Methyl orange from liquid effluents was investigated by utilizing an extract from bio-wastes of luffa skin peels and soybean hulls in 2016. The enzymatic dye elimination approach was tested with raw soybean peroxidase and luffa peroxidase extracts, which had a consistent enzymatic activity of 0.373 and 0.355 μ m/L, respectively. The dye decolorization efficiency of soybean peroxidase was 81.4% after an hour of incubation at 30 °C with 0.5 cm^3 raw soybean peroxidase, 2 mM hydrogen peroxide, and 30 mg/L Methyl orange at pH 5.0. Furthermore, luffa peroxidase provided the greatest decolorization effectiveness (75.3%) when 2 mM hydrogen peroxide, 1.5 cm^3 crude luffa peroxidase, and 10 mg/L Methyl orange at pH 3.0 were used for 40 min [11].

3.4 *Moringa Oleifera Seed and Grewia Venusta Peel*

Agbahoungbata et al. demonstrated yet another method for manufacturing bio-adsorbents from *G. venusta* peel and *M. oleifera* seeds in the year 2016. The adsorption behaviors of Methyl orange and Indigo carmine dyes were investigated using batch studies. The optimal conditions for maximal Indigo carmine elimination were pH 2, 0.5 g of *G. venusta* peel, and 150 mg/L as the initial dye solution concentration (99%). Maximum elimination rate was 85% when the identical conditions were applied with *M. oleifera* seed, at pH 5. The Freundlich model ($R^2 = 0.9968$) was applied to the equilibrium data for *M. oleifera* seed, and the Langmuir model ($R^2 = 0.9957$) for *G. venusta* peel. For the adsorption of Indigo carmine and Methyl orange dyes onto *M. oleifera* seeds and *G. venusta* peel, adsorption was a pH-dependent process, with acidic pH providing the best elimination efficiency. *M. oleifera* seeds were effective in removing both the dyes, whereas *G. venusta* peel removed only Indigo carmine [2].

3.5 *Cucumber Peel*

Reactive black 5 and Direct blue 71 as contaminants were eliminated by Sayğılı and Güzel utilizing cucumber peel as an adsorbent. The effects of the modification were seen in fourier transform infrared spectra, scanning electron microscope pictures, x-ray diffraction, thermogravimetric analysis measurements, pH_{zpc} values, and elemental analysis data. Dye binding was greatly aided by protonated hydroxyl groups. Dye binding relied heavily on protonated hydroxyl groups. At pH 2, the dyes had the most adsorption on cucumber peel powder. Bio-sorption enhanced with the rising temperature and ionic strength. Intraparticle diffusion occurred in the sorption process, and the pseudo-first-order model matched with the experimental data well. The equilibrium bio-sorption results showed to be a good fit for the Langmuir isotherm. The maximum bio-sorption capacity was 129.87 mg/g for Direct blue 71, and 95.24 mg/g for Reactive black 5. The magnitude of thermodynamic parameters like DG° and DH°, as well as the Freundlich constants, demonstrated that the physisorption occurred during bio-sorption. The adsorbed dyes were recovered using NH_3 and NaOH solutions [48].

3.6 *Peanut and Onion Peel*

A low-cost adsorbent was developed using onion and peanut peels. In low and medium quantities of mixed wastewater, onion peel had the maximum transmittance of 80.1% and 75.6%, respectively. Peanuts achieved 69.4% transmittance at high concentrations of mixed coffee and Acid black 48 effluents. The ideal pH for mixed coffee and Acid black 48 effluents in the adsorption process was approximately 5.7. According to the isothermal study, the equilibrium adsorption data was better fit for the Langmuir isotherm than Freundlich isotherm, with correlation values of 0.83 for peanut hull and 0.82 for onion peel. In the elimination of binary waste of colors and coffee waste from effluent, adsorption of low-cost adsorbents improved with increasing dosage. Onion peel adsorption was maximum till a dose of 1.2g, which steadily reduced. Both onion peel and peanut hull showed appreciable adsorption capabilities as 58.1 mg/g$^-$and 46.3 mg/g, respectively [24].

3.7 *Potato Peel*

An experimental study on the removal of Crystal violet from an aqueous solution using potato peel, an agricultural byproduct was carried out. Fourier transform infrared spectroscopy, scanning electron microscope, and energy dispersive x-ray analysis were used to characterize the potato peel, giving information related to functional groups involved in absorption of Crystal violet, and morphology of potato peel. Potato peel dose, contact time, pH, and initial Crystal violet concentration had substantial impact on Crystal violet adsorption ability on potato peel. As the initial concentration of the dye, duration, and pH increased, the Crystal violet adsorption capacity also mounted, and the ideal pH for adsorption being pH 6. The kinetics of Crystal violet removal demonstrated a 20 min optimal contact time using a two-stage adsorption kinetic profile (initial fast and subsequent slow equilibrium). Potato peel adsorbent showed a monolayer adsorption capacity of 555 mg.g^{-1} and a Langmuir constant of 0.128 L/mg, demonstrating that the adsorbent was successful in eliminating Crystal violet from an aqueous solution. The pseudo-second-order model governed the adsorption kinetics of Crystal violet elimination by potato peel [34].

Cibacron blue P-3R was eliminated from an aqueous solution using phosphoric acid activated potato peel waste which was calcined at 800 °C. The adsorption kinetics was compatible with a chemisorption adsorption approach described by a pseudo-second-order model. Langmuir isotherm was utilized to verify the single-layer, homogeneous, and chemisorption adsorption process. According to the results of the experiments, calcined potato peel had a greater adsorption capacity than potato peel and acid activated potato peel [9].

3.8 *Tapioca Peel*

Vigneshwaran et al. succeeded in producing a sulfur-tethered adsorbent of tapioca peel biochar in 2021. Biochar made from tapioca peel was used to get rid of organic dyes like Rhodamine B and Malachite green from water. Tapioca peel was utilized as a starting material for biochar synthesis. According to the isotherm study, the sulfur-tethered adsorbent of tapioca peel biochar showed adsorption efficiencies of 33.10 for Rhodamine B and 30.18 for Malachite green dye molecules. The adsorption investigation employed dye solutions at a concentration of 25 mg/L in 50 cm^3. The highest adsorption was recorded at a pH of 8, and saturation took 2 h to achieve. The Freundlich isotherm model was utilized in the adsorption technique, which followed pseudo-second-order kinetics [57]. Various vegetable peels used for the purpose of dye removal have been summarized in Table 2.

A lot of research has been published on peel-based biomaterials for the removal of dyes from synthetic dye solutions and artificial textile wastewater; however, there is less research available on actual textile wastewater treatment. In terms of peels, the majority of research has focused on orange peel and banana peel, with less work done on green pea peel, luffa fruit peel, *M. oleifera* seed and *G. venusta* peel, and tapioca peel, which gives a future scope in the mentioned area.

4 Conclusion

The current study has revealed the potential of different peels to remove various dyes from wastewater. Peels could be used to replace the existing high-cost dye removal procedures in textile effluents with a low-cost, environmentally friendly bio-sorbent. Many of the peels are transformed to activated carbon with appreciable dye adsorption capabilities. The use of peel-based biomaterials as adsorbents in textile wastewater treatment can reduce biomaterial emissions while also removing dyes from textile wastewater. Some peels such as carbonized mandarin peel have shown almost 99.77% elimination of Methylene blue from textile water.

Although much work has been carried out on the use of peels, however more focus should be laid on variations in activated carbon preparations to improve adsorptive performance, and recovering and recycling peels, as well as removing unwanted adsorbents. Additional research could be conducted using a sample of actual textile

Table 2 Removal efficiency of various vegetable peel-based biomaterials for dyes in textile wastewater

Sr. No	Peel used	Dyes removed	Adsorption capacity	References
1	Green pea peel	Methylene blue	163.94 mg.g^{-1}	Dod et al. [16]
2	Water chestnut peel waste	Rhodamine B	3.0 mg.g^{-1}	Khan et al. [31]
3	Soybean hulls and luffa skin peels	Methyl orange	Greater than 62.5% (removal efficiency)	Chiong et al. [11]
4	*Moringa oleifera* seed and *Grewia venusta* peel	Methyl orange Indigo carmine	MO = 31.25 mg.g^{-1}MOS IC = 91.74 mg.g^{-1}MOS MO = 5.78 mg.g^{-1}GVP IC = 188.68 mg.g^{-1}GVP	Agbahoungbata et al. [2]
5	Cucumber peel	Reactive black 5 Direct blue 71	95.24 mg.g^{-1} 129.87 mg.g^{-1}	Sayğılı and Güzel [48]
6	Peanut hull Onion peel	Combination of Acid black 48 and coffee wastewater	46.3 mg/g 58.1 mg/g	Hung et al. [24]
7	Potato peel	Crystal violet	555 mg.g^{-1}	Lairini et al. [34]
8	Calcined potato peel Potato peel Acid activated Potato peel	Cibacron blue P-3R	270.3 mg/g 100 mg/g 125 mg/g	Bouhadjra et al. [9]
9	Tapioca peel	Malachite green Rhodamine B	30.18 mg. g^{-1} 33.10 mg.g^{-1}	Vigneshwaran et al. [57]

wastewater effluents to examine the impact of combining and changing bio-sorbents for improved absorption capacity.

References

1. Adegoke KA, Bello OS (2015) Dye sequestration using agricultural wastes as adsorbents. Water Resour Ind 12:8–24. https://doi.org/10.1016/j.wri.2015.09.002
2. Agbahoungbata MY, Fatombi JK, Ayedoun MA, Idohou E, Sagbo EV, Osseni SA, Mama D, Aminou T (2016) Removal of reactive dyes from their aqueous solutions using *Moringa oleifera* seeds and *Grewia venusta* peel. Desalin Water Treat 57:22609–22617. https://doi.org/10.1080/19443994.2015.1132478
3. Ahmad MA, Eusoff MA, Oladoye PO, Adegoke KA, Bello OS (2021) Optimization and batch studies on adsorption of methylene blue dye using pomegranate fruit peel based adsorbent. Chem Data Collections 32:100676. https://doi.org/10.1016/j.cdc.2021.100676
4. Ahmad R, Kumar R (2010) Adsorption studies of hazardous malachite green onto treated ginger waste. J Environ Manage 91:1032–1038. https://doi.org/10.1016/j.jenvman.2009.12.016

5. Ali A, Saeed K, Mabood F (2016) Removal of chromium (VI) from aqueous medium using chemically modified banana peels as efficient low-cost adsorbent. Alex Eng J 55:2933–2942. https://doi.org/10.1016/j.aej.2016.05.011
6. Arami M, Limaee NY, Mahmoodi NM, Tabrizi NS (2005) Removal of dyes from colored textile wastewater by orange peel adsorbent: equilibrium and kinetic studies. J Colloid Interface Sci 288:371–376. https://doi.org/10.1016/j.jcis.2005.03.020
7. Ardejani FD, Badii K, Limaee NY, Mahmoodi NM, Arami M, Shafaei SZ, Mirhabibi AR (2007) Numerical modelling and laboratory studies on the removal of Direct red 23 and Direct red 80 dyes from textile effluents using orange peel, a low-cost adsorbent. Dyes Pigm 73:178–185. https://doi.org/10.1016/j.dyepig.2005.11.011
8. Bakar AA, Wan Mazlan WW, Akbar NA, Badrealam S, Ali KM (2019) Agriculture waste from banana peel as low cost adsorbent in treating methylene blue from batik textile waste water effluents. J Phys: Conf Ser 1349:1. https://doi.org/10.1088/1742-6596/1349/1/012078
9. Bouhadjra K, Lemlikchi W, Ferhati A, Mignard S (2021) Enhancing removal efficiency of anionic dye (Cibacron blue) using waste potato peels powder. Sci Rep 11:1–10. https://doi.org/10.1038/s41598-020-79069-5
10. Chakraborty S, Chowdhury S, Saha PD (2011) Adsorption of Crystal violet from aqueous solution onto NaOH-modified rice husk. Carbohydr Polym 86:1533–1541. https://doi.org/10.1016/j.carbpol.2011.06.058
11. Chiong T, Lau SY, Lek ZH, Koh BY, Danquah MK (2016) Enzymatic treatment of Methyl orange dye in synthetic wastewater by plant-based peroxidase enzymes. J Environ Chem Eng 4:2500–2509. https://doi.org/10.1016/j.jece.2016.04.030
12. Chowdhury AN, Rahim A, Ferdosi YJ, Azam MS, Hossain MM (2010) Cobalt–nickel mixed oxide surface: a promising adsorbent for the removal of PR dye from water. Appl Surf Sci 256:3718–3724. https://doi.org/10.1016/j.apsusc.2010.01.013
13. Chu H, Zhang Y, Dong B, Zhou X, Cao D, Qiang Z, Yu Z, Wang H (2012) Pretreatment of micro-polluted surface water with a biologically enhanced PAC-diatomite dynamic membrane reactor to produce drinking water. Desalin Water Treat 40:84–91. https://doi.org/10.1080/19443994.2012.671152
14. de Campos V-C, Marin-morales MA (2013) Azo dyes: characterization and toxicity—a review. Text Light Ind Sci Technol 2:85–103
15. De Gisi S, Lofrano G, Grassi M, Notarnicola M (2016) Characteristics and adsorption capacities of low-cost sorbents for wastewater treatment: a review. Sustain Mater Technol 9:10–40
16. Dod R, Banerjee G, Saini S (2012) Adsorption of Methylene blue using green pea peels (*Pisum sativum*): a cost-effective option for dye-based wastewater treatment. Biotechnol Bioprocess Eng 17:862–874. https://doi.org/10.1007/s12257-011-0614-5
17. El-Said AG, Mansour HF, Gamal AM (2012) Potential application of orange peel (op) as an eco-friendly adsorbent for textile dyeing effluents 7:1–13
18. Fernández-Sempere J, Ruiz-Beviá F, Salcedo-Díaz R, García-Algado P, García-Rodríguez M (2012) Digital holographic interferometry visualization of PEG-10000 accumulation on an acetate cellulose membrane: assessment of polarization layer and adsorption phenomenon. Desalin Water Treat 42:49–56. https://doi.org/10.1080/19443994.2012.683105
19. Freitas TKFS, Policiano Almeida CA, Manholer DD, Geraldino HCL, de Souza MTF, Garcia JC (2018) Review of utilization plant-based coagulants as alternatives to textile wastewater treatment. In: Detox Fashion. Springer, Singapore, pp 27–79. https://doi.org/10.1007/978-981-10-4780-0_2
20. Fungaro DA, Yamaura M, Carvalho TEM (2011) Adsorption of anionic dyes from aqueous solution on zeolite from fly ash-iron oxide magnetic nanocomposite. J At Mol Sci 2:305–316. https://doi.org/10.4208/jams.032211.041211a
21. Gupta VK (2009) Application of low-cost adsorbents for dye removal—a review. J Environ Manage 90:2313–2342. https://doi.org/10.1016/j.jenvman.2008.11.017
22. Huang R, Liu Q, Huo J, Yang B (2017) Adsorption of Methyl orange onto protonated cross-linked chitosan. Arab J Chem 10:24–32. https://doi.org/10.1016/j.arabjc.2013.05.017

23. Humelnicu I, Băiceanu A, Ignat ME, Dulman V (2017) The removal of Basic blue 41 textile dye from aqueous solution by adsorption onto natural zeolitic tuff: kinetics and thermodynamics. Process Saf Environ Prot 105:274–287. https://doi.org/10.1016/j.psep.2016.11.016
24. Hung YT, Pamula ASP, Paul HH (2020) Treatment of combined Acid black 48 and coffee wastewater by low-cost adsorbents. Environ Res Eng Manag 76:47–61. https://doi.org/10.5755/j01.erem.76.3.25338
25. Jawad AH, Kadhum AM, Ngoh YS (2018) Applicability of dragon fruit (*Hylocereus polyrhizus*) peels as low-cost biosorbent for adsorption of Methylene blue from aqueous solution: kinetics, equilibrium and thermodynamics studies. Desalin Water Treat 109:231–240. https://doi.org/10.5004/dwt.2018.21976
26. Jegatheesan V, Pramanik BK, Chen J, Navaratna D, Chang CY, Shu L (2016) Treatment of textile wastewater with membrane bioreactor: a critical review. Biores Technol 204:202–212. https://doi.org/10.1016/j.biortech.2016.01.006
27. Kabbout R, Taha S (2014) Biodecolorization of textile dye effluent by biosorption on fungal biomass materials. Phys Procedia 55:437–444. https://doi.org/10.1016/j.phpro.2014.07.063
28. Katheresan V, Kansedo J, Lau SY (2018) Efficiency of various recent wastewater dye removal methods: a review. J Environ Chem Eng 6:4676–4697. https://doi.org/10.1016/j.jece.2018.06.060
29. Khaled A, El Nemr A, El-Sikaily A, Abdelwahab O (2009) Treatment of artificial textile dye effluent containing Direct yellow 12 by orange peel carbon. Desalination 238:210–232
30. Khaled A, Nemr AEl, El-Sikaily A, Abdelwahab O (2009) Removal of Direct N Blue-106 from artificial textile dye effluent using activated carbon from orange peel: adsorption isotherm and kinetic studies. J Hazard Mater 165:100–110. https://doi.org/10.1016/j.jhazmat.2008.09.122
31. Khan TA, Nazir M, Khan EA (2013) Adsorptive removal of Rhodamine B from textile wastewater using water chestnut (*Trapa natans L.*) peel: adsorption dynamics and kinetic studies. Toxicol Environ Chem 95:919–931. https://doi.org/10.1080/02772248.2013.840369
32. Khanna S, Rattan VK (2017) Removal of Acid red 1 from aqueous waste streams using peel of *Cucumis sativus* fruit. Equilibrium studies. J Chem Technol Metall 52:803–811
33. Krishni RR, Foo KY, Hameed BH (2014) Food cannery effluent, pineapple peel as an effective low-cost biosorbent for removing cationic dye from aqueous solutions. Desalin Water Treat 52:6096–6103. https://doi.org/10.1080/19443994.2013.815686
34. Lairini S, El Mahtal K, Miyah Y, Tanji K, Guissi S, Boumchita S, Zerrouq F (2017) The adsorption of Crystal violet from aqueous solution by using potato peels (*Solanum tuberosum*): equilibrium and kinetic studies. J Mater Environ Sci 8:3252–3261
35. Lerch A, Siebdrath N, Berg P, Heijnen M, Gitis V, Uhl W (2012) Fouling minimised reclamation of secondary effluents with reverse osmosis (ReSeRO). Desalin Water Treat 42:181–188. https://doi.org/10.1080/19443994.2012.683139
36. Mane RS, Bhusari VN (2012) Removal of colour (dyes) from textile effluent by adsorption using orange and banana peel. Int J Eng 2:1997–2004
37. Marchis T, Avetta P, Bianco-Prevot A, Fabbri D, Viscardi G, Laurenti E (2011) Oxidative degradation of remazol turquoise blue G 133 by soybean peroxidase. J Inorg Biochem 105:321–327. https://doi.org/10.1016/j.jinorgbio.2010.11.009
38. Meyn T, Altmann J, Leiknes T (2012) In-line coagulation prior to ceramic microfiltration for surface water treatment—minimisation of flocculation pre-treatment. Desalin Water Treat 42:163–176. https://doi.org/10.5004/dwt.2012.2590
39. Nan J, He W (2012) Characteristic analysis on morphological evolution of suspended particles in water during dynamic flocculation process. Desalin Water Treat 41:44. https://doi.org/10.1080/19443994.2012.664676
40. Nayak SS, Mirgane NA, Shivankar VS, Pathade KB, Wadhawa GC (2021) Adsorption of Methylene blue dye over activated charcoal from the fruit peel of plant hydnocarpus pentandra. Mater Today: Proc 37:2302–2305. https://doi.org/10.1016/j.matpr.2020.07.728
41. Nor NM, Hadibarata T, Yusop Z, Lazim ZM (2015) Removal of brilliant green and Procion red dyes from aqueous solution by adsorption using selected agricultural wastes. Jurnal Teknologi 74. https://doi.org/10.11113/jt.v74.4880

42. Nur Dini D, Puzi SM, Khalijah S, Rozi M (2019) New application of pomegranate peel waste: the decontamination of toxic Methylene blue dye from textile wastewater. J Eng Res Educ 11:45–58
43. Pang YH, Liu J, Sun MM, Wang L, Shen XF (2020) Preparation of biomass activated carbon from pomelo peel and its adsorption of textile dye and formaldehyde. 24–29
44. Pishgar M, Farizadeh K, Souhangir M, Pishgar A (2019) Adsorptive treatment of hazardous Methylene blue dye from textile wastewater using banana peel waste as a cellulosic adsorbent : isotherm study. 11–13
45. Policiano Almeida CA, Zanela TMP, Machado C, Flores JAA, Scheibe LF, Hankins NP, Debacher NA (2016) Removal of Methylene blue by adsorption on aluminosilicate waste: equilibrium, kinetic and thermodynamic parameters. Water Sci Technol 74:2437–2445. https://doi.org/10.2166/wst.2016.408
46. Rahman A, Urabe T, Kishimoto N (2013) Color removal of reactive Procion dyes by clay adsorbents. Procedia Environ Sci 17:270–278. https://doi.org/10.1016/j.proenv.2013.02.038
47. Raman CD, Kanmani S (2016) Textile dye degradation using nano zero valent iron: a review. J Environ Manage 177:341–355. https://doi.org/10.1016/j.jenvman.2016.04.034
48. Sayğılı GA, Güzel F (2017) Chemical modification of a cellulose-based material to improve its adsorption capacity for anionic dyes. J Dispersion Sci Technol 38:381–392. https://doi.org/10.1080/01932691.2016.1170608
49. Sbizzaro Filho D, Bota GB, Borri RB, Teran FJC (2012) Electrocoagulation/flotation followed by fluidized bed anaerobic reactor applied to tannery effluent treatment. Desalin Water Treat 37:359–363. https://doi.org/10.1080/19443994.2012.661292
50. Schurer R, Janssen A, Villacorte L, Kennedy M (2012) Performance of ultrafiltration and coagulation in an Uf-RO seawater desalination demonstration plant. Desalin Water Treat 42:57–64. https://doi.org/10.1080/19443994.2012.683107
51. Sivanatham M, Venkata Rao Tata B, Anand BC (2012) Synthesis of crown ether functionalized polyacrylamide gel beads and their extraction behaviour for strontium ions. Desalin Water Treat 38:8–14. https://doi.org/10.1080/19443994.2012.664106
52. Solmaz SKA, Üstün GE, Çiner F (2011) Treatability of organized industrial district (OID) effluent for reuse in agriculture. Desalin Water Treat 33:156–163. https://doi.org/10.5004/dwt.2011.2626
53. Sud D, Mahajan G, Kaur MP (2008) Agricultural waste material as potential adsorbent for sequestering heavy metal ions from aqueous solutions—a review. Biores Technol 99:6017–6027. https://doi.org/10.1016/j.biortech.2007.11.064
54. Sumalatha B, Kumar YP, Kumar KK, Babu DJ, Narayana AV, Das KM, Venkateswarulu TC (2014) Removal of indigo carmine from aqueous solution by using activated carbon. Res J Pharm Biol Chem 5:1–12
55. Torres N, García-raurich J, Canal JM, Riba-moliner M (2021) Orange peel-based adsorbent for the removal of C. I. Disperse blue 56 from textile dye industry effluents. https://doi.org/10.20944/preprints202103.0504.v1
56. Unugul T, Nigiz FU (2020) Preparation and characterization an active carbon adsorbent from waste mandarin peel and determination of adsorption behavior on removal of synthetic dye solutions. Water Air Soil Pollution 231. https://doi.org/10.1007/s11270-020-04903-5
57. Vigneshwaran S, Sirajudheen P, Karthikeyan P, Meenakshi S (2021) Fabrication of sulfur-doped biochar derived from tapioca peel waste with superior adsorption performance for the removal of malachite green and rhodamine B dyes. Surf Interfaces 23:100920. https://doi.org/10.1016/j.surfin.2020.100920
58. Yagub MT, Sen TK, Afroze S, Ang HM (2014) Dye and its removal from aqueous solution by adsorption: a review. Adv Coll Interface Sci 209:172–184. https://doi.org/10.1016/j.cis.2014.04.002
59. Yu H, Wang T, Dai W, Yu L, Ma N (2018) Competitive adsorption of dye species onto biomass nanoporous carbon in single and bicomponent systems. Braz J Chem Eng 35:253–264. https://doi.org/10.1590/0104-6632.20180351s1s20160

Applications of Luffa Based Biomaterials in Textile Waste Water

Sanjay B. Bhagwat and Dipika Jaspal

Abstract Worldwide textile industry is one of the most dominating industries among all-existing, contributing to 14% of the total industrial production and to the economy. Even though textile industries have enriched our lives by giving us valuable products, at the same time these have also led to environmental pollution. Textile waste contains dyes along with toxic metal ions which lead to hazardos effects on the environment. Different technologies have been explored for textile waste water treatment including adsorption, coagulation, precipitation, filtration, ion exchange, electrodialysis, membrane separation and oxidation. Adsorption is considered as the most effective and economical method for the treatment of textile waste water. Even though a variety of adsorbents are available, cost of these adsorbents is a major concern and therefore the development of new innovative low-cost, natural, nontoxic and biodegradable adsorbents is the need of recent times. This chapter discusses the use of Luffa Cylindrica, a vine in cucumber family, as a bio adsorbent, for textile waste water. Recent developments related to modification of Luffa Cylindrica have been discussed including the parameters affecting adsorption process, characterization methods, kinetics model and adsorption capacity. Luffa Cylindrica has been found to be the best biomaterial on account of its ability to capture textile dyes such as Trypan blue, Reactive orange dye, Red procion, Yellow procion, Malachite green, Basic magenta and metal ions including Cr, Cu, Ni and Pb. The potential applications of Luffa Cylindrica for the removal of hazardous dyes, metal ions and toxic pollutants from aqueous environment have been reviewed. Several studies have clearly indicated that Luffa Cylindrica is an excellent bio feedstock for the preparation of activated carbon and biochar materials which has shown extraordinary performance in the removal of dyes, metal ion and toxic emerging pollutants from aqueous media. Luffa has shown promising results for capturing popular textile dyes including malachite green and Congo red and has shown adsorption capacity in the

S. B. Bhagwat · D. Jaspal (✉)
Department of Applied Science, Symbiosis Institute of Technology (SIT), Symbiosis International (Deemed University) (SIU), Lavale, Pune, Maharashtra 412115, India
e-mail: dipikaj@sitpune.edu.in

S. B. Bhagwat
School of Chemistry, Dr. Vishwanath Karad MIT World Peace University, Pune 411038, India

S. S. Muthu and A. Khadir (eds.), *Textile Wastewater Treatment*, Sustainable Textiles: Production, Processing, Manufacturing & Chemistry,
https://doi.org/10.1007/978-981-19-2852-9_5

range of 20–90 mg/g. Luffa has also proven its worth for capturing of heavy metal ions from waste water. Polyaniline-coated Luffa has shown 300 mg/g adsorption capacity towards chromium (VI), which is greater than various available adsorbent materials. Recently worldwide researchers have used Luffa sponge for the removal of various toxic pharmaceutical compounds from aqueous environment. Biochar derived from Luffa sponge has shown highest 250 mg/g adsorption capacity towards adsorption of toxic norfloxacin, proving to be a better alternative adsorbent.

Keywords Textile waste · Hazardous · Removal · Toxic pollutants · Adsorption · Azo dyes · Dye removal · Heavy metals · Lignocellulosic · Luffa Cylindrica

1 Introduction

Textile industry is globally recognized and one of the key industries due to its contribution in enhancing economy, providing employment opportunities and earning foreign exchange through export. The utilization of dyes in world is more than 10,000 tonnes per year and around 100 tonnes per year of dyes are being released into environment per year [29]. According to a new report by grand view research, the global textile market size will reach to USD 1412.5 billion by 2028. The textile market is projected to expand at a compound annual growth rate of 4.4% from 2021 to 2028. Even though textile industries have enriched our life by producing highly valuable products such as textile yarn, fibre, fabric and apparels, at the same time these have also damaged the environment and created pollution. These have contributed majorly to pollution on account of usage of at least 8000 chemicals and large amount of water. Textile industry utilizes about 1.6 million litres of water per day for the manufacturing of about 8000 kg of fabric. The effluent produced by the dying industries is almost 54% [42]. As per the World bank reports, 17–20% of textile industry water pollution has originated from textile treatments [12]. Textile waste contains highly carcinogenic halogenated solvents such as chloroform, dichloromethane and carbon tetrachloride. Textile waste also has toxic chemical compounds such as aliphatic hydrocarbons and aromatic hydrocarbons and oxygen containing compounds, salts, heavy metals, enzymes, oxidizing and reducing agents [28, 46]. Dyes in the textile effluent can cause hazardous effects on not only human life but also on the environment. Dyes are mutagenic and carcinogenic in nature and affect the whole food chain [3]. Scientific reports indicate that dyes can disturb photosynthesis in plants [43] decrease dissolved oxygen level in water, and affect the overall growth of the plants. Therefore, textile waste water must be treated properly before releasing into environment.

Various physical, chemical and biological methods have been developed to treat the textile waste water [17]. Most of these have limitations such as partial pollutant removal, excessive use of reagent, high energy requirements and generation of hazardous sludge [9]. Traditional technologies are not capable of eliminating synthetic dyes because of their extraordinary chemical stability [34]. Among the methods explored adsorption is found to be superior due the ease in mode of operation and simple mechanism. Even though activated carbon is popular adsorbent, but it has high cost therefore, some novel, environmentally friendly, efficient and economical adsorbents are needed having low cost and with efficient elimination of contaminants from waste water. One such adsorbent is Luffa Cylindrica, a lignocellulosic material, has been explored as an efficient bio adsorbent due to its inherent characteristics such as biodegradability, high stability, non-toxicity, ease of availability and low cost [35].

The present chapter highlights the use of an adsorbent Luffa Cylindrica, and modified Luffa Cylindrica as biomaterial for adsorption of dyes and heavy metals. Luffa Cylindrica has been used in a variety of forms such as fibres, sponges, activated carbon and biochar for textile waste water treatments. This chapter also summarizes the recent modifications done on the mentioned material for the exclusion of various metal ions, dyes, toxic pollutants and pharmaceutical drugs from textile waste water.

2 Hazardous Effect of Dyes on Environment

Dyes used in textile industry have been divided into two basic categories: natural dyes and synthetic dyes (Fig. 1) [22]. The focus is now shifting to the use of synthetic dyes from natural dyes for different applications. Azo dyes are known to provide varied colour intensities. Worldwide out of the total manufacturing of dyes, 50% dyes are mainly azo dyes used in textile manufacturing processes. These are found to be major contaminants in textile waste [19].

Presence of untreated dyes in water bodies has shown hazardous effects on environment. It has been reported that dyes hamper photosynthesis, as these block

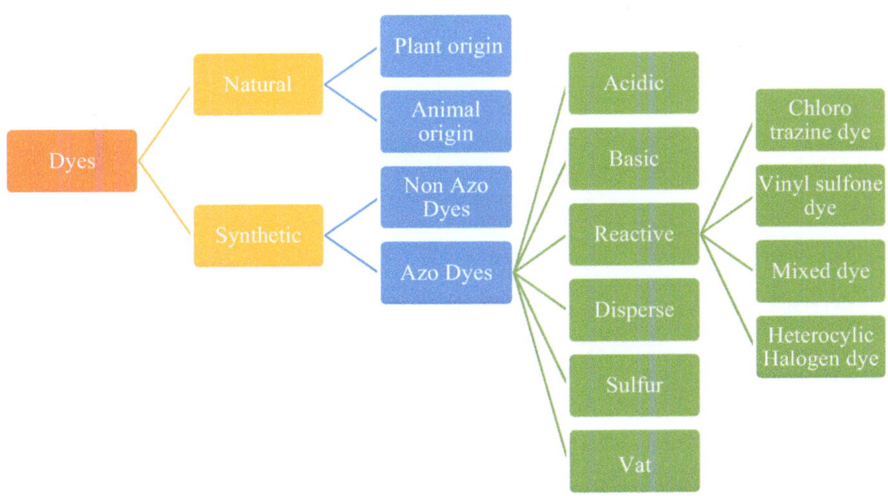

Fig. 1 Classification of dyes used in textile industries. (*Source* [22])

the radiations entering into the water streams. The biggest hurdle associated with treatment of waste water in presence of dyes is associated to degradation process.

3 Conventional Methods and Materials Used for Textile Waste Water Treatments

Textile waste water treatments are classified into three main groups physical, chemical and biological methods (Fig. 2).

3.1 Physical Methods

Membrane-based processes used in textile waste water treatment include ultrafiltration, nanofiltration and reverse osmosis. Ultrafiltration is less effective in textile waste water treatment as it is not capable of eliminating low molecular weight dyes. Combination of brackish water reverse osmosis membrane and sea water reverse osmosis membrane has been successfully used, with 95% water recovery [36]. Use of membrane-based technologies are efficient, but membrane fouling and membrane replacement are the two major problems which exist. Adsorption is the best method for textile waste water treatment due to characteristics such as high efficiency and reusability. Even though activated carbon is the best adsorbent, its high cost limits its use in textile treatments.

Recently the use of bio-based adsorbents has attracted attention as they offer extraordinary features such high abundance, low cost, biodegradable, high surface area and favourable kinetics [39]. Many plants and fruit-based materials such as

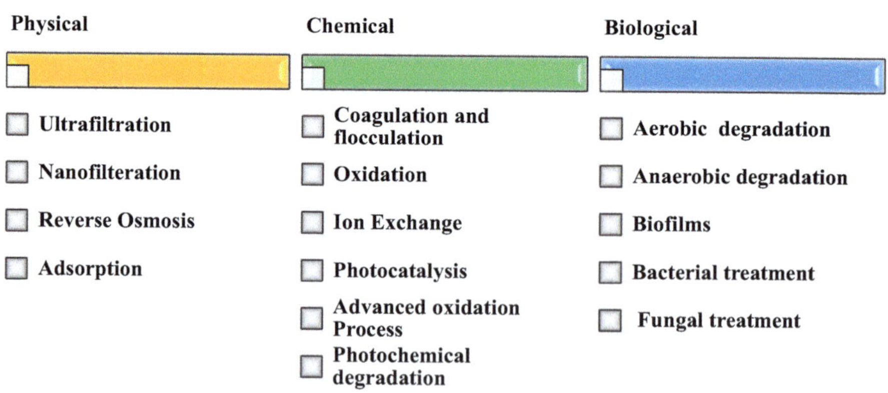

Fig. 2 Various methods used in textile waste water treatments

neem leaf powder, indica leaf powder, pineapple leaf powder, saw dust, orange peel, peel of cucumis sativa, banana trunk fibres, sugarcane fibres, chitin shell, peanut hull and rice husk have been explored, in addition to the others for contaminant removal from waste water [22].

3.2 Chemical Methods

Coagulation and flocculation are popular methods for textile effluent treatment on account of short internment time and ease to manage [51]. Most preferred chemical coagulants include alum, lime and ferric sulphate for the waste water treatment [28]. Oxidation is another chemical process in which transformation of pollutants into simple, harmless and biodegradable compounds takes place by chemical oxidizing agents such as ozone, chlorine, Fenton's reagent [40]. Ion exchange method involves the use of resins for waste water treatment. Ion exchange resins are microporous, cross-linked, long chain organic polymers with loosely bound ions for ion exchange process. The main limitations of ion exchange method are the high cost involved and secondary pollution taking place, due to resin regeneration [49]. Advanced oxidation processes usually involve production of hydroxyl radical using ozone, combination of hydrogen peroxide and ultraviolet light and other combination such as O_3 /ultraviolet, O_3/H_2O. Ozonization is also an advanced oxidation process in waste water treatment due to the formation of hydroxyl radical in presence of organic matter in large amount. Use of homogenous (Fe^{2+}, Mn^{2+}, Cu^{2+}) or heterogenous catalyst (TiO_2, Al_2O_3, MnO_2, CuO) can be very efficient in speeding up ozonization, and finally decomposition of targeted compounds [48]. Various types of advanced oxidation processes have been highlighted in Fig. 3.

Recently a blend of heterogenous photocatalysis (ZnO/ultraviolet or TiO_2/ultraviolet) and biological process have been used in treatment of synthetic and industrial textile waste water treatment. Coupling of photocatalytic (TiO_2/ultraviolet) and biological treatments have resulted in 97% of colour and 63% of total organic carbon removal, from industrial waste water respectively. For synthetic waste water, the combined photocatalytic (ZnO/UV or TiO_2/UV) and biological systems have shown extraordinary result with 98% of colour removal and more than 80% total organic carbon reduction [14].

Photochemical degradation is one of the emerging options where dye molecules are cleaved to CO_2 and water. Photochemical degradation offers an advantage as sludge is not generated [15] and appreciable reduction in foul odours has been reported [28]. Chemical methods are efficient but the high cost and generation of secondary pollutants limit their usage in textile waste water treatments.

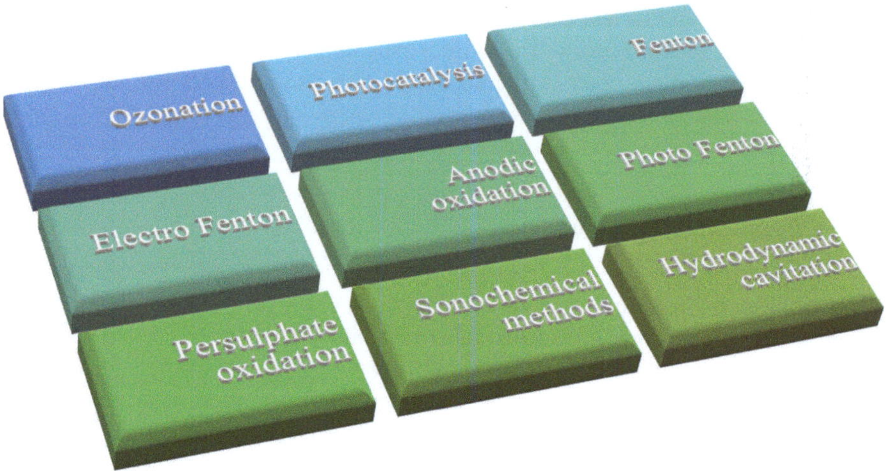

Fig. 3 Types of advanced oxidation process

3.3 Biological Methods

Biological methods are preferred for textile waste water treatment as these methods offer advantages of being eco-friendly, cost-effective with low consumption of water and less sludge production [21]. Aerobic degradation utilizes bacteria for decomposing textile waste into harmless inorganic solids. This is carried by using activated sludge and biofilms. Anaerobic degradation involves the use of anaerobic bacteria which cleaves the organic matter under anaerobic condition. Effluent release from textile industries has high amount of organic matter which can be successfully treated under anaerobic with impressive results. Biofilms are biological processes that involve the use of microorganisms which form biofilms on fixed object and flowing waste water is purified by mere exposure to biofilms [30]. Among the biological methods, use of pure fungal culture, pure bacterial culture and yeast pure culture have been found to be effective approaches for treatment of textile waste water. Many microbial sources such as fungi, bacteria, yeasts, algae and actinomycetes have been used for textile waste water treatments. Phanerochaete chrysosporium is found to be popular fungus used for decolorizing various dyes from textile waste water [32].

3.4 Adsorption for Textile Waste Water Treatment

In last several decades, adsorption has received the highest attention due to low cost and its effectiveness in removal of pollutants [28]. Adsorption refers to the accumulation of substances on surfaces and is used to remove soluble organic pollutants. The method offers numerous advantages such as regeneration at low cost, sludge-free operation and ease of recovery of sorbent material [11]. Hazardous chemicals such as pesticides, cyanides and phenols have been eliminated from aqueous environment using adsorption. Adsorption has been used to remove various dyes such as methylene blue, reactive orange dye, congo red and malachite green, red procion, yellow procion, etc., from aqueous environment. Study of influence of various parameters on adsorption includes adsorbent dosages, pH, temperature, particle size, contact time and surface area of adsorbent.

The efficiency of adsorbent for dye or metal ion removal have been determined on the basis of its adsorption capacity and reusability. Activated carbon is one the most effective adsorbents for effluent treatment [45]. Activated carbon is an effective adsorbent due to characteristics such as porous structure low cost, nontoxic, high surface area and good efficiency [38]. Activated carbon is usually prepared using carbonaceous materials such as saw dust, peanut shell, cellulosic material, maize leaves, tea waste, coconut husks, palm fruit bunches and Luffa Cylindrica.

4 Luffa Cylindrica as an Adsorbent

Luffa Cylindrica is the dried fruit of a sponge-type plant, from curcubitacea family (Fig. 4). Luffa Cylindrica growth rate is about 8000 units per hectare in Asian countries. Luffa Cylindrica can spread 100,000 dozen per year per turn. It is cultivated in countries including China, Korea, India, Japan and Central America. Luffa Cylindrica has gained considerable attention worldwide due to its inherent characteristics

Fig. 4 Luffa Cylindrica sponge. (*Source*https://www.amazon.in/)

such as biodegradability, low cost, abundance in nature, low weight and non-toxicity [37].

4.1 Composition and Characteristics of Luffa Cylindrica

Luffa fibres have inbuilt characteristics such as large surface area and fast adsorption kinetics making them potential biosorbent materials [9] (Fig. 5). As per chemical composition, Luffa Cylindrica contains 63% cellulose, 18% hemicellulose and 11.5% lignin. One of the promising features of Luffa fibres is that it has a fibrous vascular system that allows efficient exclusion of water pollutants [23].

Luffa Cylindrica is a lignocellulosic material and structurally comprises of lignin, cellulose and hemicellulose with variety of functional groups such carbonyl, carboxyl, amine and hydroxyl groups which are the main adsorptive sites. Presence of these adsorptive sites makes this bio-based material capable of capturing of metal ions, which involves phenomena such as ion exchange or complex formation or hydrogen bonding [27].

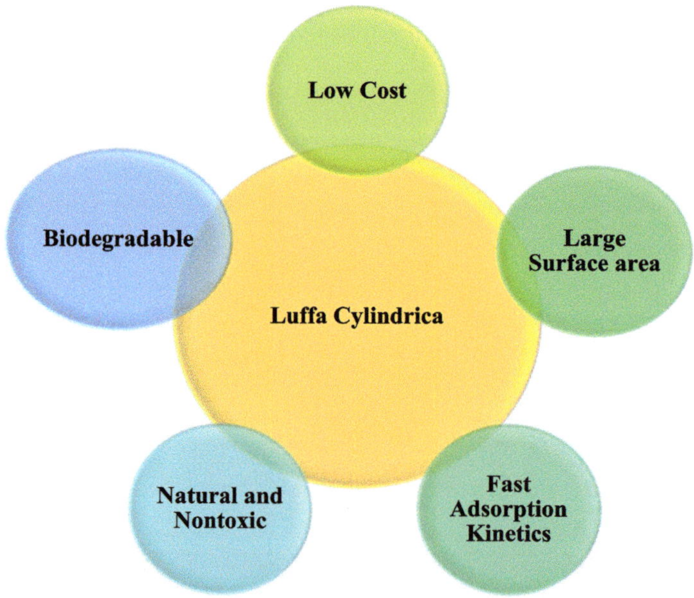

Fig. 5 Characteristics of Luffa Cylindrica as bio adsorbent

4.2 *Luffa Cylindrica for Metal Ion Adsorption*

Luffa Cylindrica has been used for removing hexavalent chromium from textile waste water. Activated carbon has been prepared from Luffa Cylindrica by carbonization and characterized by using scanning electron microscopy. Adsorption studies were carried out by means of the Langmuir and Freundlich adsorption isotherm models. Freundlich isotherm has shown better connection for Cr (VI) adsorption. Adsorbent was found to be efficient at pH 8 with contact time of 60 min and adsorption efficiency was 188.50 mg/g. The pseudo-first-order kinetic model was best fitted than the pseudo-second-order kinetic model [33]. Functional groups present on Luffa Cylindrica sponge have been found to be responsible for capturing metal ions from aqueous solution. Luffa Cylindrica sponge was used to adsorb Pb^{2+} ions from contaminated water. Surface morphology of adsorbent was studied by using scanning electron microscopy coupled with energy dispersion spectroscopy. The Brunauer–Emmett–Teller Surface area was reported as 6 m^2/g indicating that sorption tendency was due to chemical interaction of functional group with Pb^{2+} ion. Pb^{2+} ions were getting adsorbed on Luffa sponge due to ion exchange mechanism. pH 6 was found to be optimum for adsorption of Pb^{2+} ion on Luffa sponge. The highest adsorption capacity of Pb^{2+} on Luffa sponge was found to be 75.853 mg/g [4].

In an interesting research, immobilized bio sorbent was developed using Luffa sponge and microalga Chlorella sorokiniana for trapping nickel (II) from aqueous solution as well as electroplating effluent. Exclusion of nickel (II) by this adsorbent was found to be depend on various factors including pH, concentration of bio sorbent as well as metal ions. Loofa sponge-immobilized biomass of Chlorella sorokiniana showed extraordinary physical and chemical stability as no rupture of immobilized biomass was seen during its reuse in adsorption desorption cycles. Prepared adsorbent could be reproduced and can be reutilized at least for 7 cycles [5].

In a novel study, activated biochar was prepared from Luffa Cylindrica fibres eradicated copper (II) from aqueous solutions. Luffa Cylindrica fibres were thermally treated at 700 °C for 1 h to get biochar. Biochar was further activated by subjecting it to 8 M HNO_3 for 3 h at 80 °C. The activated biochar was characterized by Fourier transform infrared spectroscopy. Adsorption studies were carried out by considering factors such as mass of adsorbent, pH, initial metal ion concentration, ionic strength, contact time and temperature. Fourier transform infrared spectrum clearly indicates that oxidation of Luffa biochar by nitric acid leads to generation of carboxylic group on its surface. Fourier transform infrared spectrum was used to prove that formation of inner sphere complexes between carboxylic group and Cu (II). The adsorption capacity of the activated biochar fibres was reported to be 248 g kg^{-1} for Cu (II) [26].

Recently a research, the surface of Luffa Cylindrica, was modified by polyaniline for effective removal of chromate anion from aqueous solution. Polyaniline coating was achieved by carrying out in situ polymerization of aniline. Coating was confirmed by various characterization methods including Fourier transform infrared

Table 1 Applications of Luffa Cylindrica for capturing metal ions

Adsorbent	Targeted metal ion	Maximum monolayer adsorption capacity	References
Luffa Cylindrica activated carbon	Chromium (VI)	188.50 mg/g	Kenechi Nwosu-Obieogu et al. [33]
Luffa Cylindrica Sponge	Lead (II)	75.85 mg/g	Adewale Adewuyi et al. [4]
Luffa sponge and microalga	Nickel (II)	60.57 mg/g	Akhtar [5]
Activated biochar from Luffa Cylindrica	Cu (II)	248 g kg^{-1}	Ioanna [26]
Polyaniline coated Luffa Cylindrica	Cr (VI)	300 mg/g	Awatef [16]
Luffa-epichlorohydrin–Polyethyleneimine	U(VI)	208.3 mg/g	Shouzheng [47]
Luffa Cylindrica-poly glycidyl methacrylate-polyethyleneimine	U(VI)	438.59 mg/	Shouzheng [47]

spectroscopy, Raman spectroscopy, scanning electron microscopy and X-ray diffraction spectroscopy. In this study, impact of various important parameters such as contact time, Polyaniline/Luffa dose, pH and temperature was studied to optimize the process. Cr (VI) rejection efficiency decreased with rise in temperature and efficiency was found to be 300 mg g^{-1} [16].

In one of the chemical modifications, Luffa fibres were employed for the adsorption of U(VI) from seawater; first, Luffa fibres were treated with NaOH followed by epichlorohydrin for introduction of epoxy group on fibre. After introduction of epoxy fibre, polyethyleneimine was anchored on fibre due to a ring opening reaction between epoxy and polyethyleneimine. This Luffa-epichlorohydrin–polyethyleneimine adsorbent showed a maximum monolayer adsorption capacity of 208.3 mg/g towards uranium (VI). In this study, another adsorbent Luffa Cylindrica-poly glycidyl methacrylate–polyethyleneimine was also prepared. This adsorbent was prepared by carrying out graft polymerization of glycidyl methacrylate onto Luffa fibres and then finally reaction with polyethyleneimine. This adsorbent had shown excellent adsorption capacity of 438.59 mg/g towards uranium (VI) adsorption from sea water [47] (Table 1).

4.3 *Luffa Cylindrica for Dye Removal*

Luffa Cylindrica fibres were applied for elimination of malachite green dye from aqueous solution. Effect of several factors such as temperature, pH and contact were studied for dye removal. Fourier transform infrared spectroscopy analysis was carried out in order to check the adsorption of malachite green on to Luffa fibres. On increasing pH from 3 to 5, the amount of dye adsorbed has rapidly increased but after pH 5 it was found to be constant. Experimental data was found to be fitting with Langmuir equation, confirming monolayer sorption on a homogenous surface. Adsorption capacity for malachite green was 29.4 mg/g [7]. Reactive orange dye removal was accomplished by developing two different types of activated carbon using Luffa. Luffa carbon 1 was manufactured by soaking dried Luffa in mixture of H_3PO_4 and $ZnCl_2$ followed by carbonization to get activated carbon 1. Luffa carbon 2 was prepared by soaking in 30% HNO_3 followed by carbonization to get activated carbon 2 adsorbent. Both the type of adsorbents had shown excellent adsorption at pH 1 and at 25 °C [1]. Noureddine Boudechiche and co-workers in the year 2015 used Luffa Cylindrica in powdered form for elimination of methylene blue dye from aqueous solution. This bio sorbent was characterized by using Fourier transform infrared spectroscopy and scanning electron microscopy analysis. The amount of methylene blue adsorbed (Q) was calculated by using Eq. (1).

$$\boldsymbol{Q} = \frac{\boldsymbol{A}_0 - \boldsymbol{A}_t}{\boldsymbol{m}} \times \boldsymbol{V} \tag{1}$$

Where, Q is the amount of dye adsorbed (mg/g), A_0, the initial dye concentration (mg /litre), A_t, concentration of dye at time t, m, mass of bio sorben and V, volume of solution (L). Dye removal percentage was computed by using Eq. (2)

$$\textbf{Dye removal}\ (\ \%) = \frac{(\boldsymbol{A}_0 - \boldsymbol{A}_t)}{\boldsymbol{A}_0} \times 100 \tag{2}$$

Data were well-fitted with the Langmuir model and this adsorbent revealed an adsorption capacity of 49.46 mg g^{-1} at 20 °C [13].

The Luffa sponge blended with Zn nanoparticles was synthesized and applied for exclusion of Trypan blue dye from contaminated water. Scanning electron microscopy was used to locate the changes before and after dye adsorption. The maximum dye elimination occurred at pH 7. Luffa sponge modified with Zn nanoparticles showed an adsorption capacity of 47.3 mg/g [31]. Red procion and yellow procion dyes from aqueous solution were eradicated using Luffa Cylindrica [35]. Suitable pH for adsorption and desorption processes were found to be at 2.5 and 12 respectively. The sorption capacities were 12.2 and 13.9 mg/g for yellow procion and red procion respectively. Luffa sponge was modified by simple chemical graft method by using pyromellitic dianhydride and was used for adsorption of basic magenta and methylene blue respectively, in an investigation. This prepared Luffa sponges revealed 83.5 and 85.5 mg g^{-1} adsorption capacity towards basic magenta and methylene blue respectively [52]. In another study, Luffa Cylindrica sponge was employed to pick up Brilliant green dye from aqueous environment using batch study. The optimum elimination of dye was found at pH 8 and highest monolayer adsorption capacity was found to be 18.2 mg/g. Brilliant green dye molecules were

Table 2 Luffa Cylindrica uses for dye removal in textile waste water treatment

Adsorbent	Targeted dye	Maximum monolayer adsorption capacity	References
Luffa Cylindrica	Malachite green	29.4 mg/g	Altinişik et al. [7]
Luffa carbon 1 and Luffa carbon 2	Reactive orange dye	38.3 mg/g using Luffa carbon 1 50 mg/g using Luffa carbon 2	Abdelwahab [1]
Raw Luffa Cylindrica	Methylene blue	49.46 mg/g	Boudechiche et al. [13]
Luffa sponge modified with Zn nanoparticles	Trypan blue	47.3 mg /g	Nadaroglu et al. [31]
Luffa Cylindrica	Red procion HE-7B Yellow procion H-4R	13.9 mg/g for red procion 12.2 for yellow procion	Oliveira et al. [35]
Luffa sponge modified by using pyromellitic dianhydride	Basic magenta Methylene blue	83.5 mg/g for basic magenta 85.5 mg/g for methylene blue	Yu et al. [52]
Luffa Cylindrica sponge	Brilliant green dye	18.2 mg/g	Segun Esan et al. [44]
Luffa Cylindrica fibres	Alpacide blue	9.63 mg/g	Kesraoui et al. [23]

adsorbed on Luffa surface due to hydrogen bonding [44]. Luffa Cylindrica fibres were also used to adsorbed popular textile alpacide blue dye from aqueous solution. Optimal parameters were achieved after 2 h of treatment in a batch system using an initial dye concentration of 20 mg/L, a mass of 1 g of Luffa fibres and at pH 2 for maximum quantity of biosorption dye. Freundlich model was well-fitted for a capturing of alpacide blue dye by Luffa fibres. Luffa fibres portrayed a biosorption capacity of 9.63 mg/g towards Alpacide blue [23] (Table 2).

4.4 *Luffa Cylindrica for Toxic Pollutant Removal*

Alkali-treated Luffa sponge was employed in removal of cationic surfactant 1-cetylpyridinium bromide from aqueous environment. Adsorption process was found to be dependent on by several parameters such as initial concentration, adsorbent dosage, solution pH and contact time. Fourier transform infrared spectroscopy was used to confirm the sorption of 1-cetylpyridinium bromide on to Luffa Cylindrica. Kinetic study proved that pseudo-second-order kinetic model was followed [50]. Lignocellulosic fibres were suitable precursors in preparation of activated carbon due to high hardness and density. Lignocellulosic fibres surfaces were modified by oxidation, and number of active sites were found to increase for effective capturing of various metal ions from aqueous solution (in form of carboxylic and phenolic group). In another novel study, Luffa Cylindrica was chemically modified using 2-chloroethyldiethylamine hydrochloride and subsequently cross-linked with the help of epichlorohydrin. Quaternization of cross-linked diethyl amino ethylated Luffa was achieved by using methyl iodide and checked for trapping of sulphate, nitrate and phosphate anions. Quaternized diethyl amino ethylated Luffa showed higher adsorption capacity for phosphate anions subsequently sulphate and nitrate [20]. Phenolic compounds are one of the key raw materials for production of dyes. At high concentration, these compounds are hazardous if absorbed through skin. Luffa Cylindrica fibres were used to remove phenol from aqueous environment. Maximum adsorption of phenol occurred at pH 7 and adsorption kinetics followed pseudo-second-order model [2]. In one of the interesting modifications, Luffa-oxidized carbon from Luffa was applied to trap caffeine from aqueous environment. Adsorption of caffeine occurred due to electrostatic attraction and the maximum monolayer adsorption capacity was 59.9 mg/g [8]. Recently in 2021, Luffa Cylindrica was explored as an inexpensive bio feedstock for synthesizing biochar for elimination of 4-nitro phenol from waste water [41]. Sulphuric acid-treated Luffa biochar showed excellent monolayer adsorption capacity as compared to Luffa biochar. On the basis of functional group analysis using Fourier transform infrared spectroscopy, hydrogen bonding and pore filling were key factors for enhanced adsorption. In another study, Arana et al. in 2017 explored the potential of Luffa sponge for capturing of cyanide ion which is very hazardous to ecosystem, with 98% retention capacity [10] (Table 3).

4.5 *Luffa Cylindrica for Drug Removal*

Yu Feng et al. used Luffa for the removal of pharmaceutical antibiotic norfloxacin, due to its excellent porosity. In this study, biochar was prepared from Luffa sponge and was used for absorbing norfloxacin from waste water [18]. Infrared spectroscopic data clearly indicated that surface of biochar being full of oxygen containing functional groups such as phenol, hydroxyl, lactone and carbonyl, made the removal of norfloxacin easier. Maximum adsorption capacity of biochar was found to be 250 mg/g. Ofloxacin, a fluoroquinolone antibiotic, has very poor biodegradation ability and is also one of the contributors to environmental pollution. Activated carbon was produced from Luffa sponge and was utilized to trap ofloxacin. The surface area of activated carbon was found to be 834 $m^{2/}g$ and adsorption capacity of adsorbent was 132 mg/g [25]. Luffa Cylindrica fibres were employed for trapping two drug molecules viz. dextropropoxyphene and paracetamol, which are widely consumed in pharmaceutical products and their presence in water may cause damage to human health [6]. With the increase in pH, adsorption was found to increase and reached to maximum at pH 10 for both drug molecules. Recently, Luffa was reinforced by using a conducting polymer polypyrrole and was investigated for the removal of ibuprofen from aqueous media [24]. This composite showed good results in terms

Table 3 Application of Luffa in removal of toxic pollutants and drugs

Adsorbent	Targeted pollutant	Maximum monolayer adsorption capacity	References
Luffa Cylindrica fibres	1-cetylpyridinium bromide	37 mg /g	Ye et al. [50]
Quaternized diethylamino ethylated Luffa fibres	Sulphate, nitrate and phosphate anions	–	Hassan [20]
Luffa Cylindrica fibres	Phenol	9.25 mg/g	Abdelwahab et al. [2]
Biochar from Luffa Cylindrica sponge	Norfloxacin	250 mg/g	Feng [18]
Oxidized carbon derived from Luffa	Caffeine	59.9 mg/g	Ioannis Anastopoulos et al. [8]
Activated carbon derived from Luffa sponge	Ofloxacin	132 mg/g	Qiang [25]
Biochar derived from Luffa	4-nitrophenol	436.6 mg/g	Mohammad [41]
Luffa Cylindrica fibres	Dextropropoxyphene paracetamol	–	Amel [6]
Luffa—polypyrrole composite	Ibuprofen	19.157 mg/g	Ali [24]
Luffa sponge	Cyanide	98.2% retention capacity	Arana [10]

of reusability, as only 10% decrease in removal percentage was reported over five cycles with eightfold higher removal efficiency than raw Luffa.

5 Conclusions

Luffa Cylindrica has been found to be a potential biomaterial for adsorption of various hazardous dyes, toxic metal ions and chemical from textile waste water. Luffa Cylindrica has been found to be the best alternative to the existing adsorbent due to its characteristics such as high abundance, low cost and biodegradability. Structurally Luffa Cylindrica fibres have presence of hydroxyl (-OH) groups and hence surface modifications on these fibres offers a lot of opportunity for researcher to develop a novel low cost environmentally friendly bio-based adsorbent in future for textile waste water treatment. There is lot of development seen from last several years and new Luffa-based biomaterials have emerged as potential adsorbents for adsorption of textile pollutants. Activated carbon from Luffa Cylindrica has been used for elimination of dyes, metal ions and emerging pollutants from aqueous environment. Bio adsorbents such as Luffa Cylindrica fibres are best substituent for commercially available activated carbon for the elimination of the dyes from aqueous solution. Luffa Cylindrica has been chemically modified through epoxidation, cross-linking, in situ polymerization, Quaternization reactions and has been characterized successfully using Fourier transform infrared spectroscopy, Raman spectroscopy, X-ray diffraction, adsorption isotherms and scanning electron microscopy. Modified Luffa-based materials have shown best performance in terms of adsorption capacity and reusability, in comparison to several other adsorbents for textile waste water treatments. Researchers are now developing composites by using Luffa Cylindrica and these composites have shown good performance for removal of dyes and metal ions from aqueous environment.

References

1. Abdelwahab O (2008) Evaluation of the use of loofa activated carbons as potential adsorbents for aqueous solutions containing dye. Desalination 222:357–367. https://doi.org/10.1016/j.desal.2007.01.146
2. Abdelwahab O, Amin NK (2013) Adsorption of phenol from aqueous solutions by Luffa Cylindrica fibers: kinetics, isotherm and thermodynamic studies. Egypt J Aquat Res 39:215–223. https://doi.org/10.1016/j.ejar.2013.12.011
3. Abou Oualid H, Abdellaoui Y, Laabd M (2020) Eco-efficient green seaweed codium decorticatum biosorbent for textile dyes: characterization, mechanism, recyclability, and rsm optimization. ACS Omega 5:22192–22207. https://doi.org/10.1021/acsomega.0c02311
4. Adewuyi A, Pereira FV (2017) Underutilized Luffa Cylindrica sponge: a local bio-adsorbent for the removal of pb (ii) pollutant from water system. Beni-Suef Univ J Basic Appl Sci 6:118–126. https://doi.org/10.1016/j.bjbas.2017.02.001
5. Akhtar N, Iqbal J, Iqbal M (2004) Removal and recovery of nickel(ii) from aqueous solution by loofa sponge-immobilized biomass of chlorella sorokiniana: characterization studies. J Hazard Mater 108:85–94. https://doi.org/10.1016/j.jhazmat.2004.01.002
6. Aloui A, Bouziane N, Bendjeffal H, Bouhedja Y, Zertal A (2021) Biosorption of some pharmaceutical compounds from aqueous medium by Luffa Cylindrica fibers: application of the linear form of the redlich–peterson isotherm equation. Desalin Water Treat 214:367–378. https://doi.org/10.5004/dwt.2021.26736
7. Altinişik A, Gür E, Seki YA (2010) Natural sorbent Luffa Cylindrica for the removal of a model basic dye. J Hazard Mater 179:658–664. https://doi.org/10.1016/j.jhazmat.2010.03.053
8. Anastopoulos I, Pashalidis I (2019) The application of oxidized carbon derived from luffa Cylindrica for caffeine removal. equilibrium, thermodynamic, kinetic and mechanistic analysis. J Mol Liq 296. https://doi.org/10.1016/j.molliq.2019.112078
9. Anastopoulos I, Pashalidis I (2020) Environmental applications of luffa cylindrica-based adsorbents. J Mol Liq 319. https://doi.org/10.1016/j.molliq2020.114127
10. Arana J, González S, Navarrete L, Caicedo O (2017) Luffa Cylindrica as a natural adsorbent of cyanide ion in aqueous medium. Dyna 84:210–215. https://doi.org/10.15446/dyna.v84n201.60070
11. Bharathi KS, Ramesh ST (2013) Removal of dyes using agricultural waste as low-cost adsorbents: a review. Appl Water Sci 3:773–790. https://doi.org/10.1007/s13201-013-0117-y
12. Bhatia D, Sharma NR, Singh J, Kanwar RS (2017) Biological methods for textile dye removal from wastewater: a review. Crit Rev Environ Sci Technol 47:1836–1876. https://doi.org/10.1080/10643389.2017.1393263

13. Boudechiche N, Mokaddem H, Sadaoui Z, Trari M (2016) Biosorption of cationic dye from aqueous solutions onto lignocellulosic biomass (Luffa cylindrica): characterization, equilibrium, kinetic and thermodynamic studies. Int J Ind Chem 7:167–180. https://doi.org/10.1007/s40090-015-0066-4
14. Da Silva LS, Gonçalves MMM, Raddi de Araujo LR (2019) Combined photocatalytic and biological process for textile wastewater treatments. Water Environ Res 91:1490–1497. https://doi.org/10.1002/wer.1143
15. Dafale N, Wate S, Meshram S, Neti NR (2010) Bioremediation of wastewater containing azo dyes through sequential anaerobic-aerobic bioreactor system and its biodiversity. Environ Rev 18:21–36. https://doi.org/10.1139/A10-001
16. Dali A, Boulguemh IE, Louafi F, Mouats C (2021) Synthesis, characterization and environmental application of an original adsorbent: polyaniline-coated Luffa cylindrica. J Polym Res 28. https://doi.org/10.1007/s10965-020-02365-1
17. Deng D, Aryal N, Ofori-Boadu A, Jha MK (2018) Textiles wastewater treatment. Water Environ Res 90:1648–1662. https://doi.org/10.2175/106143018x15289915807353
18. Feng Y, Liu Q, Yu Y (2018) Norfloxacin removal from aqueous solution using biochar derived from luffa sponge. J Water Supply Res Technol—AQUA 67:703–714. https://doi.org/10.2166/aqua.2018.040
19. Gupta SK, Nayunigari MK, Misra R (2016) Synthesis and performance evaluation of a new polymeric composite for the treatment of textile wastewater. Ind Eng Chem Res 55:13–20. https://doi.org/10.1021/acs.iecr.5b03714
20. Hassan ML (2006) Quaternization and anion exchange capacity of sponge gourd (Luffa cylindrica). J Appl Polym Sci 101:2495–2503. https://doi.org/10.1002/app.23747
21. Holkar CR, Jadhav AJ, Pinjari DV (2016) A critical review on textile wastewater treatments: possible approaches. J Environ Manage 182:351–366. https://doi.org/10.1016/j.jenvman.2016.07.090
22. Hynes NRJ, Kumar JS, Kamyab H (2020) Modern enabling techniques and adsorbents-based dye removal with sustainability concerns in textile industrial sector-a comprehensive review. J Clean Prod 272. https://doi.org/10.1016/j.jclepro.2020.122636
23. Kesraoui A, Moussa A, Ben AG, Seffen M (2016) Biosorption of alpacide blue from aqueous solution by lignocellulosic biomass: Luffa Cylindrica fibers. Environ Sci Pollut Res 23:15832–15840. https://doi.org/10.1007/s11356-015-5262-4
24. Khadir A, Negarestani M, Mollahosseini A (2020) Sequestration of a non-steroidal anti-inflammatory drug from aquatic media by lignocellulosic material (Luffa cylindrica) reinforced with polypyrrole: study of parameters, kinetics, and equilibrium. J Environ Chem Eng 8. https://doi.org/10.1016/j.jece.2020.103734
25. Kong Q, He X, Shu L, Miao M sheng (2017) Ofloxacin adsorption by activated carbon derived from Luffa sponge: kinetic, isotherm, and thermodynamic analyses. Process Saf Environ Prot 112:254–264. https://doi.org/10.1016/j.psep.2017.05.011
26. Liatsou I, Constantinou P, Pashalidis I (2017) Copper binding by activated biochar fibres derived from Luffa cylindrica. Water Air Soil Pollut 228. https://doi.org/10.1007/s11270-017-3411-8
27. Lindino CA, Marciniak AA, Gonçalves AC, Strey L (2014) Adsorption of cadmium in vegetable sponge (Luffa cylindrica). Rev Ambient e Agua 9:212–223. https://doi.org/10.4136/ambi-agua.1340
28. Madhav S, Ahamad A, Singh P, Mishra PK (2018) A review of textile industry: wet processing, environmental impacts, and effluent treatment methods. Environ Qual Manag 27:31–41. https://doi.org/10.1002/tqem.21538
29. Majhi PK, Kothari R, Arora NK, Pandey VC, Tyagi VV (2021) Impact of pH on pollutional parameters of textile industry wastewater with use of chlorella pyrenoidosa at lab-scale: a green approach. Bull Environ Contam Toxicol. https://doi.org/10.1007/s00128-021-03208-5
30. Mani S, Chowdhary P, Bharagava RN (2018) Textile wastewater dyes: toxicity profile and treatment approaches. In: Emerging and eco-friendly approaches for waste management. pp 219–244. https://doi.org/10.1007/978-981-10-8669-4_11

31. Nadaroglu H, Cicek S, Gungor AA (2017) Removing trypan blue dye using nano-zn modified Luffa sponge. Spectrochim Acta—Part A Mol Biomol Spectrosc 172:2–8. https://doi.org/10.1016/j.saa.2016.08.052
32. Neifar M, Sghaier I, Guembri M (2019) Recent advances in textile wastewater treatment using microbial consortia. J Text Eng Fash Technol 5:134–146. https://doi.org/10.15406/jteft.2019.05.00194
33. Nwosu-Obieogu K, Okolo BI (2020) Biosorption of chromium (vi) from textile waste water using Luffa Cylindrica activated carbon. Environ Qual Manag 29:23–31. https://doi.org/10.1002/tqem.21687
34. Oladoja NA, Aboluwoye CO, Akinkugbe AO (2009) Evaluation of loofah as a sorbent in the decolorization of basic dye contaminated aqueous system. Ind Eng Chem Res 48(6):2786–2794. https://doi.org/10.1021/ie801207a
35. Oliveira EA, Montanher SF, Rollemberg MC (2011) Removal of textile dyes by sorption on low-cost sorbents. a case study: sorption of reactive dyes onto Luffa Cylindrica. Desalin Water Treat 25:54–64. https://doi.org/10.5004/dwt.2011.1416
36. Petrinić I, Bajraktari N, Hélix-Nielsen C (2015) Membrane technologies for water treatment and reuse in the textile industry. Adv. Membrane Technol Water Treatment: Mater Processes Appl 2015:537–550. https://doi.org/10.1016/B978-1-78242-121-4.00017-4
37. Psarra E, Papanicolaou GC (2021) Luffa Cylindrica as a durable biofiber reinforcement for epoxy systems. Compos Sci Technol 203. https://doi.org/10.1016/j.compscitech.2020.108597
38. Rashid R, Shafiq I, Akhter P, Iqbal MJ, Hussain M (2021) A state-of-the-art review on wastewater treatment techniques: the effectiveness of adsorption method. Environ Sci Pollut Res 28:9050–9066. https://doi.org/10.1007/s11356-021-12395-x
39. Sahu O, Singh N (2018) Significance of bioadsorption process on textile industry wastewater. In: The impact and prospects of green chemistry for textile technology. pp 367–416. https://doi.org/10.1016/B978-0-08-102491-1.00013-7
40. Saini RD (2017) Textile organic dyes: polluting effects and elimination methods from textile waste water. Int J Chem Eng Res 9(1):121–136
41. Salimi M, Salehi Z, Heidari H, Vahabzadeh F (2021) Production of activated biochar from Luffa Cylindrica and its application for adsorption of 4-nitrophenol. J Environ Chem Eng 9.https://doi.org/10.1016/j.jece.2021.105403
42. Samsami S, Mohamadi M, Sarrafzadeh MH, Rene ER, Firoozbahr M (2020) Recent advances in the treatment of dye-containing wastewater from textile industries: overview and perspectives. Process Saf Environ Prot 143:138–163. https://doi.org/10.1016/j.psep.2020.05.034
43. Sarayu K, Sandhya S (2012) Current technologies for biological treatment of textile wastewater-a review. Appl Biochem Biotechnol 167:645–661. https://doi.org/10.1007/s12010-012-9716-6
44. Segun Esan O, Nurudeen Abiola O, Owoyomi O, Olumuyiwa Aboluwoye C, Olubunmi Osundiya M (2014) Adsorption of brilliant green onto Luffa cylindrical sponge: equilibrium, kinetics, and thermodynamic studies. ISRN Phys Chem 2014:1–12. https://doi.org/10.1155/2014/743532
45. Senthil Kumar P, Saravanan A (2017) Sustainable wastewater treatments in textile sector. Sustainable Fibres and Textiles 2017:323–346. https://doi.org/10.1016/B978-0-08-102041-8.00011-1
46. Sivaram NM, Gopal PM, Barik D (2018) Toxic waste from textile industries. In: Energy from toxic organic waste for heat and power generation. pp 43–54. https://doi.org/10.1016/B978-0-08-102528-4.00004-3
47. Su S, Liu QI, Liu J, Zhang H, Li R, Jing X, Wang J (2018) Polyethyleneimine-functionalized Luffa cylindrica for efficient uranium extraction. J Colloid Interface Sci 530:538–546. https://doi.org/10.1016/j.jcis.2018.03.102
48. Sukanya Devi R, Dhurai B, Sundaresan S, Selvakumar A (2021) Advanced oxidation processes (aop) effective innovative treatment methods to degrade textile dye effluent.173–203. https://doi.org/10.1007/978-981-16-0065-4_7.
49. Velusamy S, Roy A, Sundaram S, Kumar Mallick T (2021) A review on heavy metal ions and containing dyes removal through graphene oxide-based adsorption strategies for textile wastewater treatment. Chem Rec 21:1570–1610. https://doi.org/10.1002/tcr.202000153

50. Ye C, Hu N, Wang Z (2013) Experimental investigation of luffa Cylindrica as a natural sorbent material for the removal of a cationic surfactant. J Taiwan Inst Chem Eng 44:74–80. https://doi.org/10.1016/j.jtice.2012.08.006
51. Yeh RYL, Hung YT, Liu RLH, Chiu HM, Thomas A (2002) Textile wastewater treatment with activated sludge and powdered activated carbon. Int J Environ Stud 59:607–622. https://doi.org/10.1080/00207230212735
52. Yu JX, Wang LY, Chi RA, Zhang YF, Xu ZG, Guo J (2013) Removal of cationic dyes: basic magenta and methylene blue from aqueous solution by adsorption on modified loofah. Res Chem Intermed 39:3775–3790. https://doi.org/10.1007/s11164-012-0880-7

Recent Trends in Textile Wastewater Treatment Using Agricultural Waste

Chijioke Elijah Onu, Christian Oluchukwu Asadu, Paschal Enyinnaya Ohale, Emeka Victor Ojukwu, Nonso Collins Ogunaobi, Chiamaka Peace Onu, and Florence Chidinma Akaeme

Abstract The need to treat textile effluent is a global challenge that has attracted serious attention among the nations of the world. The textile effluent if not treated will pose severe environmental pollution issues. This will affect both land and water bodies. Different agricultural wastes have found useful application in the handling of effluent from textile industries with varying degrees of efficiencies. The management of textile effluent using agricultural waste offered a two-fold solution namely; effective conversion of the effluent into harmless form before discharging the wastewater into water bodies and secondly, significant reduction of large amount of agricultural wastes which on their own, constitute another environmental challenge. Many treatment methods which can be chemical, physical, or biological have been applied in

C. E. Onu (✉) · P. E. Ohale · E. V. Ojukwu · F. C. Akaeme
Department of Chemical Engineering, Nnamdi Azikiwe University, P.M.B. 5025, Awka, Anambra State, Nigeria
e-mail: ce.onu@unizik.edu.ng; onuchijioke894@gmail.com

P. E. Ohale
e-mail: pe.ohale@unizik.edu.ng

E. V. Ojukwu
e-mail: ve.ojukwu@unizik.edu.ng

F. C. Akaeme
e-mail: cf.akaeme@unizik.edu.ng

C. O. Asadu
Department of Chemical Engineering, Gregory University, P.M.B. 1012, Uturu, Abia State, Nigeria
e-mail: a.christian@gregoryuniversityuturu.edu.ng

N. C. Ogunaobi
Department of Chemical Engineering, Michael Okpara University of Agriculture, P.M.B. 7267, Umudike, Abia State, Nigeria
e-mail: oguanobi.nonso@mouau.edu.ng

C. P. Onu
Department of Physics and Industrial Physics, Nnamdi Azikiwe University, P.M.B. 5025, Awka, Anambra State, Nigeria
e-mail: cp.onu@unizik.edu.ng

S. S. Muthu and A. Khadir (eds.), *Textile Wastewater Treatment*, Sustainable Textiles: Production, Processing, Manufacturing & Chemistry,
https://doi.org/10.1007/978-981-19-2852-9_6

textile effluent treatment. Adsorption method, which is a physical method, is the most widely used method because it is efficient, easy to operate and cost-effective. The equilibrium and thermodynamics of the treatment process was highlighted. Treatment of textile effluent can be enhanced by the use of artificial intelligence.

Keywords Textile wastewater · Agricultural waste · Effluent · Treatment · Dyes · Pollution · Adsorption · Physical method · Kinetics · Equilibrium

1 Introduction

Over the years, industrialization has led to improved quality of life. However, it also gave rise to the need for the development of treatment facilities that will handle the enormous waste generated by the industries [15]. Textile industries are known to discharge significant amount of highly concentrated pollutants into the land and water bodies [25, 30]. Major pollutants in the effluent of textile industries include colors, carcinogenic substances, recalcitrant hydrocarbons, varying degrees of toxic compounds, mutagenic substances, and non-biodegradable organic compounds. Many of these textile factories and industries use organic and synthetic colorants in different stages of production.

If the effluents containing dyes and other pollutants are discharged into the surrounding environment without treatment, they have undesirable and sometimes fatal effects on man and living organisms directly or indirectly [38]. In terms of treatment, dyes are the most difficult component of textile effluent. They are hazardous and can slow down photosynthesis process. They cause terminal diseases such as liver diseases, cancer, and even mutation that can be genetic in humans and animals [3]. Discharge of wastewater containing these dyes into water bodies constitutes severe danger to the aquatic ecosystem and food web. They destroy the aesthetic nature of the water body and impart toxicity to the aquatic life [52]. These dyes and colorants can absorb and reflect sunlight thereby interfering with the photosynthesis process of aquatic plants and the growth of bacteria [55]. Hence, the management and treatment of these industrial effluents are paramount before discharging. This will lead to reduction in water and soil contamination and potentially eliminate the different possible damages aquatic lives.

In contemporary times, different techniques have been employed in handling dye-containing wastewater with various degrees of efficiency. These techniques include the degradation of the dyes using photocatalysis [13], nanofiltration [40], degradation using microorganisms [11], and adsorption [4]. Adsorption have often times been preferred to other techniques of wastewater management because of its cost-effectiveness and simplicity of design. Besides, it gives high removal efficiency of the dyes with operation methods [6, 74]. In recent times, different adsorbents have been reportedly used in removing dyes and colorants from water and industrial wastewater. Activated carbon is the most common and most widely used adsorbent material for textile industrial effluent treatment. This is because of its micro-pore structures, high degree of surface reactivity, extended surface area, and high adsorption capacity. But these commercially activated carbons are very expensive and when exhausted, the regeneration of the spent activated is not cost-effective. Therefore, researchers are continuously in search of more appropriate, effective, and low cost materials that are readily available, particularly from waste materials, that can be modified and utilized as adsorbent with high efficiency.

To minimize the economic cost and the number of experiments needed, Artificial Intelligence (AI) methods, such as adaptive neuro-fuzzy inference system (ANFIS), Artificial Neural Network (ANN), Particle Swarm Optimization (PSO) among others, have been studied in stimulating and optimizing of the adsorption technique [47]. The ANN is one of the most influential tools to model and optimize any adsorption process. Recently, it has been employed in modeling the removal of dyes from wastewater in an adsorption process. Ghaedi et al. [18] used ANN to stimulate and predict the percentage removal of methyl orange dye by using gold nanoparticles activated carbon and Tamarisk. The influence of four process variables on the efficiency of the dye adsorption were investigated and modelled namely: stirring speed, dosage of adsorbent, concentration of dye solution, and contact time. The ANN predicted dye adsorption efficiency was compared with the experimental result and the result showed that the ANN is appropriate and suitable for modeling and predicting the percentage removal of the dye on gold nanoparticles in activated carbon and Tamarisk under the various conditions [18].

ANFIS is a synergetic hybrid of fuzzy inference system (FIS) and neural networks such that the neural network can monitor the flexibility of the system while the FIS can enhance the certainty and precision of the technique. As an AI technique, there is a growing interest in the use of ANFIS in stimulating, modeling, and optimization of various industrial processes.

This book chapter focused on the evaluation of the efficiency of agricultural wastes as a precursor that can be converted into adsorbent for the effective removal of dyes from textile effluent. Equilibrium, kinetics, and thermodynamics studies were reviewed to fathom and comprehend the technicality of the adsorption process. Influence of basic adsorption variables such as contact time, solution pH, stirring speed, dye concentration, dosage or amount of adsorbent, and temperature, time were equally investigated. The use of different artificial intelligence models in dye removal was also considered.

2 Typical Characteristics of Textile Effluents

The textile effluents constitute a mixture of dangerous and toxic pollutants such as colorants/dyes, organic and inorganic compounds, heavy metals and other pollutants. The dyes are grouped into synthetic and natural dyes. Most of the dyes used in the industries are synthetic dyes. Synthetic dyes are usually characterized by their fastness, ability to be made into any desirable color, easy to produce on account of advanced technology, etc. These properties made them more widely acceptable

Table 1 Textile waste water and their characteristics 7—u

References	Roy et al. [63]	Gilpavas, et al. [22]	Islam and Mostapha [29]	Karam et al. [32]	Kehinde and Aziz [35]	Kalra et al. [31]
Temp (°C)	41	–	–	–	21–62	35–45
pH (–)	9.4	9.3	6.70–8.10	1–12	6.95–11.8	6–10
Color (Pt-Co)	1374	–	–	50–350	50–2500	50–2500
BOD (mg/l)	358	115	–	–	80–6000	100–4000
COD (mg/l)	721	720	340–784	–	150–30,000	150–12,000
TS (mg/l)	–	–	–	–	6000–7000	–
TSS (mg/l)	913		223–401	–	15–8000	100–5000
TDS (mg/l)	2250	–	2937.5–3778	–	2900–3100	1800–6000
DO (mg/l)	2.84	–	–	–	–	–
TOC (mg/l)	–	164	–	–	–	–

and applied than the natural dyes [75]. Synthetic dyes can be further classified depending on their chemical structure or their mode of application. Based on their structure, synthetic dyes are grouped into azo, anthraquinone, sulfur, phthalocyanine, and triarylmethane. With regards to their mode of application, synthetic dyes are classified into basic, direct, vat, disperse, and reactive dyes [60]. The textile effluents are prominent in coloring dyes, chemical oxygen demand (COD), metals, temperature, salts, biochemical oxygen demand (BOD), suspended solids, etc. [75, 67, 69]. Different properties of textile effluent as reported by different authors are tabulated in Table 1. Hence, there is need for adequate treatment of the textile waste water prior to discharge into water bodies.

Textile effluents are broadly classified into four with unique characteristics that require different method of prevention and treatment.

2.1 *Hard to Treat Effluent*

These textile effluents are made up of those that are difficult to treat, those that are persistent in nature, and those that hinder the efficient operation of wastewater treatment plants. Non-biodegradable waste materials (both organic and inorganic) are the primary sources of these wastes. They contain certain surfactants, color, toxic organic compounds, metals, phenols, phosphates, pesticides among others. The identification and source elimination of these wastes present the best way of eliminating the problem since they are difficult to treat. These methods can be called preventive methods. These methods can be in the form of process control and optimization, chemical, or process substitution and better work practices.

2.2 *Hazardous or Toxic Effluent*

These textile effluents are very toxic, irrespective of their quantity or volume. They include heavy metals, non-biodegradable materials such as dyes, chlorinated solvents, volatile organic materials, etc. Some of these wastes are from materials used for non-process applications such as machine cleaning. Because of their drastic impact on man and the environment, they are treated as a separate class.

2.3 *High Volume Effluent*

These are textile effluents that are more than 2000 L by volume. Large volume of wastes is a very serious problem for the textile processing industries. The most common large-volume waste is the direct effluent form the dyeing process. These are wastewater from both the preparation and the continuous dyeing processes. Also included are alkaline wastes from preparatory processes. These wastes may be eliminated or reduced by equipment or process modification, reuse or recycling of the waste materials.

2.4 *Dispersible Wastes*

The dispersible wastes include:

1. The wastes obtained from continuous operations such as preparatory, printing, dyeing, etc.
2. Lint (dyeing, preparatory, washing operations, etc.)
3. Print paste (the cleaning of drum, screen printing, and cleaning of the drum)
4. Solvents from machine cleaning
5. Foam from coating operations
6. Still bottoms from solvent recovery (dry cleaning operation)
7. Batch dumps of unused processing (finishing mixes)

3 Agricultural Wastes

Agricultural wastes are by-products of different agricultural processes. They usually comprise of materials that are potentially valuable to man but which may involve high cost of collection and processing before it can be converted to useful product for man. Sometimes, the total cost of processing the agricultural waste into useful products may exceed the economic value of the product.

Agricultural waste otherwise called agro-waste is comprised of crop waste (rice husks, corn cobs and stalks, culls and drops from vegetables and fruits, prunings, sugarcane bagasse, etc.), waste from food processing techniques (e.g., only 20% of maize is canned while the rest 80% is waste in forms of stalks, cobs, leaves, etc.), animal waste (animal carcasses, manure), and hazardous agricultural processing waste (germicides, pesticides, insecticides and herbicides, etc.). Agricultural wastes in most cases present a disposal challenge and have no economic value unless they are further processed and converted into useful products.

Some of these agricultural products have been converted in activated carbon (bio adsorbents) and utilized in treating diverse textile waste water where the emphasis is on removal of dyes. Chitosan flakcs was proccssed and used for removal of methylene blue dye [44], waste pea shells was processed and utilized in removing malachite green [38], rice husk was converted to activated carbon for removal of bromoscreosol green dye [57], teak tree bark powder was modified for removal of methylene blue [66], pomegranate activated carbon was used in treating methyl reactive remazol brilliant blue dye-contaminated wastewater [49], Enteromorpha prolifera was employed in the removal of reactive dyes [12], kola nut shell was processed into activated carbon and used to remove orange G dye from textile effluent [54], surfactant modified corn straw was used for the removal of organic dyes [73], and activated teak leaf powder was employed in the removal of cargo red dye [16] among others. Some of these agricultural waste products, utilized for the decoloration of contaminated solutions at varying operating conditions are presented in Table 2.

4 Treatment Processes of Textile Industry Wastewaters

Removing of dyes/colorants from textile effluents can be broadly classified into three methods: biological method, chemical method, and physical method.

4.1 Biological Methods

Biological treatment method involves the use of microorganisms in the handling of textile effluents. This is possible because most of the microoraginsms like algae, bacteria, fungi, yeast, etc., have the ability to accumulate in large quantity over a short period of time. This is in addition to their ability to degrade and vitiate different types of dyes. Examples of biological methods are bioremediation, microbial, or fungal decolorization, adsorption by microbial biomass (both living and dead), etc. From economic perspective, biological treatment method is better in comparison with chemical and physical methods. However, biological method demands large area for effectiveness. Besides, with the existing biodegradation process, it is not very effective in the elimination of dyes and colorants [33].

Based on oxygen requirement of each process, biological methods can be grouped into anaerobic (not requiring oxygen) and aerobic (requiring oxygen) methods. Aerobic treatment method is a common and established biological method on account

Table 2 Agricultural wastes used in textile effluent treatment

Agro-waste	Dye	Adsorption capacity (mg/g)	References
Rice husk	Bromocresol green	41.5	Onu et al. [57]
Pomegranate activated carbon	Remazol Brilliant Blue	370.86	Mohd et al. [49]
Kola nut shell	Orange G	93.5	Okpe et al. [54]
Activated teak leaf powder	Cargo red	33.33	Gedam et al. [16]
Activated carbon from Enteromorpha prolifera	Reactive dyes (RR23, RB171 and RB4)	59.88, 71.94 and 131.93	Deshuai et al. [12]
kola nut shell activated carbon	Phenol red	48.3	Nwabanne et al. [51]
Surfactant modified corn straw	Organic dyes (B21 and Y20)	96.15 and 434.78	Umpuch and Juharat [73]
Bentonite clay	Reactive red dyes	95.15	Hajira et al. [23]
Rice husk activated carbon	Phenol		Iheanacho et al. [27]
Teak tree bark powder	Methylene blue	333.3	Satish et al. [66]
Natural clay	Methyl orange	1.2 and 0.65	Karukh and Bakhtyar [34]
Activated carbon	Methylene blue		Sachin et al. [64]
Acid modified bentonite	Crystal violet	62.4	Oguanobi et al. [53]
Peanut hull	Reactive dye	55.5	Tanyildizi [72]
Citrullus lanatus rind	Crystal violet	11.9	Bharathi and Ramesh [8]

of its broad and extensive application with higher efficiency than anaerobic method in most cases. Example, isolated white rot fungi (schizophyllum commune and lenzite-seximia) was utilized in the aerobic treatment of textile effluent and removal of azo dyes where methyl orange, congo red, and Erichrome black T dyes were successfully decolorized by these fungi [68].

4.2 *Chemical Methods*

Chemical methods utilized in textile effluent treatment include precipitation, electro floatation, irradiation or electrochemical processes, oxidation, electrokinetic coagulation, etc. These chemical methods are very efficient as they remove most of the dyes from the wastewater. But the methods are very costly and equally result to accumulation of sludge over time which in turn creates environmental pollution and a disposal challenge. Excessive use of chemicals generates another pollution problem on its own. Coagulation method involves the use of appropriate chemical substances in a chemical reaction with the pollutants that will form an end product that is usually insoluble. The insoluble substances formed can then be removed through some separation techniques. Through this process, the harmful substances like dyes and colorants can be removed from textile effluents [33]. Generally, coagulants in use include, alum, ferric chloride, etc. Electrokinetic coagulation method is not suitable because it can only remove little quantity of colloidal particles. In recent times, new advanced technique, known as the advanced oxidation process, involving the generation of strong oxidizing agents like hydroxyl radicals, is used in textile wastewater treatment with great success. However, like most chemical methods, the advanced oxidation process is very costly and hence, commercially unattractive. In addition, the large volume of chemical reagents consumed and the high electrical energy demand for this method makes it unattractive for commercial purposes. Examples of the chemical methods are electrochemical oxidation, photocatalytic degradation, among others.

4.2.1 Electrochemical Oxidation

This method involves the breaking down of the dye compounds through an oxidation process powered by current. This is a substitute treatment process for color removal in textile effluent. Electrochemical oxidation is quite adept in eliminating the chromophore groups and can induce complete decolorization of the dyes in the industrial wastewater within a short period of treatment [46]. Electrochemical oxidation through hydroxyl radicals is another way to vitiate and degrade reactive dyes. This method is a novel method that institutes a new technology with very small energy requirement in treatment of different kinds of reactive dyes from textile effluents.

4.2.2 Photocatalytic Degradation

Photocatalysis is another relatively effective advanced oxidation processes for textile effluent treatment. The primary advantage of photocatalytic degradation with respect to other treatment techniques is that there is no need for additional disposal system [33]. It gives a high efficient pollutant removal. Rameshwar carried out efficient photocatalytic degradation of methylene blue using Calcium oxide [62]. Kirupavasam and Allen reported on photocatalytic degradation of Amido black-10B using nano photocatalyst with high efficiency [39]. Furthermore, prepared Nano titanium dioxide (TiO_2) was reported to be effective in removing methylene blue dye [5].

4.3 *Physical Methods*

This method include coagulation, eletrocoagulation, membrane filtration processes (electrodialysis, reverse osmosis, nanofiltration), flocculation, or a combination floatation and filtration, and adsorption techniques. Puasa et al. [61] reported on the use of Micellar-enhanced ultrafiltration (MEUF) to decolorize solution containing C.I Reactive Orange 16 and C.I Reactive Black 5. The MEUF utilized was Cetylpyridinium chloride (CPC). Mohammad studied the elimination dyes such as reactive blue, reactive black and acid red using nanofiltration membrane [48]. Onu et al. [56] investigated the application of adsorption technique for efficient removal of eriochrome black T dye from wastewater.

4.3.1 Electrocoagulation

This method is one of the most commonly used physical methods in decolorizing solutions of dyes. It is a simple and efficient technique treating aqueous waste containing dyes and colorants. Electrocoagulation is a systematic method that involves the creation of flocks of hydroxide of different metals inside the textile effluent by electro dissolution of soluble iron (Fe) or aluminum (Al) electrodes. Currently, several researches are being concentrated on the application of electrocoagulation in textile effluent treatment because of its simplicity and efficiency. Hussein et al. [26] studied the use of iron electrodes in an electrocoagulation experiment to remove Reactive Blue 19 from industrial textile. Stercipoulous et al. [71] also reported on the effective elimination of indigo carmine from textile effluent via electrochemical decolorization.

4.3.2 Adsorption

Adsorption is the most common technique for removing color and dyes from textile wastewater. Adsorption technique provides a cost-effective alternative for handling wastewater from textile. It is desirable if the adsorbent material is cheap with little or no required additional pretreatment before usage. It gives a high quality and clean treated effluent as the end product. Several researchers have reported the results of the decolorization of textile dye solutions from the wastewater using different kinds of prepared adsorbents. Aadil et al. [1] used modified peanut shell as sorbent to study the removal of brilliant green and Congo red in a biosorption process. Karukh and Bakhtyar [34] employed natural germak clay as an adsorbent in investigating the elimination of methyl orange dye from solutions. Hajira et al. [23] studied the characterization of modified bentonite clay and its effect in the decolorization of waste water containing reactive dyes. Umpuch and Jutarat [73] used surfactant modified corn straw to investigate the adsorption of organic dyes from solution. Deshuai et al. [12] reported on the conversion of *Enteromorpha prolifera* into activated carbon and its application on the adsorption of reactive dyes. Satish et al. [66] modified teak tree park powder and studied the effects of process variables like solution pH, temperature, initial dye concentration, agitation speed, adsorbent dosage, particles size, and contact time on the adsorption of Methylene blue. Sachin et al. [64] applied hyacinth and activated carbon as adsorbents in the removal of methylene blue dye with significant result. The adsorption process can provide an appealing substitute in the decolorization and handling of textile contaminated waste water, especially if the process of preparing the adsorbent material is simple and cost-effective.

5 Factors Affecting the Treatment Process

Some process variables that influence the efficiency of dye removal are contact time, dye concentration, pH, adsorbent dosage among others. Optimization of these variables will enhance the full-scale textile effluent treatment on an industrial scale.

5.1 Contact Time

The influence of time is significant in the treatment any textile waste water. Contact time is the duration or length of time when the adsorbent is in contact with the adsorbate. An increase in contact time will usually lead to considerably increase in the rate of removal of the dye. However, after some time, the rate of dye removal will decrease because of the clogging of the deposited dye molecules on the vacant adsorption site of the adsorbent material. Additional increase in time may not necessarily result to increase in dye uptake from the solution [50]. At this condition, dynamic equilibrium is established between the quantity of being adsorbed and the quantity of dye being desorbed from the adsorbent. The time taken to reach this equilibrium state is called the equilibrium time while the equilibrium adsorption capacity is the amount of dye (mg/g) adsorbed at that equilibrium time. The equilibrium adsorption capacity is the maximum adsorption capacity of the adsorbent [50]. The contact time between the adsorbate and the adsorbent influences the dye removal process significantly [14]. Example Heibati et al. [24] reported that there is a direct relationship between the contact time of walnut activated carbon with acid dye and the removal efficiency.

5.2 Adsorbent Dosage

Different weights of adsorbents can be used to determine the influence of the adsorbent dose on the dye removal process by adding different masses of the adsorbents to the solutions containing constant initial dye concentration [65]. Usually, the percentage removal of the dye increased as the amount of the adsorbent dosage increased. At the initial time, as the amount of the adsorbent dosage was increased, there is a fast and instantaneous increase in the percent dye removal which subsequently slowed down as the adsorbent dosage increased. This trend is because at lesser dosage of the adsorbent, the quantity adsorbed per unit adsorbent is higher due to the adsorbate being readily accessible. With increase in dosage, less comparable quantity adsorbed per unit adsorbent will be observed leading to the unsaturation of several adsorption sites during the adsorption process [8]. The removal percentage continues to decrease till when it becomes insignificant when compared with the increase in dosage. Hence, the quantity of dye adsorbed per unit mass of the adsorbent (q_t) reduces with increase in the adsorbent dosage while the percentage removal increases [8].

Sharma and Uma [70] investigated the effect of using different dosages of rice husk activated carbon (0.40–0.60 g) for methylene blue adsorption. The result indicated that the removal percentage rose from 86.75 to 99.83% as the adsorbent dosage was increased from 0.4 to 0.60 g. Heibati et al. [24] reported similar trend on the effect of dosage on the adsorption of acid red 18 dye using poplar wood (ACP) and walnut (ACW) prepared activated carbon.

5.3 *Solution pH*

The effect of solution pH is critical in removal of dyes from textile effluents. A medium pH will regulate the extent of the electrostatic charge which is influenced by ionizable dye molecules and usually leads to variation in the rate adsorption of the dyes from solutions [10]. Generally, for cationic dyes, the removal percentage will increase with increase in pH while the removal percentage decreases for anionic dyes with increase in pH [65].

There is a considerably decrease in the electrostatic repulsion between the sorbent surface and the positively charged dye as the surface charge density reduces. Most times this leads to increase in the adsorption percentage [8]. Mohd et al. [49] studied the influence of pH on the removal of remazol dye by activated carbon prepared from pomegranate.

5.4 *Initial Dye Concentration*

The influence of initial concentration of the dye in an effluent is also vital because a fixed quantity of adsorbent can only adsorb a set quantity of dye from solution [8]. The effect of the effluent dye concentration is investigated by using various solutions of different initial dye concentrations at constant adsorbent dosage till equilibrium is achieved [65]. The influence of the dye concentration is dependent on the synergy between the concentration of the dye and the vacant binding sites on the surface of the adsorbent [65]. Generally, an increase in dye concentration will result to decrease on the dye removal percentage. This is on account of a fixed amount of dye particles that can be removed by a particular quantity of adsorbent [8]. Isah et al. [28] investigated the influence of initial dye concentration on adsorption of reactive blue dye. Heibati et al. [24] studied the influence of initial acid dye concentration on the removal efficiency using poplar activated carbon and walnut activated carbon.

5.5 *Agitation Speed*

The effect of agitation speed is significant in the elimination of dyes from textile effluent. This is because the agitation enhances the contact between the dye molecules and the adsorbents for effective removal. The effect of the agitation speed can be investigated by varying the speed of the agitation while maintaining other process variables at constant values. The adsorption mechanism of the treatment process consists of three main steps: (1) the migration of effluent substances from the wastewater solution to the adsorbent surface; (2) diffusion in to the adsorbent surface; and (3) intra-particle diffusion of the adsorbate into the inner parts of the adsorbent. If the speed of agitation increases, there will be simultaneous reduction in the resistance of the boundary layer to the transfer of adsorbate to the adsorbent surface from the bulk solution. According to Patel and Vashi [59], increase in agitation speed will lead to significant increase in the percentage of crystal violet adsorption.

5.6 *Temperature*

The influence of temperature gives an indication on the nature of adsorption especially as regards to the endothermic or exothermic nature of the dye adsorption process [65]. The treatment process is an endothermic process if as the temperature increased, there is a corresponding increase in the quantity adsorbed per unit mass of adsorbent. This is because of the increased mobility of the dye molecules with increasing temperature [8]. Giannakoudakis et al. [21] and Miao et al. [45] reported the influence of temperature on the removal of cephalecin and reactive dye using activated carbon prepared from alligator weed and commercial activated carbon. Isah et al. [28] investigated the effect of temperature on reactive blue 19 removal using activated carbon prepared from coconut shell.

6 Kinetics of the Treatment Process

The kinetics of the treatment process is based on the effect of time on the quantity of adsorbate removed. It depends on the adsorbate–adsorbent interaction. Two vital components necessary for any adsorption technique are the reaction rate and reaction mechanism. It explains the kinetics for solid–liquid adsorption system [8]. The removal kinetics is very identical to the adsorption equilibrium. The kinetics data is usually tested by standard kinetic model equations like such as intra-particle diffusion, pseudo-first-order, elovich, pseudo-second-order models among others.

Adsorption capacity according to Khan et al. [37] can be calculated by Eq. (1),

$$q_{\mathrm{e}} = \frac{(C_{\mathrm{i}} - C_{\mathrm{e}})V}{m} \tag{1}$$

where q_{e} = adsorption capacity (mg/g); C_{i} and C_{e} are the initial and equilibrium dye concentration (mg/L), respectively; V − volume of metal ion solution (L); m − adsorbent dose (g).

7 Equilibrium Isotherms of the Treatment Processes

The equilibrium study is characterized by changes in the concentration of the solutions. It is usually expressed by an adsorption isotherm, delineated by peculiar constants whose calculated values indicate the characteristics and properties of the sorbent sorption. The equilibrium data is usually fitted to different isotherm models so as to determine the appropriate model suitable for design purposes [8]. The isotherm also gives an insight into the adsorption capacity of the sorbent material and is very important in the description of the interactive relation between the adsorbate and the adsorbent. The adsorbent surface can either be a monolayer surface or multilayer surface [65]. Some of the equilibrium models used to describe the treatment process are Langmuir, Freundlich, Temkin, and Dubinin Radushkevich isotherm models.

7.1 Langmuir Model

In this type of isotherm, the active sites of the absorbent materials have the same affinity for the adsorbate molecule and hence constant activation energy and enthalpies. The linear Langmuir model of adsorption is given by Eq. (2).

$$\frac{C_{\mathrm{e}}}{q_{\mathrm{e}}} = \frac{C_{\mathrm{e}}}{q_{\mathrm{m}}} + \frac{1}{Kq_{\mathrm{m}}} \tag{2}$$

where q_e is adsorption capacity of adsorbent at equilibrium (mg g^{-1}), C_e is the adsorbate concentration at equilibrium (mg L^{-1}), q_m is Langmuir model adsorption capacity (mg g^{-1}), and K is the Langmuir equilibrium constant. K is connected to the adsorption energy of the process.

The constants q_m and K can be evaluated from the slope and intercept of the linear plot of C_e/q_e against C_e [37].

7.2 Freundlich Model

This isotherm investigates the non-ideality, reversibility, and the multilayer nature of the adsorption process [37]. The model pays particular attention to the heterogeneous nature of the adsorbent surface [37]. The linear form is as shown in Eq. (3) according to Onu and Nwabanne [55]

$$\log q_{\mathrm{e}} = \log K_{\mathrm{f}} + \frac{1}{n}\log C_{\mathrm{e}} \tag{3}$$

where q_{e} is the dye adsorption capacity at equilibrium (mg/g), C_{e} is the equilibrium concentration (mg/L). n and K_{f} are Freundlich constants connected to the adsorption intensity and adsorption capacity of the adsorbent, respectively. The Freundlich constants can be evaluated from the plot of $\log q_{\mathrm{e}}$ versus $\log C_{\mathrm{e}}$ [37].

7.3 Temkin Model

This model examines the effect of adsorbent-adsorbent interaction in the adsorption process. This model assumes that the heat of absorption of all molecules in the layer show a constant decreases with surface coverage [37].

The Temkin model is expressed in linear form as in Eq. (4) according to Nwabanne et al. [51];

$$q_{\mathrm{e}} = B\mathrm{In}K_{\mathrm{KT}} + B\mathrm{In}C_{\mathrm{e}} \tag{4}$$

where $B = \mathrm{RT}/b$, is the Temkin constant linked to the adsorption heat of the process, K_{KT} is isotherm binding constant (L/mg) related to maximum binding energy [37]. The values of these constants can be determined from the slope and intercept obtained from the linear plot of q_{e} against $\mathrm{In}C_{\mathrm{e}}$ [37].

The data generated from the adsorption process will be fitted into any of the above isotherms and the calculated R^2 value used to check the degree of fitness.

8 Thermodynamic Parameters

Thermodynamic parameters give a better insight of the temperature effect on any dye adsorption process. Thermodynamic properties such as change in entropy (ΔS), enthalpy changes (ΔH), and change in Gibbs free energy https://www.sciencedirect.com/topics/earth-and-planetary-sciences/gibbs-free-energy (ΔG) are determined and used to describe the process. A positive value of enthalpy change indicates that the adsorption process is endothermic while a negative value suggests that the adsorption is exothermic nature. This is due to the increase or decrease of the rate of diffusion of adsorbate particles across the internal pores and external boundary layer of the adsorbents. Positive ΔS value is implies that there is increased degree of freedom at the solid–liquid interface during the adsorption process. Furthermore, the positive value suggests increased randomness at the interface of the liquid and solid. It also shows the level of the affinity between the adsorbent and the adsorbate [49]. The negative ΔG value is an indication of the spontaneous nature of the adsorption process. The isotherm, kinetics, and thermodynamics of some selected adsorbents used for the treatment of textile wastewater are shown in Table 3.

9 Hybrid Application of Agricultural Waste in Textile Water Treatment

This is a means of achieving increased elimination of dyes and colorants from textile wastewater. It involves the combination of more than one approach to the decolorization of dyes-contaminated wastewater. The hybrid application may involve a combination of adsorption, bioremediation, electrocoagulation, microbial degradation, photo Fenton-like oxidation methods among others [7]. The removal of methylene blue via adsorption and *Rhodococcus* strain UCC 0003 in banana peels was investigated by Maniyam et al. [42]. In this hybrid, the bacteria strain was supported with banana peel while the strain itself removed the dyes efficiently under optimized conditions of 35 °C, pH value of 7 and 1.0 g/L banana peel concentration [42].

Similarly, Bener et al. [7] reported that combination of corn cob support material on $BiNiO_3$ catalyst effective for color removal in a photo Fenton-like oxidation method. The optimum processing parameters were pH of between 7 and 7.5 and dosage of 0.75 g/L $BiNiO_3$/corncob based catalyst. This gave a result of 49% COD, 87% TOC, 96% turbidity, with about 90% color removal. Furthermore, 95–97% removal of suspended solids was obtained. Hybrid approach affords the improvement of the dye removal from textile wastewater far more than can ever be obtained from any single method.

Table 3 Kinetics, isotherm, and thermodynamics properties of various agricultural materials for textile effluent treatment

S. No	Dyes	Adsorbent used	pH	Kinetic model	Isotherm model	Thermodynamic	References
1	Bromoscresol green dye	Rice husk	4	Pseudo second order	Langmuir and Harkin-Jura	Exothermic and spontaneous	Onu et al. [57]
2	Methylene blue	Teak tree bark powder	3–11	Lagergen Pseudo second order	Langmuir and Freundlich	Endothermic and spontaneous	Satish et al. [66]
3	Reactive red dye	Bentonite clay	2–12	Pseudo-second order	Langmuir	Endothermic and spontaneous	Hajira et al. [23]
4	Orange G dye	Kolanut shell	2	Pseudo second order	Freundlich	Exothermic and spontaneous	Nwabanne et al. [51]
5	Organic dyes	Surfactant modified corn straw	–	Pseudo second order	Langmuir and freundlich	–	Umpuch and Jutarat [73]
6	Methylene blue	Activated carbon	5–8	langmuir	Langmuir and freundlich	–	Sachin et al. [64]
7	Eriochrome black T dye	Activated bentonite	2	–	Freundlich and Jovanoic	Endothermic and spontaneous	Onu et al. [56]
8	Reactive dyes	Activated carbon from Enteromorph aprolifera	4.5–6.0	Pseudo second order	freundlich	Endothermic and spontaneous	Deshuai et al. [12]
9	Cargo red	Activated teak leaf powder	–	Elovich and intra-particle diffusion	Langmuir	Endothermic and spontaneous	Gedam et al. [16]
10	Crystal violet dye	Modified kalonite	6	Pseudo second order	Dubinin Radushkevich	Endothermic and spontaneous	Oguanobi et al. [53]
11	Remazol brilliant blue	Pomegranate activated carbon	2	Pseudo second order	Freundlich	Exothermic and spontaneous	Ahmad et al. (2014)

10 Application of Artificial Intelligence in Textile Wastewater Treatment

The goal of any industrial process is to optimize the process parameters and the output. Several process variables such as temperature, dye concentration, particle size, contact time, solution pH, adsorbent dosage, stirring speed, among others affect the quality of the treated effluent. The application of only statistical models in stimulating and optimizing the different process variables in dye removal processes is difficult and not effective since the relationship between the process variables is nonlinear [20]. Hence, the need for modeling, optimization, and simulation of the treatment process. This will enhance the industrial treatment process, save time, minimize losses, reduces stress, and increase the efficiency of the process.

One of the most effective tool for modeling of treatment textile waste water is the Artificial Intelligence (AI) techniques, such as adaptive neuro-fuzzy inference system (ANFIS), Genetic Algorithm (GA), Artificial Neural Network (ANN), etc. Since tremendous advances have been made on AI in recent times in diverse industrial processes, it can also be investigated as a potential modern technique that can enhance the treatment of textile effluent, model and optimize the process for greater efficiency.

Mahmoodi et al. [41] investigated the use of ANN in stimulating and predicting dye removal via both single and multicomponent systems using bio waste as adsorbent. Bozkurt and Geyik [9] reported on the use of GA and ANN for the prediction of the azo-metal complex dye adsorption onto lentil straw. Ghaedi et al. [19] reported on the use of ANN and GA optimization tools in the adsorption of malachite green dye onto activated carbon prepared from copper nanowires. Khajeh et al. [36] studied the use of silver nanoparticles in the removal of methylene blue dye from aqueous solutions via PSO-ANN and RSM models. Ghaedi et al. [17] equally reported on the use of hybrid ANN and PSO for the elimination of brilliant green dye from textile effluent with the aid of zinc sulfide nanoparticle loaded on activated carbon.

Furthermore, Marjan et al. [43] employed ANN to optimize the adsorption of methyl orange onto polyaniline and found that the used model well-estimated the performance of the adsorption technique. Similar results were also reported by Ghaedi and Vafaei [20]. Maniyam et al. [42] reported a tripled-fold increase in the removal of 2 g/L methylene blue by the optimization of the 3 significant process variables namely pH, temperature and concentration of banana peel via response surface methodology (RSM). The optimization enhanced the stability and regeneration/reusability of the spent banana peel up to 12 cycles before it experienced a significant reduction in efficiency.

11 Conclusion

Untreated or inadequately treated textile effluents are harmful to man and the environment. Adequate treatment of such waste is required to reduce adverse impact on man, protect the environmental protection, reuse purposes among others. Several treatment methods categorized under physical, chemical, and biological methods are being applied for the treatment of dye-contaminated waste water with various degrees of successes, before discharging the effluent into the environment and water bodies.

Different methods and factors were reviewed in this chapter for the removal of dyes from solutions. Adsorption was observed to be the most economical and very efficient among all methods. The influence of various parameters that affect the textile effluent treatment process namely; contact time, the mass of adsorbent, agitation speed, solution pH, temperature and concentration of dye were reviewed. The kinetics, isotherm, and thermodynamics study of the dye treatment process were equally reviewed in the work. It was observed that artificial intelligence techniques like artificial neural network (ANN), adaptive neuro-fuzzy inference system (ANFIS), Genetic Algorithm (GA), etc., models can be used to model and optimized dye adsorption processes effectively. It is concluded that the treatment of textile effluent using agricultural waste offered a twofold solution involving removal of pollutants and waste reduction.

Acknowledgement The authors will like to thank the department of Chemical Engineering, Nnamdi Azikiwe University Awka for the conducive environment they provided which facilitated this work.

References

1. Aadil A, Shahzad M, Kashif S, Muhammad M, Rabta A, Saba A (2012) Comparative study of adsorptive removal of congo red and brilliant green dyes from water using peanut shell. Middle East J Sci Res 11(6):828–832
2. Ahmad MA, Nur AAP, Bello SO (2014) Kinetic, Equilibrium and thermody-namic studies of synthetic dye removal using pomegranate peel activated carbon prepared by microwave-induced koh activation. Water Resources and Industry. http://dx.doi.org/10.1016/j.wri.2014.06.002
3. Ali I, Al-Othman ZA, Alwarthan A (2016) Molecular uptake of congo red dye from water on iron composite nanoparticles. J Mol Liq 224:171–176
4. Ali I, Asim M, Khan TA (2012) Low-cost adsorbents for the removal of organic pollutants from wastewater. J Environ Manag 113:170–183. https://doi.org/10.1016/j.jenvman.2012.08.028
5. Balachandran K, Sivaraj R, Venckatesh R (2013) Removal of methylene blue using prepared nano titanium dioxide(TiO_2) photocatalyst. J Environ Nanotech 2(1):32–39. https://doi.org/10.13074/jent.2013.02.121027
6. Bapat S, Jaspal D (2016) Parthenium hysterophorus: novel adsorbent for the removal of heavy metals and dyes. Glob J Environ Sci Manag 2:135–144. https://doi.org/10.7508/gjesm.2016.02.004
7. Bener S, Atalay S, Ersöz G (2020) The hybrid process with eco-friendly materials for the treatment of the real textile industry wastewater. Ecol Eng 148:105789. https://doi.org/10.1016/j.ecoleng.2020.105789
8. Bharathi KS, Ramesh ST (2013) Removal of dyes using agricultural waste as low-cost adsorbent: a review. Appl water sci 3:773–790. https://doi.org/10.1007/s13201.013.0117y

9. Bozkurt ÇA, Geyik F (2013) Use of artificial neural networks and genetic algorithms for prediction of sorption of an azo-metal complex dye onto lentil straw. Bioresour Technol 129:396–401. https://doi.org/10.1016/j.biortech.2012.11.085
10. Brahim OI, Belmedani M, Belgacem A, Hadoun H, Sadaoui Z (2014) Discoloration of azo dye solutions by adsorption on activated carbon prepared from the cryogenic grinding of used tires. Chem Eng Transactions 38:121-126. http://dx.doi.org/10.3303.CET1438021
11. Dellamatrice PM, Silva-Stenico ME, de Moraes LAB, Fiore MF, Monteiro RTR (2017) Degradation of textile dyes by cyanobacteria. Braz J Microbiol 48:25–31. https://doi.org/10.1016/j.bjm.2016.09.012
12. Deshuai S, Zhongyizhang M, Wang Y (2013) Adsorption of reactive dyes on activated carbon developed from, *Enteromorpha prolifera.* Am J Anal Chem 4:17–26. https://doi.org/10.4236/ajac.2013.47A003
13. Ding J, Yang Z, He C, Tong X, Li Y, Niu X, Zhang H (2017) UiO-66 (Zr) coupled with Bi 2 MoO 6 as photocatalyst for visible-light promoted dye degradation. J Colloid Interface Sci 497:126–133. https://doi.org/10.1016/j.jcis.2017.02.060
14. El-Sayed GO, Yehia MM, Asaad AA (2014) Assessment of activated carbon prepared from corncob by chemical activation with phosphoric acid. Water Res Ind 7–8:66–75. https://doi.org/10.1016/j.wr.2014.10.001
15. Emembolu LN, Nwabanne JT, Onu CE, Obiakor IO (2017) Simulation of kinetic model for anaerobic digestion of pharmaceutical wastewater. Chem Sci Int J 20(3):1–13. https://doi.org/10.9734/CSJI/2017/34050
16. Gedam VV, Raut P, Chahande A et al (2019) Kinetic, thermodynamics and equilibrium studies on the removal of congo red dye using activated teak leaf powder. Appl Water Sci 9:55. https://doi.org/10.1007/s13201-019-0933-9
17. Ghaedi M, Ansari A, Bahari F, Ghaedi AM, Vafaei A (2015) A hybrid artificial neural network and particle swarm optimization for prediction of removal of hazardous dye brilliant green from aqueous solution using zinc sulphide nanoparticle loaded on activated carbon. Spectrochim Acta A 137:1004–1015. https://doi.org/10.1016/j.saa.2014.08.011
18. Ghaedi MA, Ansari A, Mohammadi F, Vafaei A (2014) Artificial neural network and particle swarm optimization for removal of methyl orange by gold nanoparticles loaded on activated carbon and tamarisk. Spectrochim Acta A Mol Biomol Spectrosc 132:639–654. https://doi.org/10.1016/j.saa.2014.04.175
19. Ghaedi M, Shojaeipour E, Ghaedi AM, Sahraei R (2015) Isotherm and kinetics study of malachite green adsorption onto copper nanowires loaded on activated carbon: artificial neural network modelling and genetic algorithm optimization. Spectrochim Acta A 142:135–149. https://doi.org/10.1016/j.saa.2015.01.086
20. Ghaedi M, Vafaei A (2017) Application of artificial neural network for adsorption removal of dyes from aqueous solution: a review. https://doi.org/10.1016/j.cis.2017.04.015
21. Giannakoudakis DA, Kyzas GZ, Avranas A, Lazaridis NK (2016) Multi-parametric adsorption effects of the reactive dye removal with commercial activated carbons. J Mol Liq 213:381–389. https://doi.org/10.1016/j.molliq.2015.07.010
22. GilPavas E, Dobrosz-Gómez I, Gómez-García MÁ (2018) Optimization of sequential chemical coagulation-electro-oxidation process for the treatment of an industrial textile wastewater. J Water Process Eng 22:73–79
23. Hajira T, Muhammad S, Zainab Q (2013) Physiochemical modification and characterization of bentonite clay and its application for the removal of reactive dyes. Int J Chem 5(3):19–32. https://doi.org/10.5539/ijc.v5n3p19
24. Heibati B, Rodriguez-Couto S, Al-Ghouti MA, Asif M, Tyagi I, Agarwal S, Gupta VK (2015) Kinetics and thermodynamics of enhanced adsorption of the dye AR 18 using activated carbons prepared from walnut and poplar woods. J Mol Liq 208:99–105. https://doi.org/10.1016/j.molliq.2015.03.057
25. Holkar CR, Jadhav AJ, Pinjari DV, Mahamuni NM, Pandit ABA (2016) Critical review on textile wastewater treatments: possible approaches. J Environ Manag 182:351–366. https://doi.org/10.1016/j.jenvman.2016.07.090

26. Hussein HS, Sabry R, Hassan N, Morsi MS, Sharrawy HH (2014) Removal of reactive blue 19 from textile wastewater using iron electrodes. Res J Pharm Biol Chem Sci 5(3):2091–2105
27. Iheanacho CO, Nwabanne JT, Onu CE (2019) Optimum process parameters for activated carbon production from rice husk for phenol adsorption. Curr J Appl Sci Technol 36(6):1–11. https://doi.org/10.9734/CJAST/2019/v36i630264
28. Isah U, Abdulraheem AG, Bala S, Muhammad S, Abdullahi M (2015) Kinetics, equilibrium and thermodynamics studies of C.I. reactive blue 19 dye adsorption on coconut shell based activated carbon. Int Biodeterior Biodegradation 102:265–273. https://doi.org/10.1016/j.ibiod.2015.04.006
29. Islam MR, Mostafa MG (2020) Characterization of textile dyeing effluent and its treatment using polyaluminum chloride. Appl Water Sci 10(5):1–10
30. Jegatheesan V, Pramanik BK, Chen J, Navaratna D, Chang CY, Shu L (2016) Treatment of textile wastewater with membrane bioreactor: a critical review. Biores Technol 204:202–212. https://doi.org/10.1016/j.biortech.2016.01.006
31. Kalra SS, Mohan S, Sinha A, Singh G (2011) Advanced oxidation processes for treatment of textile and dye wastewater: a review. In: 2nd international conference on environmental science and development IPCBEE. vol 4. IACSIT Press, Singapore
32. Karam A, Zaher K, Mahmoud AS (2020) Comparative studies of using nanozerovalent iron, activated carbon, and green synthesized nanozerovalent iron for textile wastewater color removal using artificial intelligence, regression analysis, adsorption isotherm, and kinetic studies. Air Soil Water Res 13:1178622120908273
33. Karthik V, Saravanan K, Bharathi P, Dharanya VC (2014) An overview of treatments for the removal of textile dyes. J Chem Pharm Sci 7(4):301–307
34. Karuk AB, Bakhtyar KA (2019) Adsorptive removal of methyl orange from aqueous solutions with natural germak clay as a cheap and efficient adsorbent in batch and continuous systems. https://doi.org/10.17656/j2s.10767
35. Kehinde FO, Aziz HA (2014) Textile waste water and the advance doxidative treatment process, an overview. Int J Innovat Res Sci Eng Technol 3:15310–15317
36. Khajeh M, Kaykhaii M, Sharafi A (2013) Application of PSO-artificial neural network and response surface methodology for the removal of methylene blue using silver nanoparticles from water samples. J Ind Eng Chem 19:1624–1630. https://doi.org/10.1016/j.jiec.2013.01.033
37. Khan AMA, Anwar N, Naeem M (2020) Comparative study of congo red dye on rice husk, rice husk char and chemically modified rice hush char from aqueous solution. Bull Chem Soc Ethiop 34(1):41–54. https://doi.org/10.4314/bcse.v34i1.4
38. Khan TA, Rahman RI, Ali EA, Khan EA, Mukhlif AA (2014) Removal of malachite green from aqueous solution using waste pea shells as low-cost adsorbent–adsorption isotherms and dynamics. Toxicol Environ Chem 96:569–578
39. Kirupavasam EK, Allen GRG (2012) Photocatalytic degradation of amido black-10B using nano photocatalyst. J Chem Pharm Res 4(6):2979–2987
40. Liu M, Chen Q, Lu K, Huang W, Lü Z, Zhou C, Yu S, Gao C (2016) High efficient removal of dyes from aqueous solution through nanofiltration using diethanolamine-modified polyamide thin-film composite membrane. Sep Purif Technol 173:135–143. https://doi.org/10.1016/j.seppur.2016.09.023
41. Mahmoodi NM, Taghizadeh K, Taghizadeh KA (2018) Mesoporous activated carbons of low-cost agricultural bio-wastes with high adsorption capacity: preparation and artificial neural network modelling of dye removal from single and multicomponent (binary and ternary) systems. J Mol Liq 269. https://doi.org/10.1016/j.molliq.2018.07.108
42. Maniyam MN, Hari M, Yaacob NS (2020) Enhanced methylene blue decolourization by *Rhodococcus* strain UCC 0003 grown in banana peel agricultural waste through response surface methodology. Biocatal Agric Biotechnol 23:101486. https://doi.org/10.1016/j.bcab.2019.101486
43. Marjan T, Sayyed HH, Asieh DK, Martin O (2017) Artificial neural network optimization for methyl orange adsorption onto polyaniline nano-adsorbent: kinetic, isotherm and thermodynamics studies. https://doi.org/10.1016/j.molliq.2017.08.112

44. Marrakchi F, Ahmed MJ, Khanday WA, Asif M, Hameed BH (2017) Mesoporous-activated carbon prepared from chitosan flakes via single-step sodium hydroxide activation for the adsorption of methylene blue. Int J Biol Macromol 98:233–239. https://doi.org/10.1016/j.ijbiomac.2017.01.119
45. Miao MS, Liu Q, Shu L, Wang Z, Liu YZ, Kong Q (2016) Removal of cephalexin from effluent by activated carbon prepared from alligator weed: kinetics, isotherms and thermodynamic analyses. Process Saf Environ Prot. https://doi.org/10.1016/j.psep.2016.03.017
46. Milica J, Dalibor S, Dragan M, Ivan A, Andelija M, Biljana D, Goran R (2013) Study of the electrochemical oxidation of reactive textile dyes using a platinum electrode. Int J Electrochem Sci 8:168–183
47. Mingyi F, Jiwei H, Rensheng C, Kangning X, Xionghui W (2017) Modeling and prediction of copper removal from aqueous solutions by nZV1/rGO magnetic nanocomposites using ANN-GA and ANN-PSO. Sci Rep 7:18040. https://doi.org/10.1038/s41598-017-18223-y
48. Mohammad FA, Mumtaz AZ, Abeer MAA (2012) Experimental study of dye removal from industrial wastewater by membrane technologies of reverse osmosis and nano-filtration. Iran J Environ Health Sci Eng 9(17):1–9. https://doi.org/10.1186/1735-2746-9-17
49. Mohd AA, Nur AAP, Olugbenga SB (2014) Kinetic, equilibrium and thermodynamic studies of synthetic dye removal using pomegranate peel activated carbon prepared by microwave-induced KOH activation. Water Resour Ind. https://doi.org/10.1016/j.wri.2014.06.002
50. Mohd A, Mohammad R, Mimi A, Mohd H, Rafidah H (2017) Factors affecting textile dye removal using adsorbent from activated carbon: a review. https://doi.org/10.10051/mateconf/2017030615
51. Nwabanne JT, Okpe EC, Igbokwe PK, Asadu CC, Onu CE (2016) Isotherm and kinetic modelling of adsorption of dyestuffs onto kola nut (*Cola acuminate*) shell activated carbon. J Chem Technol Metall 51(2):188–201
52. Nwobasi VN, Igbokwe PK, Onu CE (2020) Removal of methylene blue dye from aqueous solution using modified ngbo clay. J Mater Sci Res Rev 5(2):33–46
53. Oguanobi NC, Onu CE, Onukwuli OD (2019) Adsorption of a dye (crystal violet) on an acid modified non-conventional adsorbent. J Chem Technol Metall 54(1):95–110
54. Okpe EC, Asadu CO, Onu CE (2018) Statistical analysis for orange G adsorption using kola nut shell activated carbon. J Chin Adv Mater Soc. https://doi.org/10.1080/22243682.2018.1534607
55. Onu CE, Nwabanne JT (2014) Adsorption kinetics for malachite green removal from aqueous solution using nteje clay. J Environ Hum 1(2):133–150. https://doi.org/10.15764/EH.2014.02015
56. Onu CE, Nwabanne JT, Ohale PE, Asadu CO (2021) Comparative analysis of RSM, ANN and ANFIS and the mechanistic modeling in eriochrome black-T dye adsorption using modified clay. S Afr J Chem Eng 36:24–42. https://doi.org/10.1016/j.sajce.2020.12.003
57. Onu CE, Oguanobi NC, Okonkwo CO, Nnamdi-Bejie J (2020) Application of modified agricultural waste in the adsorption of bromocresol green dye. Asian J Chem Sci 7(1):15–24. https://doi.org/10.9734/AJOCS/2020/v7i119011
58. Onu CE, Nwabanne JT (2014b) Adsorption studies on the removal of Eriochrome Black-T from aqueous solution using Nteje clay. SOP Transactions on Applied Chemistry. 1 (2) 14 – 25.:https://doi.org/10.15764/STAC.2014.02003
59. Patel H, Vashi RT (2010) Adsorption of crystal violet dye onto tamarind seed powder. E-J Chem 73:975–984. https://doi.org/10.1155/2010/143439
60. Popli S, Patel UD (2015) Destruction of azo dyes by anaerobic-aerobic sequential biological treatment: a review. Int J Environ Sci Technol 12:405–420. https://doi.org/10.1007/s13762-014-0499-x
61. Puasa SW, Ruzitah MS, Sharifah ASA (2012) Competitive removal of reactive black 5/reactive orange 16 from aqueous solution via micellar-enhanced ultrafiltration. Int J Chem Eng Appl 3(5). https://doi.org/10.7763/IJCEA.2012.V3.217
62. Rameshwar A, Dileep K, Priyanka J (2014) Photocatalytic degradation of methylene blue using calcium oxide. Acta Chim Pharm Indica 4(1):20–28

63. Roy CK, Jahan MAA, Rahman SS (2018) Characterization and treatment of textile wastewater by aquatic plants (macrophytes) and algae. Eur J Sustain Dev Res 2(3):29. https://doi.org/10.20897/ejosdr/85933
64. Sachin MK, Gaikwad RW (2011) Removal of methylene blue from effluent by using activated carbon and water hyacinth as adsorbent. Int J Chem Eng Appl 2(5):317–399. https://doi.org/10.7763/IJCEA.2011.v2.126
65. Salleh MAM, Mahmoud DK, Karim WAWA, Idris A (2011) Cationic and anionic dye adsorption by agricultural solid wastes: a comprehensive review. Desalination 280:1–13. https://doi.org/10.1016/j.desal.2011.07.019
66. Satish P, Sameer R, Naseema P (2011) Removal of methylene blue, a basic dye from aqueous solutions by adsorption using teak tree (*Tectona grandis*) bark powder. Int J Environ Sci 1(5). 10.1.1.225.337
67. Sekomo CB, Rousseau DPL, Saleh SA, Lens PNL (2012) Heavy metal removal in duckweed and algae ponds as a polishing step for textile wastewater treatment. Ecol Eng 44:102–110. https://doi.org/10.1016/j.ecoleng.2012.03.003
68. Selvam K, Shanmuga PM (2012) Biological treatment of azo dyes & textile industry effluent by newly isolated white rot fungi schizophyllum commune and lenziteseximia. Int J Environ Sci 2(4):1926–1935. https://doi.org/10.6088/ijes.00202030076
69. Shah MP, Patel KA, Nair SS, Darji AM (2013) Optimization of environmental parameters on microbial degradation of reactive black dye. J Biomed Biodeg 4:183. https://doi.org/10.4172/2155-6199.1000183
70. Sharma CY, Uma (2010) Optimization of Parameters for Adsorption of Methylene Blue on a Low-Cost Activated Carbon. J Chem Eng Data 55(1):435-439. https://doi.org/10.1021/je900408s
71. Stergiopoulos D, Dermentzis K, Giannagoudakis P, Sotiropoulos S (2014) Decolourization and removal of indigo carmine textile dye from wastewater. Glob NEST J 16(3):499–506. https://doi.org/10.30955/gnj.001330
72. Tanyildizi MS (2011) Modeling of adsorption isotherms and kinetics of reactive dye from aqueous solution by peanut hull. Chem Eng J 168:1234–1240. https://doi.org/10.1016/j.cej.2011.02.021
73. Umpuch C, Jutarat B (2013) Adsorption of organic dyes from aqueous solution by surfactant modified corn straw. Int J Chem Eng Appl 4(3):134–139. https://doi.org/10.7763/JCEA2013.V4.279
74. Xie M, Zeng L, Zhang Q, Kang Y, Xiao H, Peng Y, Chen X, Luo J (2015) Synthesis and adsorption behaviour of magnetic microspheres based on chitosan/organic rectorite for low-concentration heavy metal removal. J Alloy Compd 647:892–905
75. Yaseen DA, Scholz M (2018) Textile dye wastewater characteristics and constituents of synthetic effluents: a critical review. https://doi.org/10.1007/s13762-018-2130-z2018

Application of Agricultural Waste in Anionic Dyes Removal from Wastewater

Titus Chinedu Egbosiuba

Abstract Environmental pollution and public health is a serious concern posed by effluents from textile industries. The effluents from textile industries, which are primarily contaminated by dyes, are due to large use of dyes and water. The dye contaminants are primarily classified into anionic, cationic and non-ionic dyes. The anionic dyes are essentially composed of acid, direct and reactive dyes containing chromophore groups characteristic for azo, triarylmethane, phthalocyanine and anthraquinone. Therefore, the presence of the anionic dyes in textile effluents may cause a number of diseases to humans such as damages to the liver, kidney, intestines and highly mutagenic and carcinogenic. In the environment, the presence of these contaminants also obstructs the penetration of sunlight, thereby causing reduction in the rate of photosynthesis. Thus, these contaminants also decrease the amount of dissolved oxygen available to the ecosystem. In view of these, there is a great concern for appropriate treatment method to ensure effective removal of the dyes. We reviewed (1) textile dyes classification, (2) toxicity of dyes, (3) treatment technologies for dyes, (4) agricultural waste materials and (5) the removal of anionic dyes using agricultural waste. The removal of the anionic dyes by agricultural wastes were studied based on adsorption due to its advantages of ease of operation, simplicity, lack of chemical use and effectiveness of the process.

Keywords Agricultural waste · Toxicity of dyes · Anionic dyes · Direct dyes · Acid dyes · Reactive dyes · Wastewater treatment methods · Adsorption · Isotherm · Kinetics

T. Chinedu Egbosiuba (✉)
Chemical Engineering Department, Chukwuemeka Odumegwu Ojukwu University, Uli Campus, Anambra, Nigeria
e-mail: egbosiubachinedu@gmail.com

S. S. Muthu and A. Khadir (eds.), *Textile Wastewater Treatment*, Sustainable Textiles: Production, Processing, Manufacturing & Chemistry,
https://doi.org/10.1007/978-981-19-2852-9_7

1 Introduction

Water has been acknowledged as a major resource for the sustainability of the environment, but had been continually threatened in the recent times due to rapid expansion of human population and increasing industrialization. The sustained human activities in the industries to satisfy increasing human needs culminated to water contamination by different pollutants such as dyes, heavy metals, municipal wastes, alkalis, acids, oils, surfactants, salts, chemicals, solvents and other hazardous materials [55, 76, 130]. Among these pollutants, dyes are discharged into the water bodies from industries such as textile, printing, cosmetics, paper, plastic, food processing, dye manufacturing, leather and rubber industries [49, 165]. At the moment, textile industries have been ranked as the highest source of dye contaminated wastewater globally due to the application of over 100,000 different types of commercially available dyes with the production capacity above 700,000 tons yearly [9, 19, 31, 45, 150, 163]. Sequel to that, about 10–15% of the dyes are lost annually during dyeing and finishing processes in textile industry which are discharged into the water bodies without treatment [19, 76, 116]. Consequently, these dyes have been identified to be toxic, carcinogenic and harmful to animals, aquatic and human lives [66, 163]. Since these dyes remain over a lengthened period and can only be removed through further treatment, there is a need to remove the dye effluents prior to discharge to the environment or reuse in the industry.

In recent times, a number of conventional physical, chemical and biological treatment technologies such as adsorption [8, 48, 50, 51,87, 90], filtration [40, 94, 96], coagulation and flocculation [47, 70, 117, 172], sedimentation/flotation [3, 20, 136, 152], ion exchange [22, 53, 98], membrane processes [69, 134, 138, 170], advanced oxidation [30, 85, 110, 127], electrochemical [54, 105, 137], photocatalysis [5, 12, 13, 78, 88], phytoremediation [65, 76, 151] and phycoremediation [37, 81, 94, 133, 156] have been used to eliminate dyes from textile wastewater. Compared with other conventional methods, adsorption has been the most preferred due to its versatility, effective dye removal, economical procedure, low energy usage and simplicity of the treatment process [9, 19, 45].

Before now, numerous materials have been used as adsorbent toward the removal of dyes from textile effluents but activated carbon has been reported to be more effective [45]. Although commercial activated carbon had been widely reported in the treatment of dye contaminated textile effluent, they are relatively expensive to be applied in large-scale media. Consequently, alternative naturally abundant, low-cost and effective materials are highly desired. In view of that, different researchers have explored their interest on the use of wastes such as agricultural wastes which are abundant, cheap and possessing good chemical stability. The application of agricultural wastes to remove dyes from textile wastewater enhances the use of green chemistry principles in line with the global environmental regulations against indiscriminate waste disposal and industrial effluents discharge to the water bodies [27]. In the meantime, researchers have reported the use of cheap adsorbents for anionic dyes removal such as barley straw [71], corn stalks [145], tea residue [73], pineapple peels [159], wheat husk [22], oil palm [21], pine cone [31], avocado seed [108], litchi seed [48], cattail root [68], cocoa pod husks [115], date pits [62], coffee waste [165], coconut residual powder [126], sugarcane bagasse [130], bamboo powder [171], banana peel [104], walnut shell [113], orange peel [107], cabbage [164], guava leaf [112], garlic peel [16], jojoba residue [6], shrimp shell [123], wheat [91], cucumber and potato [149].

In this review, studies on the adsorption of anionic dyes from textile industries using agricultural waste are selectively reported. Research on the removal of dyes using adsorbents had been on for decades, but remains highly interesting due to increasing scientific contributions to knowledge in this field in the recent times. Although agricultural wastes are prepared into activated carbon, some are also modified into multifunctional adsorbents and composite materials for the adsorption of anionic dyes from textile effluents. Therefore, it is paramount to carry out reviews on the progress made so far. In this chapter, detailed information on the current methods of exploiting different agricultural wastes to remove anionic dyes and the various adsorption capacities are reported.

2 Textile Dyes and Classification

A dye can be described as any coloring substance used for the provision and modification of an existing color of different materials. Dyes are ionizable, bright and aromatic organic compound that are principally used in aqueous solution, displaying affinity toward the substrate of its application [17]. The colors of dyes are due to the presence of chromophores and auxochromes molecules which absorb light at specified wavelength within the visible spectrum of 400–700 nm [102]. Primarily, dyes are classified into natural and synthetic dyes depending on their source of origin, chemical structure, chromophore color and the application through the proposed color index of global society of colorists and American association of textile chemists as shown in Table 1 [11, 102, 143].

Natural dyes can be acquired from natural coloring sources of origin including organic (plants and animals) and inorganic (minerals, earths, clays, metallic salts and clays) dyestuffs [109]. Overall, different plant parts ranging from the leaves, stems, roots, fruit, bark and seeds have been explored and established to contain extractive coloring agents [61]. Additionally, botanical sources such as insects, shellfish and lichens can be utilized to obtain natural dyes. On the other hand, synthetic dyes are developed to satisfy the needs for modernized physically and biologically resistant coloring agents. Synthetic dyes are classified according to the chemical groups as azo, triphenylmethane, phthalocyanine, anthraquinone, nitro, indigoid and sulfur [14, 36].

However, synthetic dyes are categorized based on their ionic forms, application, chemical structure and color as anionic (acid, direct and reactive), cationic (basic) and non-ionic (disperse and vat) dyes based on the particle charge [7, 19, 44, 55, 90]. While anionic dyes are defined as acidic and negatively charged chromophore

Table 1 Textile dyes classification

Dyes	Chemical group	Solubility	Substrate	Example
Anionic				
Acid	a. Azo b. Anthraquinone and c. Triphenylmethane d. Azine e. Xanthene f. Nitro g. Nitroso	Soluble in water	1. Paper 2. Leather 3. Wool 4. Nylon and 5. Ink	i. Acid blue 92 ii. Methyl orange iii. Congo red iv. Acid brown 354 v. Eosin yellow vi. Acid red vii. Acid orange etc
Direct	a. Azo b. Phthalocyanine c. Stilbene and d. Oxazine	Soluble in water	1. Cotton 2. Rayon 3. Nylon 4. Leather and 5. Paper	i. Direct blue 6 ii. Direct red 5B iii. Direct black 22 etc
Reactive	a. Azo b. Phthalocyanine c. Anthraquinone d. Oxazine e. Formazan and f. Basic	Soluble in water	1. Wool 2. Cotton 3. Nylon and 4. Silk	i. Reactive blue 4 ii. Reactive yellow 145 iii. Reactive blue 19 iv. Reactive red 218 v. Reactive black 5 etc
Cationic				
Basic	a. Azo b. Anthraquinone c. Oxazine d. Cyanine e. Triarylmethane f. Acridine g. Hemicyaninem h. Azine i. Xanthenes j. Diazahemicyanine k. Diphenylmethane	Soluble in water	1. Inks 2. Nylon 3. Paper 4. Polyester and 5. Polyacrylonitrile	i. Methylene blue ii. Crystal violet ii. Malachite green iii. Basic yellow 28 iv. Rhodamine B v. Brilliant green vi. Safranin etc
Non-ionic				
Disperse	a. Azo b. Anthraquinone c. Benzodifuranone d. Nitro(e)Styryl	Insoluble in water	1. Plastic 2. Acrylic 3. Acetate 4. Polyester and 5. Polyamide	i. Disperse orange 1 ii. Disperse red 9 iii. Disperse red 60 iv. Disperse yellow 26 v. Disperse brown etc
Vat	a. Indigoids b. Anthraquinone	Insoluble in water	1. Wool and 2. Cotton	i. Indigo blue ii. Alizarin red iii. Vat red 10 iv. Vat orange 1 etc

dyes, cationic dyes are basic and positively charged chromophore dyes [74, 149]. Among the chemical groups, azo dyes are the most generally used colorants, thereby constituting over 70% of dyes in textile industry usage, followed by anthraquinone with nearly 15% [17, 45]. This is due to the coverage of larger color range and additional characteristics of production cost-effectiveness, availability of low-cost starting materials and superior fastness. Azo dyes show stability in light and resists microbial degradation. Notably, azo dyes are remarkably the largest group of dyes with a functional group of N=N as a chromophore in aromatic structure. Essentially, chromophores are composed of functional groups such as $-N = N$, $-C = O$, $-NO_2$ and $O = (C_6H_4) = O$, while auxochromes contains $-OH$, $-NH_3$, $-CO_2H$ and $-SO_3H$ [17, 36].

3 Toxicity of Dyes

Large volumes of wastewater containing dyes are discharged into the water bodies from textile industries. The large volume of textile dyes that are lost and discharged into water bodies through dyeing procedures activates health issues of toxicity, carcinogenicity and mutagenicity in human beings [73, 119, 122] Particularly, the health hazards posed by these dyes could be linked to their nature and non-biodegradable by-products [14, 62, 87].

Correspondingly, the quantity of dissolved oxygen (DO) and biochemical oxygen demand (BOD) in water is significantly reduced by discharged anionic dyes through the formation of a thin layer on the receiving water bodies' surface. Moreover, anionic dyes in wastewater greatly increase the chemical oxygen demand (COD) and changes the turbidity level, pH and temperature, thereby indicating an elevated level of environmental pollution [28, 126, 171]. Although dyeing effluents are severely hazardous to aquatic life by damaging the intestines, liver, gills and aquatic populations in the ecosystem, the utilization of the wastewater for irrigation purposes has negative environmental impact including reduction in seed germination, soil fertility, plants protein content and chlorophyll [17, 66, 149]. Also, the increasing concentration of color from the dyes in the water bodies obstructs the penetration of sunlight and causes reduction in receiving water bodies' oxygenated potential, thereby influencing photosynthetic activity [17, 168].

Furthermore, azo dyes consisting of amine groups have been reported to be responsible for different health challenges such as immune suppression, contact dermatitis, chemosis, skin irritation, vomiting gastritis, exophthalmos, hypertension, central nervous system disorders, respiratory and circulatory diseases, permanent blindness, leukemia, lung edema, eye infection, diarrhea, insomnia, salivation, paralysis, cancer and damage to the brain, reproductive system, kidney and liver [14, 21] (Darwesh et al. 2021b; [48, 166, 167, 169]).

4 Treatment Technologies for Dyes

Over the literature, several conventional treatment techniques have been utilized to achieve a cost-efficient and renewable removal of anionic dyes from textile wastewater and these methods are important to achieve a sustainable safe environment and public health [15, 106]. To achieve efficient removal of anionic dyes from textile wastewater, various chemical processes, biological processes, physical processes and integrated processes have been applied. As shown in Fig. 1, the treatment techniques under the chemical processes are irradiation, oxidative process, photocatalysis, coagulation-flocculation, precipitation (electrocoagulation), Fentons process and ozonation [5, 13, 38, 59, 109, 131], while the biological treatment methods include enzyme degradation, algae degradation, fungi decolorization and biosorption by microbial biomass [65, 138, 151, 157]. In the physical treatment process, adsorption, membrane method, filtration, ion exchange and reverse osmosis are the characteristic treatment technologies for anionic dyes elimination from textile effluents [1, 25, 37, 56, 70, 81, 101, 103].

4.1 Chemical Treatment

Chemical treatment methods for dye are those techniques that use chemicals to achieve the removal of dye. The conventional chemical methods for the removal of

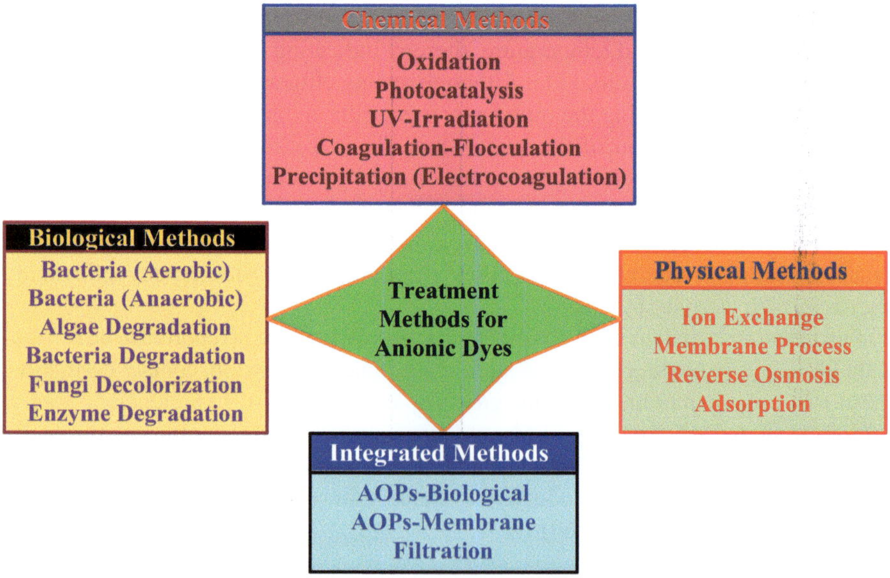

Fig. 1 Treatment methods for dyes removal from textile wastewater

dye include oxidation, photocatalysis, UV-irradiation, coagulation-flocculation and precipitation (electrocoagulation) as shown in Fig. 1. A number of the chemical techniques for dye removal are more expensive compared to biological and physical treatment techniques. Chemical methods require specific equipment that demands high electrical energy, involve the input of large amount of chemicals and generate secondary pollutants [36, 42, 76, 111].

4.1.1 Chemical Oxidation Processes

In chemical processes of anionic dye elimination, chemical oxidation indicates the removal of anionic dyes using chemical agents of oxidation. In mild oxidation environments such as in biological systems, anionic dyes are very resistant. Hence, in order to achieve efficient removal of anionic dyes from textile wastewater, potent oxidizing agents such as Fentons reagent, ozone, hydrogen peroxide and sodium hypochlorite are predominantly applied [106]. In Fentons process, a solution of iron ions (usually iron (II) sulfate) and peroxides (mostly hydrogen peroxide (H_2O_2)) are typically employed as a catalyst having active oxygen species to oxidize the dyes in wastewater [30, 85, 139, 161]. In the Fentons procedure, simultaneous oxidation of ferrous ion to ferric ion and transformation of hydrogen peroxide to hydroxyl radicals and hydroxide ion are achieved, while radical oxidation and precipitation of colored compounds with ferric ions are attained [61]. Fentons process is highly limited due to the formation of large volume of sludge. Ozonation method involves the treatment of dye contaminated colored wastewater using ozone, which is a strong oxidant and readily available commercially. Importantly, ozone had been reported to effectively oxidize unsaturated bonds and chromophore group responsible for color [111, 127, 137]. A key advantage of ozonation is that it does not support increase in wastewater and sludge volume, while its short half-life (usually 20 min) due to use of gaseous ozone, high installation costs of ozonation plant and destabilization by pH, salt and temperature are the major disadvantages of ozonation process [106]. Notably, peroxidases technique uses hydrogen peroxide to oxidize and decolorize aromatic spectrum of dyes from textile effluent, while oxidation using sodium hypochlorite to degrade the amino group contained in dye molecule using Cl^+ initiates and facilitates the cleavage of azo bond [56, 154, 160]. These chemical techniques consume a large amount of chemicals involving high cost of operation and generate toxic by-products [14].

4.1.2 Photocatalysis

Photocatalysis can be described as the process in which a reaction by a catalyst is accelerated in the presence of light. On absorption of light waves by the photocatalysts with a greater energy than the bandgap of the semiconductor pairs of electrons and holes are produced in the conduction and the semiconductor valence band [146]. In all, semiconductor catalyst including Fe_2O_3, TiO_2, ZnO, ZnS, CeO_2, CdS and

GaP are mostly utilized for the decomposition of dyes and other organic contaminants in textile wastewater into biodegradable compounds and subsequent conversion to nontoxic carbon dioxide and water [23, 80, 82, 95, 140]. The advantages of heterogeneous photocatalysis include low cost of operation, major components and intermediate compound mineralization without producing secondary pollutants and ambient application of pressure and temperature [146].

4.1.3 UV-Irradiation

Chemical approach involving irradiation makes use of electron beams or gamma rays to remove and disinfect anionic dyes from textile wastewater [166]. In irradiation technique, a large amount of dissolved oxygen and its continuous supply is essential for efficient degradation of anionic dyes due to the rapid consumption of the dissolved oxygen [37, 77, 154].

4.1.4 Coagulation-Flocculation

Coagulation-flocculation is a complex technique involving the use of large quantity of coagulants such as iron sulfate (Fe_2SO_4), iron chloride ($FeCl_3$) and alum (aluminum sulfate, Al_2SO_4) [146]. Generally, small particles are converted to larger aggregates (flocs), while the dissolved organic materials are adsorbed by the particulate aggregates [33, 37, 154, 155]. Subsequently, the contaminants are eliminated in the sedimentation/flotation and filtration stage. Hence, three steps involving the formation of coagulant, destabilization of colloids/particles and aggregation of particles are basic in coagulant method [101]. Specifically, the formation of coagulant and destabilization of colloids/particles are promoted at the fast mixing stage, while the aggregation of particles (formation of flocs) is achieved at the flocculation step, where large particles of the flocs are produced through inter-particle collisions [40, 57, 63, 94, 172]. Herein, coagulation describes the initial process involving the destabilization of colloid dispersion due to charge neutralization of the particles to be removed as they coagulate and precipitate, while flocculation refers to the aggregation and floc formation of the destabilized colloids [40, 47, 101, 103, 144]. The efficiency of the coagulation-flocculation process depends on the parameters such as time, mixing speed and suspension pH [61]. Coagulation-flocculation method is a less effective process requiring large addition of alkaline substances to achieve optimum pH [142].

4.1.5 Electrocoagulation

Electrocoagulation is an important electrochemical method applied to remove pollutants from wastewater by electricity instead of chemicals that are very expensive and also toxic [3, 20]. Before now, electrocoagulation has been used efficiently to eliminate contaminants from different industrial effluents such as textile, dye, food, electroplating, tanneries and cosmetics that contain dyes, organic materials, suspended solids and other pollutants [84, 100, 105]. Through the enhanced influence of electric field that drives the movement of charged particles to achieve greater coagulation due to higher destabilization, electrocoagulation has shown greater efficiency to eliminate smaller colloidal particles than the conventional coagulation-flocculation method. An advantage of electrocoagulation includes the reduction in processing times and the generation of lesser secondary pollutants through wastewater treatment [34, 141].

In electrocoagulation, coagulants ions such as Fe^{3+} and Al^{3+} are produced from the sacrificial electrodes of required current density in the processing tank. The dissolved metal ions in the electrodes change to polymetallic hydroxides at higher pH according to electrolysis and initial coagulation. Thereafter, the polymetallic hydroxides obtained facilitate the elimination of organic substances by coagulation and conversion of the colloidal particles to bigger flocs through aggregation [61]. However, as the metal cations are produced at the anode, the released hydrogen gas at the cathode may aid the elimination of the flocs by electroflotation method [137].

4.2 *Biological Methods*

Biological technologies for wastewater is an environmentally friendly and cheap treatment process producing small amount of sludge due to less use of chemicals and complete mineralization of dyes [121, 173]. Over the years, researchers have explored the application of microorganisms such as some bacteria, fungi, algae, yeast and enzymes to achieve highly efficient dyes removal and degradation from textile wastewater [36, 53]. The advantages of biological elimination of dyes from textile wastewater include dyes conversion to harmless products such as carbon dioxide, biomass and water [129, 132]. Generally, two basic dye decolorization processes involved in biological method are adsorption onto microbial biomass and biodegradation of dyes through biocatalyst (microbial cells). Biological treatment method facilitates the reduction of BOD and COD through contact with the textile wastewater and the microbial cells [17]. Biological method is highly efficient with effective reusability characteristics [166]. The disadvantages of the biological technique include difficulty to control the microorganisms due to its high sensitivity, damage of the microbial cells by the intermediates, highly selective and toxic [36, 142, 166]. The different materials used in the biological treatment methods are described in the following sections.

4.2.1 Aerobic Treatment

Aerobic treatment occurs in the presence of oxygen or air using microorganisms called aerobes that utilize molecular-free oxygen for the conversion of organic materials to carbon dioxide, biomass and water [17]. In aerobic treatment, bacteria and fungi are the predominantly used microorganisms to degrade dyes from textile wastewater. These microorganisms can be used over a lengthened period for complete degradation of dyes and have been regarded as a cheap method that enhance complete mineralization of dyes without toxic by-products [83, 99]. On the other hand, aesthetic issues in receiving waters are attributed to the inefficiency of aerobic technologies in decolorization due to resistance of commercial dyes. Additionally, difficulty of the dye ions to acclimate onto the cells of the microorganisms pose another challenge to the aerobic biological process [61, 128].

4.2.2 Anaerobic Treatment

Anaerobic removal method of dyes from textile wastewater is a procedure whereby anaerobic bacteria is utilized to achieve the decomposition of organic substances in an oxygen-free environment to carbon dioxide and methane [61]. Also, ammonia and hydrogen sulfide are obtained as product of anaerobic treatment if nitrogen and sulfide are contained in the pollutants. In anaerobic method, significant quantity of heat generated and methane produced are commonly for different intensive energy applications [35, 64]. Transfer of electrons from the oxidized substrate by microorganism referred to as electron donor toward the dye effluents as the electron acceptor have been responsible for the decolorization of dyes through the reductive cleavage of the chromophore [36].

4.2.3 Algae Degradation

The degradation of dyes from textile wastewater by algae which requires no preservation due to the dependence of its growth on carbon dioxide and sunlight have been widely researched because it does not produce secondary pollutants [76, 158]. Algae degradation operates a simplified procedure in a photobioreactor at standardized atmospheric conditions, thereby making it an environmentally friendly and low-cost process than other conventional treatment techniques [65, 138]. Previously, algae biomass has shown effectiveness as an important biosorbent attributable to its binding sites and enhanced surface area that improved the removal capacity toward the dyes contaminants in the wastewater [61, 151]. Above all, three steps are involved in the degradation of dyes by algae. These steps include the use of chromophores for the algal biomass harvest by the algae, the formation of non-chromophore materials from the alteration of the chromophore materials and the absorption of the chromophore by the algal biomass [132].

4.2.4 Bacteria Degradation

The removal of dyes from textile wastewater has revealed the effectiveness of bacterial species than microbes and the efficiency of the bacteria largely depend on their ability to adapt and function actively under a given environmental condition [76]. Numerous bacteria, for instance *Escherichia, Paenibacillus, Rhodococcus, Sphingomonas, Nocardia, Klebsiella, Streptomyces, Clostridium, Eromonas, Shigella, Bacillus, Aeromonassp, Micrococcus, Pseudomonas* and *Citrobacter* have shown great capacity to efficiently degrade azo dyes [79, 132]. The cultivation ease, better rate of growth compared to other microorganisms and versatility as biocatalysts in the mineralization of azo dyes contained in the textile wastewater are the primary advantages of bacteria degradation. In microbial degradation of azo dyes, the mechanism of bacteria degradation involves the reductive cleavage of azo bonds (–N = N–) that occurs typically in the presence of azo reductase enzymes [61]. For the removal of dyes from textile wastewater using bacteria, parameters such as pH, dye concentration, mixing speed and temperature can be varied to make the process faster and more effective [36].

4.2.5 Fungi Degradation

Dyes can be removed from textile wastewater through fungal culture to biosorb, bioaccumulate, biodegrade and treat with enzymes using laccase, manganese peroxidase and lignin peroxidase to facilitate metabolic activities [121, 153]. In fungi treatment method, hazardous dye molecules in wastewater are degraded and detoxified using fungal strains [4, 162]. Fungi functions usually with intra- and extra-cellular enzymatic system to modify aromatic materials from textile, pesticides, paper and pulp industries [132]. At the moment, researchers have explored the use of fungal strains such as white-rot fungi (*Phenocyte, Polaria, Coriolis, Ketomium, Pyulo-spora* and *Hypoxylon*), string-shaped soil fungi (*Fusarium*) and *Phanerochaete chrysosporium* to eliminate color from wastewater [61]. The xenobiotic characteristics of azo dyes having a minimum of a chromophore chemical group facilitate its difficulty to decolorize in contaminants, except by azo bond cleavage using filamentous fungi and bacteria species [76]. An advantage of fungi degradation lies on the flexibility and ability to remove a number of dyes simultaneously, while the disadvantages include unstable system, large reactors requirements for dye removal, extended growth phase and need for a nitrogen restricted zone for growth [79].

4.2.6 Enzyme Degradation

Enzymes are expensive in pure form and are not usually the primary choice in wastewater treatment. On the other hand, it has been widely reported that industrial enzymes are cheap, effective, highly regenerative and exist in liquid form [17, 79]. In enzymatic degradation, complex contaminants are converted to products, while precipitation technique facilitates the removal from wastewater [76]. From the literature, azo reductases and lactases have indicated efficient capacity toward enzyme decolorization and azo dye degradation [132]. The advantages of enzyme degradation method include high efficiency, non-toxicity, highly reusable, cheap and capacity to use enzymes in the degradation of dyes, while the production of unreliable quantity of enzymes lies the disadvantage of this treatment process [79].

4.3 Physical Methods

Physical techniques of dye elimination from textile wastewater represents direct method that are governed by the mechanisms of mass transfer. A variety of physical technologies such as ion exchange, membrane filtration, reverse osmosis and adsorption have been applied extensively in the removal of dyes from textile wastewater because of low operation cost and enhanced dye uptake capacity [36]. The physical methods are shown in Fig. 1. Compared to the chemical and biological treatment methods, physical methods are predominantly used due to its simplicity, low chemical usage, non-utilization of living organisms and effectiveness in dye removal capacity [79]. In physical methods, adsorbents such as agricultural wastes, biochar, activated carbon, wood, coal, clay, silica, metallic materials and composites have been widely used to treat textile wastewater [2, 61, 125].

4.3.1 Ion Exchange

Ion exchange is highly recommended for dyes removal from textile wastewater, but it has not been widely utilized due to lack of accommodation of wide variety of dyes by the ion exchangers [61]. In ion exchange, a reversible interchange of ions takes place between the liquid and solid phase. Dyes from the solution are removed by the insoluble material (resin), while discharging other ions of similar charges without the structural internal resin modification. In the treatment of dyes contaminated wastewater, the wastewater is passed through the ion exchange resin pending the saturation of the existing exchange sites [22, 53]. Ion exchange technique is appropriate for the effective removal of both the anionic and cationic dyes. The advantages of ion exchange include removal of soluble dyes, selectivity, lack of adsorbent loss through regeneration and solvent recovery after use [79]. Additionally, anionic exchange resin, cationic exchange resin and chelating exchange resin are the classifications of ion exchange resin according to their functional groups.

4.3.2 Membrane Processes

Membrane process method occurs without chemical addition to the feed stream and phase change and can serve as an alternative to other processes such as adsorption, coagulation/flocculation, precipitation, ion exchange, distillation and biological treatment. Membrane process is highly attractive due to its improved product quality, higher separation effectiveness and low consumption of energy. In membrane process, dyes contaminated wastewater is passed through a membrane, thereby separating the dyes from the clean water [146]. The advantage of membrane process includes the effectiveness of water recovery, environmental friendliness, simplicity, low energy consumption and reusability, while the disadvantages include production of concentrated sludge, membrane fouling, high initial cost of investment and high cost for maintenance [129]. There are four basic membrane separation process: membrane filtration, microfiltration, nanofiltration and ultrafiltration.

Membrane filtration is utilized for the removal of dissolved materials and fine particles from solution and can be applied continuously for the concentration and separation of dyes from effluents [76]. The use of membrane filtration is highly attractive due to its resistance to increase in temperature, microbial attack and adverse chemical environment. Membrane filtration process is best suited for effluent treatment in which the concentration of dye is low and for the reusing of textile wastewater [61].

Microfiltration has a limited area of application and shows similarity to the conventional filtration process. The effectiveness of microfiltration is usually at low pressure within 2 bar differences, while the pore sizes is in the range 0.1–10 mm. Microfiltration process is predominantly used for elimination of suspended particles and colloid dyes emanating from dye bath and rinsing bath discharge [41, 89].

In nanofiltration, pressure is the principal driving force accountable for the separation process. Nanofiltration as a new technique in membrane technology is applied for different water treatment and purifications. Nanofiltration is a separation process that is driven by pressure in the range of 4–20 MPa and molar mass of particles from 350 to 1000 Da, whereby the retainment of dissolved molecules can be achieved [142].

Ultrafiltration is a membrane-based technique applied to separate colloids from solution and macromolecules which have limited use for treatment of textile wastewater. This may be attributed to the molar masses of the dyes in the highly colored textile wastewater and are lower when compared to molar mass cut-off from ultrafiltration. Generally, ultrafiltration is utilized as a pre-treatment for reverse osmosis or used alongside other biological methods and removal of metal hydroxides [17].

4.3.3 Reverse Osmosis

Reverse osmosis is a reverse technique of osmosis whereby solvent is passed through a membrane from a high solute concentration region to a low solute concentration region with the pressure in the range 7–100 bar [61]. Reverse osmosis effectively removes dyes, color, hardness, viruses and bacteria. Reverse osmosis shows high sensitivity to fouling and as such, the influent liquid must be pre-treated carefully. If the pore size of reverse osmosis is small, the cost of operation, capital investment and the pressure of operation will be higher [10, 89]. The disadvantages of reverse osmosis include production of acidic water, unhealthy demineralized water, slow flow rates and inefficiency in the removal of volatile organic compound, chemicals, chloramines and pharmaceuticals [142].

4.3.4 Adsorption Process

In the recent times, adsorption process have been ranked as an efficient and effective technique for the elimination of dissolved dyes from textile industrial effluents [79]. The preference for adsorption process over other treatment methods may be attributed to its effectiveness in the removal of dyes, environmentally friendly, insensitivity of toxic materials, cheap, ease of operation and highly reusable [36, 51, 52, 166]. Specifically in adsorption process, adsorbate refers to the adsorbing species, while adsorbent represents the adsorbing material or phase [61]. Adsorption method is defined as a surface phenomenon whereby molecules, ions and atoms from a dissolved solid, liquid and gas termed adsorbate adhere to the active/binding sites of a solid material known as the adsorbent [146]. Adsorption is also a physical technique in which soluble dyes contained in the wastewater are transferred to the surface of a highly porous solid material. The efficiency and type of interaction between the adsorbate and the adsorbent determines the nature of the adsorption process. As a result, adsorption process is of two types, physical (physisorption) and chemical (chemisorption). Physisorption is a reversible process that occurs when the adsorbate adheres on the adsorbent surface by weak intermolecular forces like Van der Waals and is depicted by a low adsorption heat in the range of −20 to 40 kJ/mol [61]. On the other hand, chemisorption is an irreversible process that occurs when adsorbate adheres to the adsorbent surface by chemical bonds and is depicted by a high heat of adsorption above 40 kJ/mol [79, 109]. Desorption is the term describing the removal of adsorbate from the adsorbent surface. The adsorption degree is dependent on the adsorbate nature-like solution concentration, polarity, molecular size, molecular structure and molecular weight. The surface properties of the adsorbent such as surface area, surface charge and particle size also plays an important role on the extent of adsorption.

In all, parameters effecting the extent of adsorption include pH of the adsorbate, contact time, adsorbent dosage, temperature, salts present in the adsorbate, adsorbent type, nature of the adsorbent, adsorbent surface properties and agitation speed [109]. The isotherm shape gives information on the adsorbate affinity and the possible adsorption mechanism. Generally, Langmuir and Freundlich are the commonly used adsorption isotherms to evaluate the adsorption data. Kinetic investigation also defines the effectiveness of the adsorption process and the possibility to scale-up the operation. In kinetic study, pseudo-first order and pseudo-second order are the most commonly used adsorption kinetics. For the adsorption of dyes from textile industry wastewater, different adsorbents such as corn stalk [145], waste tea residue [73], red dragon fruit peel [90], walnut shell [113], sugarcane bagasse [130], green coffee residue [97], cocoa pod husks [115], ackee apple [24], jujube seeds [147], garlic peel [16], quinoa husk [39], cucumber, banana and potato peel [149] and activated carbon [124] have been previously applied. Among these adsorbent materials, activated carbon is the most commonly used adsorbent in dyes adsorption from textile wastewater. Activated carbon is produced by physical technique though extended time calcination at a higher temperature and chemically through the treatment using chemical agents [18, 93]. Activated carbon consists of a highly porous structure and a surface area range of 500–2000 m^2/g.

4.4 Integrated Methods

The major methods of wastewater treatment such as chemical, biological and physical have been studied previously by different researchers and some of the methods do not establish effective dye removal. Also, recalcitrant compounds remaining in dye contaminated wastewater makes integrated processes important alternative to their treatment. In integrated methods approach, dye contaminated wastewater can be treated by combining two or more processes from the above methods. For instance, Fentons reagent had been used with ultraviolet light for the removal of dyes, but the produced sludge has a foul odor [79]. The improved influence of AOPs and adsorption processes catalyzed by $BiNiO_3$ catalyst have been reported [26]. The dye removal capacity was over 90%, while the turbidity and total solids removal were over 96% at the end of the integrated processes [26]. Although AOPs degrade textile dyes, their mineralization effectiveness is low. Therefore, an integrated system was developed using ozonation followed by biodegradation for the treatment of acid orange producing an overall 98.7% removal at optimum conditions [86]. In another study, UV/TiO_2/electroflotation was developed for the removal of dyes from textile wastewater and the result indicated 90% removal efficiency compared to UV/TiO_2 and UV [152]. The effectiveness of integrated process using UV/H_2O_2 revealed over 91% removal efficiency [29]. The application of ozonation coupled with $MnFe_2O_4$@CA catalyst in the degradation of textile wastewater enhanced the removal efficiency up to 99% [92].

5 Agricultural Waste Materials

Over the literature, several attempts have been made to find alternative adsorbents that are low-cost and readily available for the removal of dyes from textile wastewater. In the recent times, agricultural waste has been identified as cost-effective and always available materials for adsorbent purposes owing to the large quantities produced globally. Agricultural wastes which are commonly from the residues, fruits and vegetable peels are lignocellulosic materials consisting of the three structural components of lignin, cellulose and hemicellulose [142]. Agricultural wastes contain a number of functional groups such as carboxyl, carbonyl, ketone, phenol, aldehyde and alcohol [76]. In the past, a number of agricultural wastes such as pineapple peels [159], olive cake waste [157, 158], barley straw [71], wheat husk [22], oil palm waste [21], avocado seed powder [108], ash seed [60], switch grass [60], raw date pits [62], eucalyptus leaf [87], coconut fiber [126], litchi seed powder [48] and guava leaf [112] have been reported for anionic dyes removal from textile industry wastewater.

6 Removal of Anionic Dyes Using Agricultural Waste

6.1 Acid Dye

Acid dyes have been established to negatively affect the environment, the ecosystem and highly carcinogenic to human health [32]. Hence, there is a need for the treatment of dyes contaminated wastewater from textile effluent prior to reuse or discharge to the environment. In this study, agricultural waste was used for the treatment of the anionic dyes contaminated textile wastewater. Waste tea residue was modified into a good functional, surface area, morphological and particle size for the removal of Acid Blue 25 [91]. In the study, a maximum adsorption capacity of 127.14 mg/g was reported, while the kinetic and isotherm data fitted pseudo-second order and Redlich-Peterson model as shown in Table 2. Oil palm waste was also utilized for the treatment of Acid orange 10 contaminated textile wastewater and a maximum adsorption capacity of 18.76 mg/g obtained for the investigation [21]. In another study, avocado seed powder was used as an adsorbent for the removal of Acid yellow 17 from wastewater, whereby a maximum adsorption capacity of 42.70 was reported and additional kinetic and isotherm fitting to pseudo-second order and Langmuir [108].

Furthermore, Erichrome black T was adsorbed from wastewater using modified date palm ash [32]. The authors reported successful removal of the anionic acid dye with an adsorption capacity of 425.16 mg/g. The experimental data were also reported by the authors to be best suited to the pseudo-second-order kinetic and Langmuir isotherm model. According to Table 2, jojoba seeds residues were also applied to remove Erichrome black T and the adsorbent contained oxygen functional groups and good porous surface morphology [6]. A maximum adsorption capacity of 88.96 was reported by the authors, while Freundlich and pseudo-second-order model fitted the isotherm and kinetic data, respectively. Green coffee waste was also reported for the treatment of Congo red from wastewater with an adsorption capacity of 20.04 mg/g [97]. In their work, the experimental data were best fitted to the Langmuir isotherm and pseudo-second-order model. Additionally, banana peel was effectively used to adsorb Congo red and an adsorption capacity of 1.73 mg/g was reported [104]. The

Table 2 Studies of acid dyes removal using agricultural waste

Adsorbent	Dye	Isotherm	Kinetics	Adsorption capacity (mg/g)	References
Waste tea residue	Acid blue 25	Redlich Peterson	Pseudo-second order	127.14	[73]
Oil palm wastes	Acid orange 10	Langmuir	Pseudo-second order	18.76	[21]
Avocado seed powder	Acid yellow 17	Langmuir	Pseudo-second order	42.70	[108]
Date palm ash	Erichrome black-T	Langmuir	Pseudo-second order	425.16	[32]
Jojoba seed residue	Erichrome black-T	Freundlich	Pseudo-second order	88.96	[6]
Sugarcane bagasse	Congo red	Freundlich	Intra-particle diffusion	4.08	[130]
Haloxylon recurvum plant	Acid brown	Langmuir	Pseudo-second order	10.01	[66]
Pineapple peels	Eosin yellow	Langmuir		11.76	[159]
Corn stalks	Acid red	Langmuir	Pseudo-second order	31.06	[145]
Green coffee waste	Congo red	Langmuir	Pseudo-first order	20.04	[97]
Coconut residual fiber	Congo red	Freundlich	Pseudo-second order	128.94	[126]
Walnut shell	Congo red	Langmuir	Pseudo-second order	40.00	[113]
Banana peel	Congo red	Langmuir	Pseudo-second order	1.73	[104]
Almond leaves	Congo red	Freundlich	Pseudo-second order	7.77	[114]
Coffee waste	Congo red	Langmuir	Pseudo-second order	34.36	[165]
Cabbage waste	Congo red	Langmuir	Pseudo-second order	2.31	[164]
Guava leaf	Congo red	Freundlich	Pseudo-second order	47.62	[112]
Cocoa pod husk	Congo red	Langmuir	Pseudo-second order	43.67	[115]
Litchi seed powder	Congo red	Freundlich	Pseudo-second order	20.49	[48]
Orange peel powder	Congo red	Langmuir	Pseudo-second order	163.00	[107]

(continued)

Table 2 (continued)

Adsorbent	Dye	Isotherm	Kinetics	Adsorption capacity (mg/g)	References
Rice husk	Methyl orange	Langmuir	Pseudo-second order	254.00	[67]
Wheat straw	Methyl orange	Langmuir	Pseudo-second order	506.00	[91]
Quinoa husk	Methyl green	Freundlich	Pseudo-second order	1365.10	[39]

isotherm and kinetic model was better predicted using Langmuir and pseudo-second order model. Other studies from the literature on the removal of acid dyes using agricultural waste are presented in Table 2.

6.2 *Direct Dye*

Over the years, researchers have explored the use of agricultural wastes to remove direct dyes from textile wastewater using different experimentations as indicated in Table 3. Among them, Patra et al. [120] reported the use of *Pongamia pinnata* shells to remove Direct blue 6 from adsorptive systems. The authors reported a

Table 3 Studies of direct dyes removal using agricultural waste

Adsorbent	Dye	Isotherm	Kinetics	Adsorption capacity (mg/g)	References
Pongamia pinnata shells	Direct blue 6	Langmuir	Pseudo-second order	625.00	[120]
Waste peanut shell	Direct blue 86	Langmuir	Pseudo-second order	21.60	[58]
Thevetia neriifolia Juss wood	Direct orange 102	Langmuir	Pseudo-second order	33.00	[148]
Mushroom waste	Direct red 5B	Langmuir	Pseudo-second order	18.00	[7]
Mushroom waste	Direct black 22	Langmuir	Pseudo-second order	15.46	[7]
Mushroom waste	Direct black 71	Langmuir	Pseudo-second order	20.19	[7]
Garlic peel	Direct red	Langmuir	Pseudo-second order	37.96	[16]

high adsorption capacity of 625.00 mg/g, compared to other studies and the experimental data suited Langmuir and pseudo-second-order models effectively. In another study, waste peanut shell was used to remove Direct blue 86 from aqueous solution and a maximum adsorption capacity of 21.60 mg/g was achieved under Langmuir monolayer isotherm and pseudo-second-order kinetic model [58]. Similarly, Direct orange 102 was successfully removed using activated carbon produced from *Thevetia neriifolia Juss* wood with an adsorption capacity of 33.00 mg/g [148]. The authors also reported that the experimental data was appropriately fitted by Langmuir and pseudo-second-order kinetic model.

An eco-friendly mushroom waste was applied to treat anionic dyes such as Direct red 5B, Direct black 22 and Direct black 71 [7]. The authors reported maximum adsorption capacities of 18.00, 15.46 and 20.19 mg/g for the removal of Direct red 5B, Direct black 22 and Direct black 71 using the mushroom waste. Also, the authors reported Langmuir isotherm and pseudo-second-order kinetic model as the appropriate model. Another author, Asfaram et al. [16] studied the adsorption of Direct red from wastewater using garlic peel with a maximum adsorption capacity of 37.96 mg/g, while the equilibrium data fitted effectively to Langmuir and the kinetic data to pseudo-second-order model.

6.2.1 Reactive Dye

In general, researchers have made concerted efforts to study the removal of reactive dyes from textile industry wastewater using various kinds of agricultural waste as shown in Table 4. In all, the studies have shown promising capacities of agricultural waste in the effective and sustained adsorption of reactive dyes to ensure the treated water are safe for reuse or discharge into the environment. In view of that, date palm rachis prepared by basic activation at 800 °C in inert environment was found useful toward the adsorption of reactive red [43]. A high adsorption capacity of 128.21 mg/g was achieved, while the equilibrium and kinetic data fitted well to Langmuir and pseudo-second-order model. Hence, date palm rachis indicated efficiency in reactive red dyes elimination from real wastewater. A study on the removal of Reactive red 218 and Reactive blue 18 was conducted using corn silk and obtained maximum adsorption capacities of 63.30 and 71.60 mg/g, respectively [45].

The authors revealed that the experimental data for the removal of Reactive red 218 and Reactive blue 18 were best suited by Temkin and Freundlich isotherm, while the kinetic data was best represented by pseudo-second-order model. A cheap adsorbent, jute fiber was treated using NaOH prior to use in the adsorption of Reactive red 195 and the result revealed effective removal of the anionic dye with the adsorption capacity of 32.21 mg/g [46]. Another study was also carried out using Buckwheat husks to remove Reactive black 5 from wastewater and the adsorption capacity of 85.18 mg/g showed that the dye was removed appreciably [75]. In a different investigation on anionic dye adsorption using agricultural waste, bottle gourd peel was utilized to treat water contaminated with Reactive Red 195 producing an adsorption capacity of 245.50 mg/g [118]. Equally, the equilibrium and kinetic data were suitably

Table 4 Studies of reactive dyes removal using agricultural waste

Adsorbent	Dye	Isotherm	Kinetics	Adsorption capacity (mg/g)	References
Date palm rachis	Reactive red	Langmuir	Pseudo-second order	128.21	[43]
Corn silk	Reactive red 218	Temkin	Pseudo-second order	63.30	[45]
Corn silk	Reactive blue 18	Freundlich	Pseudo-second order	71.60	[45]
Jute fiber	Reactive red 195	Langmuir	Pseudo-second order	32.21	[46]
Buckwheat husks	Reactive black 5	Langmuir	–	85.18	[75]
Borassus Flabellifer shell	Reactive yellow 85	Langmuir	Pseudo-second order	40.00	[72]
Delonix regia seed	Reactive yellow	Freundlich	Pseudo-second order	24.15	[135]
Bottle gourd peel	Reactive red 195	Temkin	Pseudo-second order	245.50	[118]
Jujube stones	Reactive	Langmuir	Pseudo-second order	28.49	[43]
Eggshell membrane	Reactive red 120	Langmuir	Pseudo-second order	191.50	[134]
Coffee waste	Reactive black	Langmuir	Pseudo-second order	77.52	[165]

fitted by Temkin and pseudo-second-order kinetic model. Studies on the utilization of eggshell membrane for the removal of Reactive red 120 had been explored and a maximum adsorption capacity of 191.50 mg/g was obtained for the effectiveness of the anionic dye removal using the low-cost agricultural waste [134].

7 Conclusion

In this review, the classification of dyes was discussed extensively and their toxic effects described. The various treatment techniques for the removal of dyes contaminated wastewater were discussed and adsorption process identified to have gained more interests and widely applied due to its ease of operation, simplicity, low cost and effective removal of contaminants. The present review focused on the application of agricultural waste that is low-cost and readily available to sustain green chemistry removal of anionic dyes from textile effluents wastewater. The isotherm model was better fitted to Langmuir isotherm compared to others, while pseudo-second order was more appropriate for the kinetic data fitting. In all, agricultural waste showed convincing potentials for effective removal of anionic dyes from textile wastewater.

References

1. Aboelfetoh EF, Aboubaraka AE, Ebeid EZM (2021) Binary coagulation system (graphene oxide/chitosan) for polluted surface water treatment. J Environ Managem 288:112481. https://doi.org/10.1016/J.JENVMAN.2021.112481
2. Adegoke KA, Bello OS (2015) Dye sequestration using agricultural wastes as adsorbents. Water Resources and Industry 12:8–24. https://doi.org/10.1016/j.wri.2015.09.002
3. Adjeroud N, Elabbas S, Merzouk B, Hammoui Y, Felkai-Haddache L, Remini H, Leclerc JP, Madani K (2018) Effect of Opuntia ficus indica mucilage on copper removal from water by electrocoagulation-electroflotation technique. J Electroanal Chem 811:26–36. https://doi.org/10.1016/J.JELECHEM.2017.12.081
4. Ahmed SF, Mofijur M, Nuzhat S, Chowdhury AT, Rafa N, Uddin MA, Inayat A, Mahlia TMI, Ong HC, Chia WY, Show PL (2021) Recent developments in physical, biological, chemical, and hybrid treatment techniques for removing emerging contaminants from wastewater. J Hazardous Mater 416:125912. https://doi.org/10.1016/j.jhazmat.2021.125912
5. Ajiboye TO, Oyewo OA, Onwudiwe DC (2021) Simultaneous removal of organics and heavy metals from industrial wastewater: a review. Chemosphere 262:128379. https://doi.org/10.1016/j.chemosphere.2020.128379
6. Al-Zoubi H, Zubair M, Manzar MS, Manda AA, Blaisi NI, Qureshi A, Matani A (2020) Comparative adsorption of anionic dyes (Eriochrome Black T and Congo Red) onto Jojoba Residues: Isotherm, Kinetics and Thermodynamic Studies. Arab J Sci Eng 45(9):7275–7287. https://doi.org/10.1007/s13369-020-04418-5
7. Alhujaily A, Yu H, Zhang X, Ma F (2020) Adsorptive removal of anionic dyes from aqueous solutions using spent mushroom waste. Appl Water Sci 10(7):1–12. https://doi.org/10.1007/s13201-020-01268-2
8. Ali AY, Hadj E, N'diaye AD, Fahmi D, Kankou MS, Stitou M (2021) Adsorption of Congo Red from aqueous solution using Typha australis leaves as a low cost adsorbent. J Environ Treatment Techniques 9:534–539. https://doi.org/10.47277/JETT/9(2)539
9. Amin M, Chetpattananondh P, Khan MN (2020) Ultrasound assisted adsorption of reactive dye-145 by biochars from marine Chlorella sp. extracted solid waste pyrolyzed at various temperatures. J Environ Chem Eng 8(6):104403. https://doi.org/10.1016/j.jece.2020.104403
10. Amiri MJ, Faraji A, Azizi M, Nejad BG, Arshadi M (2021) Recycling bone waste and cobalt-wastewater into a highly stable and efficient activator of peroxymonosulfate for dye and HEPES degradation. Process Saf Environ Prot 147:626–641. https://doi.org/10.1016/J.PSEP.2020.12.039
11. Amirza MAR, Adib MMR, Hamdan R (2017) Application of agricultural wastes activated carbon for dye removal—an overview. MATEC Web of Conferences 103:32–34. https://doi.org/10.1051/matecconf/201710306013
12. Aravind M, Amalanathan M (2020) Structural, morphological, and optical properties of country egg shell derived activated carbon for dye removal. Mater Today: Proceed 43(xxxx):1491–1495. https://doi.org/10.1016/j.matpr.2020.09.311
13. Arkaan MF, Ekaputri RF, Fatimah I, Kamari A (2020) Physicochemical and photocatalytic activity of hematite/biochar nanocomposite prepared from Salacca skin waste. Sustain Chem Pharmacy 16:100261. https://doi.org/10.1016/j.scp.2020.100261
14. Aruna Bagotia N, Sharma AK, Kumar S (2021) A review on modified sugarcane bagasse biosorbent for removal of dyes. Chemosphere 268:129309. https://doi.org/10.1016/j.chemosphere.2020.129309

15. Aryee AA, Mpatani FM, Kani AN, Dovi E, Han R, Li Z, Qu L (2021) A review on functionalized adsorbents based on peanut husk for the sequestration of pollutants in wastewater: modification methods and adsorption study. J Cleaner Prod 310:127502. https://doi.org/10.1016/j.jclepro.2021.127502
16. Asfaram A, Fathi MR, Khodadoust S, Naraki M (2014) Removal of Direct Red 12B by garlic peel as a cheap adsorbent: kinetics, thermodynamic and equilibrium isotherms study of removal. Spectrochimica Acta—Part A: Molecular and Biomolecular Spectroscopy 127:415–421. https://doi.org/10.1016/j.saa.2014.02.092
17. Ayele A, Getachew D, Kamaraj M, Suresh A (2021) Phycoremediation of synthetic dyes: an effective and eco-friendly algal technology for the dye abatement. J Chem 2021. https://doi.org/10.1155/2021/9923643
18. Azzaz AA, Khiari B, Jellali S, Ghimbeu CM, Jeguirim M (2020) Hydrochars production, characterization and application for wastewater treatment: a review. Renew Sustain Energy Rev 127. https://doi.org/10.1016/j.rser.2020.109882
19. Baidya SK, Kumar U (2021) Adsorption of brilliant green dye from aqueous solution onto chemically modified areca nut husk. South African J Chem Eng 35:33–43. https://doi.org/10.1016/j.sajce.2020.11.001
20. Balla W, Essadki AH, Gourich B, Dassaa A, Chenik H, Azzi M (2010) Electrocoagulation/electroflotation of reactive, disperse and mixture dyes in an external-loop airlift reactor. J Hazard Mater 184(1–3):710–716. https://doi.org/10.1016/J.JHAZMAT.2010.08.097
21. Baloo L, Isa MH, Sapari NB, Jagaba AH, Wei LJ, Yavari S, Razali R, Vasu R (2021) Adsorptive removal of methylene blue and acid orange 10 dyes from aqueous solutions using oil palm wastes-derived activated carbons. Alex Eng J 60(6):5611–5629. https://doi.org/10.1016/j.aej.2021.04.044
22. Banerjee S, Gautam RK, Jaiswal A, Gautam PK, Chattopadhyaya MC (2016) Study on adsorption behavior of acid orange 10 onto modified wheat husk. Desalin Water Treat 57(26):12302–12315. https://doi.org/10.1080/19443994.2015.1046151
23. Behera M, Nayak J, Banerjee S, Chakrabortty S, Tripathy SK (2021) A review on the treatment of textile industry waste effluents towards the development of efficient mitigation strategy: an integrated system design approach. J Environ Chem Eng 9(4):105277. https://doi.org/10.1016/J.JECE.2021.105277
24. Bello OS, Auta M, Ayodele OB (2013) Ackee apple (Blighia sapida) seeds: a novel adsorbent for the removal of Congo Red dye from aqueous solutions. Chem Ecol 29(1):58–71. https://doi.org/10.1080/02757540.2012.686606
25. Beluci NdeCL, Mateus GAP, Miyashiro CS, Homem NC, Gomes RG, Fagundes-Klen MR, Bergamasco R, Vieira AMS (2019) Hybrid treatment of coagulation/flocculation process followed by ultrafiltration in TIO_2-modified membranes to improve the removal of reactive black 5 dye. Sci The Total Environ 664:222–229. https://doi.org/10.1016/J.SCITOTENV.2019.01.199
26. Bener S, Atalay S, Ersöz G (2020) The hybrid process with eco-friendly materials for the treatment of the real textile industry wastewater. Ecol Eng 148. https://doi.org/10.1016/j.ecoleng.2020.105789
27. Bensalah H, Younssi SA, Ouammou M, Gurlo A, Bekheet MF (2020) Azo dye adsorption on an industrial waste-transformed hydroxyapatite adsorbent: Kinetics, isotherms, mechanism and regeneration studies. J Environ Chem Eng 8(3):103807. https://doi.org/10.1016/j.jece.2020.103807
28. Berkane N, Meziane S, Aziri S (2020) Optimization of Congo red removal from aqueous solution using Taguchi experimental design. Separation Sci Technol (Philadelphia) 55(2):278–288. https://doi.org/10.1080/01496395.2019.1577442
29. Bezerra KCH, Fiaschitello TR, Labuto G, Freeman HS, Fragoso WD, Da Costa SM, Da Costa SA (2021) Reuse of water from real reactive monochromic and trichromic wastewater for new cotton dyes after efficient treatment using H_2O_2 catalyzed by UV light. J Environ Chem Eng 9(4):1–11. https://doi.org/10.1016/j.jece.2021.105731

30. Bhat AP, Gogate PR (2021) Degradation of nitrogen-containing hazardous compounds using advanced oxidation processes: a review on aliphatic and aromatic amines, dyes, and pesticides. J Hazardous Mater 403:123657. https://doi.org/10.1016/J.JHAZMAT.2020.123657
31. Bhomick PC, Supong A, Baruah M, Pongener C, Sinha D (2018) Pine Cone biomass as an efficient precursor for the synthesis of activated biocarbon for adsorption of anionic dye from aqueous solution: isotherm, kinetic, thermodynamic and regeneration studies. Sustain Chem Pharmacy 10:41–49. https://doi.org/10.1016/j.scp.2018.09.001
32. Blaisi NI, Zubair M, Ihsanullah Ali S, Kazeem TS, Manzar MS, Al-Kutti W, Al Harthi MA (2018) Date palm ash-MgAl-layered double hydroxide composite: sustainable adsorbent for effective removal of methyl orange and eriochrome black-T from aqueous phase. Environ Sci Pollution Res 25(34):34319–34331. https://doi.org/10.1007/s11356-018-3367-2
33. Bouchareb R, Derbal K, Özay Y, Bilici Z, Dizge N (2020) Combined natural/chemical coagulation and membrane filtration for wood processing wastewater treatment. J Water Process Eng 37:101521. https://doi.org/10.1016/J.JWPE.2020.101521
34. Brillas E, Martínez-Huitle CA (2015) Decontamination of wastewaters containing synthetic organic dyes by electrochemical methods. an updated review. Appl Catal B 166–167:603–643. https://doi.org/10.1016/j.apcatb.2014.11.016
35. Bulgariu L, Escudero LB, Bello OS, Iqbal M, Nisar J, Adegoke KA, Alakhras F, Kornaros M, Anastopoulos I (2019) The utilization of leaf-based adsorbents for dyes removal: a review. J Mol Liq 276:728–747. https://doi.org/10.1016/j.molliq.2018.12.001
36. Bushra R, Mohamad S, Alias Y, Jin Y, Ahmad M (2021) Current approaches and methodologies to explore the perceptive adsorption mechanism of dyes on low-cost agricultural waste: a review. Microporous and Mesoporous Mater 319. https://doi.org/10.1016/j.micromeso.2021.111040
37. Bustos-Terrones YA, Hermosillo-Nevárez JJ, Ramírez-Pereda B, Vaca M, Rangel-Peraza JG, Bustos-Terrones V, Rojas-Valencia MN (2021) Removal of BB9 textile dye by biological, physical, chemical, and electrochemical treatments. J Taiwan Inst Chem Eng 121:29–37
38. Byrne C, Subramanian G, Pillai SC (2018) Recent advances in photocatalysis for environmental applications. J Environ Chem Eng 6(3):3531–3555. https://doi.org/10.1016/j.jece.2017.07.080
39. Chen S, Tang S, Sun Y, Wang G, Chen H, Yu X, Su Y, Chen G (2018) Preparation of a highly porous carbon material based on quinoa husk and its application for removal of dyes by adsorption. Materials 11(8). https://doi.org/10.3390/ma11081407
40. Ćurić I, Dolar D, Bošnjak J (2021) Reuse of textile wastewater for dyeing cotton knitted fabric with hybrid treatment: coagulation/sand filtration/UF/NF-RO. J Environ Managem 295:113133. https://doi.org/10.1016/J.JENVMAN.2021.113133
41. da Rocha HD, Reis ES, Ratkovski GP, da Silva RJ, Gorza FDS, Pedro GC, de Melo CP (2020) Use of PMMA/(rice husk ash)/polypyrrole membranes for the removal of dyes and heavy metal ions. J Taiwan Inst Chem Eng 110:8–20. https://doi.org/10.1016/J.JTICE.2020.03.003
42. Dai Y, Sun Q, Wang W, Lu L, Liu M, Li J, Yang S, Sun Y, Zhang K, Xu J, Zheng W, Hu Z, Yang Y, Gao Y, Chen Y, Zhang X, Gao F, Zhang Y (2018) Utilizations of agricultural waste as adsorbent for the removal of contaminants: a review. Chemosphere 211:235–253. https://doi.org/10.1016/j.chemosphere.2018.06.179
43. Daoud M, Benturki O, Kecira Z, Girods P, Donnot A (2017) Removal of reactive dye (BEZAKTIV Red S-MAX) from aqueous solution by adsorption onto activated carbons prepared from date palm rachis and jujube stones. J Mol Liq 243:799–809. https://doi.org/10.1016/j.molliq.2017.08.093
44. Darwesh OM, Abd El-Latief AH, Abuarab ME, Kasem MA (2021) Enhancing the efficiency of some agricultural wastes as low-cost absorbents to remove textile dyes from their contaminated solutions. Biomass Conversion and Biorefinery.https://doi.org/10.1007/s13399-020-01142-w
45. Değermenci GD, Değermenci N, Ayvaoğlu V, Durmaz E, Çakır D, Akan E (2019) Adsorption of reactive dyes on lignocellulosic waste; characterization, equilibrium, kinetic and thermodynamic studies. J Clean Prod 225:1220–1229. https://doi.org/10.1016/j.jclepro.2019.03.260

46. Dey AK, Dey A (2021) Selection of optimal processing condition during removal of reactive red 195 by NaOH treated jute fibre using adsorption. Groundwater Sustain Developm 12:100522. https://doi.org/10.1016/j.gsd.2020.100522
47. Dotto J, Fagundes-Klen MR, Veit MT, Palácio SM, Bergamasco R (2019) Performance of different coagulants in the coagulation/flocculation process of textile wastewater. J Clean Prod 208:656–665. https://doi.org/10.1016/J.JCLEPRO.2018.10.112
48. Edokpayi JN, Makete E (2021) Removal of congo red dye from aqueous media using litchi seeds powder: equilibrium, kinetics and thermodynamics. Phys Chem Earth 123:103007. https://doi.org/10.1016/j.pce.2021.103007
49. Egbosiuba TC, Abdulkareem AS, Kovo AS, Afolabi EA, Tijani JO, Auta M, Roos WD (2020) Ultrasonic enhanced adsorption of methylene blue onto the optimized surface area of activated carbon: adsorption isotherm, kinetics and thermodynamics. Chem Eng Res Des 153:315–336. https://doi.org/10.1016/j.cherd.2019.10.016
50. Egbosiuba TC, Abdulkareem AS, Kovo AS, Afolabi EA, Tijani JO, Roos WD (2020) Enhanced adsorption of As(V) and Mn(VII) from industrial wastewater using multi-walled carbon nanotubes and carboxylated multi-walled carbon nanotubes. Chemosphere 254:126780. https://doi.org/10.1016/j.chemosphere.2020.126780
51. Egbosiuba TC, Abdulkareem AS, Kovo AS, Afolabi EA, Tijani JO, Bankole MT, Bo S, Roos WD (2021) Adsorption of Cr(VI), Ni(II), Fe(II) and Cd(II) ions by KIAgNPs decorated MWCNTs in a batch and fixed bed process. Sci Rep 11(1):1–20. https://doi.org/10.1038/s41598-020-79857-z
52. Egbosiuba TC, Abdulkareem AS, Tijani JO, Ani JI, Krikstolaityte V, Srinivasan M, Veksha A, Lisak G (2021) Taguchi optimization design of diameter-controlled synthesis of multi walled carbon nanotubes for the adsorption of Pb(II) and Ni(II) from chemical industry wastewater. Chemosphere 266(Ii):128937. https://doi.org/10.1016/j.chemosphere.2020.128937
53. Elgarahy AM, Elwakeel KZ, Mohammad SH, Elshoubaky GA (2021) A critical review of biosorption of dyes, heavy metals and metalloids from wastewater as an efficient and green process. Cleaner Eng Technol 100209. https://doi.org/10.1016/J.CLET.2021.100209
54. Emamjomeh MM, Sivakumar M (2009) Review of pollutants removed by electrocoagulation and electrocoagulation/flotation processes. J Environ Manage 90(5):1663–1679. https://doi.org/10.1016/j.jenvman.2008.12.011
55. Felista MM, Wanyonyi WC, Ongera G (2020) Adsorption of anionic dye (Reactive black 5) using macadamia seed husks: kinetics and equilibrium studies. Scientif African 7:e00283. https://doi.org/10.1016/j.sciaf.2020.e00283
56. Fortunato L, Elcik H, Blankert B, Ghaffour N, Vrouwenvelder J (2021) Textile dye wastewater treatment by direct contact membrane distillation: Membrane performance and detailed fouling analysis. J Membrane Sci 636:119552. https://doi.org/10.1016/J.MEMSCI.2021.119552
57. Furlan FR, de Melo da Silva LG, Morgado AF, de Souza AAU, Guelli Ulson de Souza SMA (2010) Removal of reactive dyes from aqueous solutions using combined coagulation/flocculation and adsorption on activated carbon. Resourc Conserv Recycling 54(5):283–290. https://doi.org/10.1016/J.RESCONREC.2009.09.001
58. Garg D, Majumder CB, Kumar S, Sarkar B (2019) Removal of direct blue-86 dye from aqueous solution using alginate encapsulated activated carbon (PnsAC-alginate) prepared from waste peanut shell. J Environ Chem Eng 7(5):103365. https://doi.org/10.1016/j.jece.2019.103365
59. Genázio Pereira PC, Reimão RV, Pavesi T, Saggioro EM, Moreira JC, Veríssimo Correia F (2017) Lethal and sub-lethal evaluation of Indigo Carmine dye and byproducts after TiO2 photocatalysis in the immune system of Eisenia andrei earthworms. Ecotoxicol Environ Saf 143(May):275–282. https://doi.org/10.1016/j.ecoenv.2017.05.043
60. Grabi H, Lemlikchi W, Derridj F, Lemlikchi S, Trari M (2021) Efficient native biosorbent derived from agricultural waste precursor for anionic dye adsorption in synthetic wastewater. Biomass Conversion and Biorefinery.https://doi.org/10.1007/s13399-021-01280-9
61. Gürses A, Güneş K, Şahin E (2021) Removal of dyes and pigments from industrial effluents. In: Green chemistry and water remediation: research and applications. https://doi.org/10.1016/b978-0-12-817742-6.00005-0

62. Hachani R, Sabir H, Sana N, Zohra KF, Nesrine NM (2017) Performance study of a low-cost adsorbent—raw date pits—for removal of azo dye in aqueous solution. Water Environ Res 89(9):827–839. https://doi.org/10.2175/106143017x14902968254863
63. Han G, Liang CZ, Chung TS, Weber M, Staudt C, Maletzko C (2016) Combination of forward osmosis (FO) process with coagulation/flocculation (CF) for potential treatment of textile wastewater. Water Res 91:361–370. https://doi.org/10.1016/J.WATRES.2016.01.031
64. Hansen É, Monteiro de Aquim P, Gutterres M (2021) Current technologies for post-tanning wastewater treatment: a review. J Environ Managem 294. https://doi.org/10.1016/j.jenvman.2021.113003
65. Hashem AH, Saied E, Hasanin MS (2020) Green and ecofriendly bio-removal of methylene blue dye from aqueous solution using biologically activated banana peel waste. Sustain Chem Pharmacy 18:00333. https://doi.org/10.1016/J.SCP.2020.100333
66. Hassan W, Noureen S, Mustaqeem M, Saleh TA, Zafar S (2020) Efficient adsorbent derived from Haloxylon recurvum plant for the adsorption of acid brown dye: kinetics, isotherm and thermodynamic optimization. Surf Interf 20. https://doi.org/10.1016/j.surfin.2020.100510
67. Hosseinzadeh H, Mohammadi S (2016) Biosorption of anionic dyes from aqueous solutions using a novel magnetic nanocomposite adsorbent based on rice husk ash. Separation Sci Technol (Philadelphia) 51(6):939–953. https://doi.org/10.1080/01496395.2016.1142564
68. Hu Z, Chen H, Ji F, Yuan S (2010) Removal of Congo Red from aqueous solution by cattail root. J Hazard Mater 173(1–3):292–297. https://doi.org/10.1016/j.jhazmat.2009.08.082
69. Hubadillah SK, Othman MHD, Jamalludin MR, Naim R, Sheikh Abdul Kadir SH, Puteh MH, Budiman Pauzan MA, Mohd Sobri FA (2021) Fabrication and characterisation of superhydrophobic bio-ceramic hollow fibre membranes prepared from cow bone waste. Ceramics Int 47(3):4178–4186. https://doi.org/10.1016/J.CERAMINT.2020.09.295
70. Hussein TK, Jasim NA (2021) A comparison study between chemical coagulation and electro-coagulation processes for the treatment of wastewater containing reactive blue dye. Mater Today: Proceed 42:1946–1950. https://doi.org/10.1016/J.MATPR.2020.12.240
71. Ibrahim S, Fatimah I, Ang HM, Wang S (2010) Adsorption of anionic dyes in aqueous solution using chemically modified Barley Straw. Water Sci Technol 62(5):1177–1182. https://doi.org/10.2166/wst.2010.388
72. Indhu S, Muthukumaran K (2018) Removal and recovery of reactive yellow 84 dye from wastewater and regeneration of functionalised Borassus flabellifer activated carbon. J Environ Chem Eng 6(2):3111–3121. https://doi.org/10.1016/j.jece.2018.04.027
73. Jain SN, Tamboli SR, Sutar DS, Jadhav SR, Marathe JV, Shaikh AA, Prajapati AA (2020) Batch and continuous studies for adsorption of anionic dye onto waste tea residue: kinetic, equilibrium, breakthrough and reusability studies. J Cleaner Prod 252:119778. https://doi.org/10.1016/j.jclepro.2019.119778
74. Jiang Z, Hu D (2019) Molecular mechanism of anionic dyes adsorption on cationized rice husk cellulose from agricultural wastes. J Mol Liq 276:105–114. https://doi.org/10.1016/j.molliq.2018.11.153
75. Józwiak T, Filipkowska U, Kowalkowska A, Struk-Sokoowska J, Werbowy D (2021) The influence of amination of sorbent based on buckwheat (Fagopyrum esculentum) husks on the sorption effectiveness of reactive black 5 dye. J Environ Chem Eng 9(2). https://doi.org/10.1016/j.jece.2021.105092
76. Kadhom M, Albayati N, Alalwan H, Al-Furaiji M (2020) Removal of dyes by agricultural waste. Sustain Chem Pharmacy 16:100259. https://doi.org/10.1016/j.scp.2020.100259
77. Kalra A, Gupta A (2019) Recent advances in decolourization of dyes using iron nanoparticles: a mini review. Mater Today: Proceed 36:689–696
78. Kanakaraju D, Bin Ya MH, Lim YC, Pace A (2020) Combined Adsorption/Photocatalytic dye removal by copper-titania-fly ash composite. Surf Interfaces 19:100534. https://doi.org/10.1016/J.SURFIN.2020.100534
79. Katheresan V, Kansedo J, Lau SY (2018) Efficiency of various recent wastewater dye removal methods: a review. J Environ Chem Eng 6(4):4676–4697. https://doi.org/10.1016/j.jece.2018.06.060

80. Katwal R, Kothari R, Pathania D (2021) An overview on degradation kinetics of organic dyes by photocatalysis using nanostructured electrocatalyst. Delivering Low-Carbon Biofuels with Bioproduct Recovery 195–213. https://doi.org/10.1016/B978-0-12-821841-9.00005-0
81. Kaur Y, Jasrotia T, Kumar R, Chaudhary GR, Chaudhary S (2021) Adsorptive removal of eriochrome black T (EBT) dye by using surface active low cost zinc oxide nanoparticles: a comparative overview. Chemosphere 278
82. Khan FSA, Mubarak NM, Tan YH, Khalid M, Karri RR, Walvekar R, Abdullah EC, Nizamuddin S, Mazari SA (2021) A comprehensive review on magnetic carbon nanotubes and carbon nanotube-based buckypaper for removal of heavy metals and dyes. J Hazardous Mater 413:125375. https://doi.org/10.1016/j.jhazmat.2021.125375
83. Khan S, Khan A, Ali N, Ahmad S, Ahmad W, Malik S, Ali N, Khan H, Shah S, Bilal M (2021) Degradation of Congo red dye using ternary metal selenide-chitosan microspheres as robust and reusable catalysts. Environ Technol Innov 22:101402. https://doi.org/10.1016/J.ETI.2021.101402
84. Khandegar V, Saroha AK (2013) Electrocoagulation for the treatment of textile industry effluent—a review. J Environ Manage 128:949–963. https://doi.org/10.1016/j.jenvman.2013.06.043
85. Khare P, Patel RK, Sharan S, Shankar R (2021) Recent trends in advanced oxidation process for treatment of recalcitrant industrial effluents. In: Advanced oxidation processes for effluent treatment plants. Elsevier Inc. https://doi.org/10.1016/b978-0-12-821011-6.00008-6
86. Kumar Sonwani R, Pandey S, Kumar Yadav S, Shekhar Giri B, Katiyar V, Sharan Singh R, Nath Rai B (2021) Construction of integrated system for the treatment of acid orange 7 dye from wastewater: Optimization and growth kinetic study. Bioresource Technol 337:125478. https://doi.org/10.1016/j.biortech.2021.125478
87. Kumari R, Mohanta J, Dey B, Dey S (2020) Eucalyptus leaf powder as an efficient scavenger for Congo red from water: comprehensive batch and column investigation. Separation Sci Technol (Philadelphia) 55(17):3047–3059. https://doi.org/10.1080/01496395.2019.1670208
88. Leichtweis J, Silvestri S, Carissimi E (2020) New composite of pecan nutshells biochar-ZnO for sequential removal of acid red 97 by adsorption and photocatalysis. Biomass and Bioenergy 140:105648. https://doi.org/10.1016/J.BIOMBIOE.2020.105648
89. Li L, Wu M, Song C, Liu L, Gong W, Ding Y, Yao J (2021) Efficient removal of cationic dyes via activated carbon with ultrahigh specific surface derived from vinasse wastes. Bioresource Technol 322:124540. https://doi.org/10.1016/J.BIORTECH.2020.124540
90. Lim LBL, Priyantha N, Latip SAA, Lu Y (2021) Sequestration of toxic congo red dye through the utilization of red dragon fruit peel: linear versus nonlinear regression analyses of isotherm and kinetics. Desalin Water Treat 218:409–422. https://doi.org/10.5004/dwt.2021.26977
91. Lin Q, Wang K, Gao M, Bai Y, Chen L, Ma H (2017) Effectively removal of cationic and anionic dyes by pH-sensitive amphoteric adsorbent derived from agricultural waste-wheat straw. J Taiwan Inst Chem Eng 76:65–72. https://doi.org/10.1016/j.jtice.2017.04.010
92. Liu L, Chen Z, Zhang J, Shan D, Wu Y, Bai L, Wang B (2021) Treatment of industrial dye wastewater and pharmaceutical residue wastewater by advanced oxidation processes and its combination with nanocatalysts: a review. J Water Process Eng 42(22):102122. https://doi.org/10.1016/j.jwpe.2021.102122
93. Lobato-Peralta DR, Duque-Brito E, Ayala-Cortés A, Arias DM, Longoria A, Cuentas-Gallegos AK, Sebastian PJ, Okoye PU (2021) Advances in activated carbon modification, surface heteroatom configuration, reactor strategies, and regeneration methods for enhanced wastewater treatment. J Environ Chem Eng 9(4). https://doi.org/10.1016/j.jece.2021.105626
94. Long Y, You X, Chen Y, Hong H, Liao BQ, Lin H (2020) Filtration behaviors and fouling mechanisms of ultrafiltration process with polyacrylamide flocculation for water treatment. Sci Total Environ 703
95. Lum PT, Foo KY, Zakaria NA, Palaniandy P (2020) Ash based nanocomposites for photocatalytic degradation of textile dye pollutants: a review. Mater Chem Phys 241:122405. https://doi.org/10.1016/J.MATCHEMPHYS.2019.122405

96. Mansor ES, Ali H, Abdel-Karim A (2020) Efficient and reusable polyethylene oxide/polyaniline composite membrane for dye adsorption and filtration. Colloid Interface Sci Commun 39:100314. https://doi.org/10.1016/J.COLCOM.2020.100314
97. Manzar MS, Zubair M, Khan NA, Husain Khan A, Baig U, Aziz MA, Blaisi NI, Abdel-Magid HIM (2020) Adsorption behaviour of green coffee residues for decolourization of hazardous congo red and eriochrome black T dyes from aqueous solutions. Int J Environ Anal Chem 00(00):1–17. https://doi.org/10.1080/03067319.2020.1811260
98. Marin NM, Pascu LF, Demba A, Nita-Lazar M, Badea IA, Aboul-Enein HY (2019) Removal of the acid orange 10 by ion exchange and microbiological methods. Int J Environ Sci Technol 16(10):6357–6366. https://doi.org/10.1007/S13762-018-2164-2
99. Mashkoor F, Nasar A (2020) Magsorbents: potential candidates in wastewater treatment technology—a review on the removal of methylene blue dye. J Magnet Magnet Mater 500:166408. https://doi.org/10.1016/j.jmmm.2020.166408
100. Mbacké MK, Kane C, Diallo NO, Diop CM, Chauvet F, Comtat M, Tzedakis T (2016) Electrocoagulation process applied on pollutants treatment- experimental optimization and fundamental investigation of the crystal violet dye removal. J Environ Chem Eng 4(4):4001–4011. https://doi.org/10.1016/j.jece.2016.09.002
101. Mcyotto F, Wei Q, Macharia DK, Huang M, Shen C, Chow CWK (2021) Effect of dye structure on color removal efficiency by coagulation. Chem Eng J 405:126674. https://doi.org/10.1016/J.CEJ.2020.126674
102. Mishra S, Cheng L, Maiti A (2021) J Environ Chem Eng The utilization of agro-biomass / byproducts for effective bio-removal of dyes from dyeing wastewater : a comprehensive review. J Environ Chem Eng 9(1):104901. https://doi.org/10.1016/j.jece.2020.104901
103. Miyashiro CS, Mateus GAP, dos Santos TRT, Paludo MP, Bergamasco R, Fagundes-Klen MR (2021) Synthesis and performance evaluation of a magnetic biocoagulant in the removal of reactive black 5 dye in aqueous medium. Mater Sci Eng: C 119:111523. https://doi.org/10.1016/J.MSEC.2020.111523
104. Mondal NK, Kar S (2018) Potentiality of banana peel for removal of Congo red dye from aqueous solution: isotherm, kinetics and thermodynamics studies. Appl Water Sci 8(6). https://doi.org/10.1007/s13201-018-0811-x
105. Mook WT, Aroua MK, Issabayeva G (2014) Prospective applications of renewable energy based electrochemical systems in wastewater treatment: a review. Renew Sustain Energy Rev 38:36–46. https://doi.org/10.1016/J.RSER.2014.05.042
106. Mpatani FM, Han R, Aryee AA, Kani AN, Li Z, Qu L (2021) Adsorption performance of modified agricultural waste materials for removal of emerging micro-contaminant bisphenol A: a comprehensive review. Sci Total Environ 780:146629. https://doi.org/10.1016/j.scitotenv.2021.146629
107. Munagapati VS, Kim DS (2016) Adsorption of anionic azo dye Congo Red from aqueous solution by Cationic modified orange peel powder. J Mol Liq 220:540–548. https://doi.org/10.1016/j.molliq.2016.04.119
108. Munagapati VS, Wen HY, Vijaya Y, Wen JC, Wen JH, Tian Z, Reddy GM, Raul Garcia J (2021) Removal of anionic (Acid Yellow 17 and Amaranth) dyes using aminated avocado (Persea americana) seed powder: adsorption/desorption, kinetics, isotherms, thermodynamics, and recycling studies. Int J Phytoremediation 0(0):1–13. https://doi.org/10.1080/15226514.2020.1866491
109. Natarajan S, Bajaj HC, Tayade RJ (2018) Recent advances based on the synergetic effect of adsorption for removal of dyes from waste water using photocatalytic process. J Environ Sci (China) 65:201–222. https://doi.org/10.1016/j.jes.2017.03.011
110. Nidheesh PV, Gopinath A, Ranjith N, Praveen Akre A, Sreedharan V, Suresh Kumar M (2021) Potential role of biochar in advanced oxidation processes: a sustainable approach. Chem Eng J 405:126582. https://doi.org/10.1016/J.CEJ.2020.126582
111. Nidheesh PV, Gopinath A, Ranjith N, Praveen A, Sreedharan V, Kumar MS (2021) Potential role of biochar in advanced oxidation processes: a sustainable approach. Chem Eng J 405:126582. https://doi.org/10.1016/j.cej.2020.126582

112. Ojedokun AT, Bello OS (2017) Kinetic modeling of liquid-phase adsorption of Congo red dye using guava leaf-based activated carbon. Appl Water Sci 7(4):1965–1977. https://doi.org/10.1007/s13201-015-0375-y
113. Ojo TA, Ojedokun AT, Bello OS (2019) Functionalization of powdered walnut shell with orthophosphoric acid for Congo red dye removal. Part Sci Technol 37(1):74–85. https://doi.org/10.1080/02726351.2017.1340914
114. Okorocha NJ, Okoji JJ, Osuji C (2016) Adsorption of basic and acidic dyes onto agricultural wastes. Int Letters of Chem Phys Astronomy 70:12–26. https://doi.org/10.18052/www.scipress.com/ilcpa.70.12
115. Olakunle MO, Inyinbor AA, Dada AO, Bello OS (2018) Combating dye pollution using cocoa pod husks: a sustainable approach. Int J Sustain Eng 11(1):4–15. https://doi.org/10.1080/19397038.2017.1393023
116. Oloo CM, Onyari JM, Wanyonyi WC, Wabomba JN, Muinde VM (2020) Adsorptive removal of hazardous crystal violet dye form aqueous solution using Rhizophora mucronata stem-barks: Equilibrium and kinetics studies. Environ Chem Ecotoxicol 2:64–72. https://doi.org/10.1016/j.enceco.2020.05.001
117. Onukwuli O, Nnaji P, Menkiti M, Anadebe VC, Oke E, Ude CN, Ude CJ, Okafor NA (2021) Dual-purpose optimization of dye-polluted wastewater decontamination using bio-coagulants from multiple processing techniques via neural intelligence algorithm and response surface methodology. J Taiwan Institute of Chem Eng.https://doi.org/10.1016/J.JTICE.2021.06.030
118. Palamthodi S, Lele SS (2016) Optimization and evaluation of reactive dye adsorption on bottle gourd peel. J Environ Chem Eng 4(4):4299–4309. https://doi.org/10.1016/j.jece.2016.09.032
119. Park JH, Wang JJ, Meng Y, Wei Z, DeLaune RD, Seo DC (2019) Adsorption/desorption behavior of cationic and anionic dyes by biochars prepared at normal and high pyrolysis temperatures. Colloids Surf, A 572:274–282. https://doi.org/10.1016/j.colsurfa.2019.04.029
120. Patra C, Gupta R, Bedadeep D, Narayanasamy S (2020) Surface treated acid-activated carbon for adsorption of anionic azo dyes from single and binary adsorptive systems: a detail insight. Environ Pollution 266:115102. https://doi.org/10.1016/j.envpol.2020.115102
121. Pavithra KG, Senthil Kumar P, Jaikumar V, Sundar Rajan P (2019) Removal of colorants from wastewater: a review on sources and treatment strategies. J Indus Eng Chem 75:1–19. https://doi.org/10.1016/j.jiec.2019.02.011
122. Quansah JO, Hlaing T, Lyonga FN, Kyi PP, Hong SH, Lee CG, Park SJ (2020) Nascent rice husk as an adsorbent for removing cationic dyes from textile wastewater. Appl Sci (Switzerland) 10(10). https://doi.org/10.3390/app10103437
123. Ramadhani P, Chaidir Z, Zilfa Tomi ZB, Rahmiarti D, Zein R (2020) Shrimp shell (Metapenaeus monoceros) waste as a low-cost adsorbent for metanil yellow dye removal in aqueous solution. Desalination and Water Treatment 197:413–423. https://doi.org/10.5004/dwt.2020.25963
124. Ramathilagam G, Ananthakumar K (2018) Removal of orange G from aqueous solutions by activated carbon obtained from agricultural wastes: isotherm and kinetic studies. Asian J Chem 30(8):1857–1862. https://doi.org/10.14233/ajchem.2018.21350
125. Rangabhashiyam S, Anu N, Selvaraju N (2013) Sequestration of dye from textile industry wastewater using agricultural waste products as adsorbents. J Environ Chem Eng 1(4):629–641. https://doi.org/10.1016/j.jece.2013.07.014
126. Rani KC, Naik A, Chaurasiya RS, Raghavarao KSMS (2017) Removal of toxic Congo red dye from water employing low-cost coconut residual fiber. Water Sci Technol 75(9):2225–2236. https://doi.org/10.2166/wst.2017.109
127. Rostam AB, Taghizadeh M (2020) Advanced oxidation processes integrated by membrane reactors and bioreactors for various wastewater treatments: a critical review. J Environ Chem Eng 8(6):104566. https://doi.org/10.1016/j.jece.2020.104566
128. Roy M, Saha R (2021) Dyes and their removal technologies from wastewater: a critical review. In: Intelligent environmental data monitoring for pollution management. Elsevier Inc. https://doi.org/10.1016/b978-0-12-819671-7.00006-3

129. Saad MS, Wirzal MDH, Putra ZA (2021) Review on current approach for treatment of palm oil mill effluent: integrated system. J Environ Managem 286:112209. https://doi.org/10.1016/j.jenvman.2021.112209
130. Said AEAA, Aly AAM, Goda MN, Abd El-Aal M, Abdelazim M (2020) Adsorptive remediation of congo red dye in aqueous solutions using acid pretreated sugarcane bagasse. J Polym Environ 28(4):1129–1137. https://doi.org/10.1007/s10924-020-01665-3
131. Salgado BCB, Cardeal RA, Valentini A (2019) Photocatalysis and photodegradation of pollutants. In: Nanomaterials applications for environmental matrices: water, soil and air. Elsevier Inc. https://doi.org/10.1016/B978-0-12-814829-7.00015-X
132. Samsami S, Mohamadi M, Sarrafzadeh MH, Rene ER, Firoozbahr M (2020) Recent advances in the treatment of dye-containing wastewater from textile industries: overview and perspectives. Process Saf Environ Prot 143:138–163. https://doi.org/10.1016/j.psep.2020.05.034
133. Santos DHS, Duarte JLS, Tavares MGR, Tavares MG, Friedrich LC, Meili L, Pimentel WRO, Tonholo J, Zanta CLPS (2020) Electrochemical degradation and toxicity evaluation of reactive dyes mixture and real textile effluent over DSA® electrodes. Chem Eng Process—Process Intensification 153
134. Saratale RG, Sun Q, Munagapati VS, Saratale GD, Park J, Kim DS (2021) The use of eggshell membrane for the treatment of dye-containing wastewater: batch, kinetics and reusability studies. Chemosphere 281:130777. https://doi.org/10.1016/j.chemosphere.2021.130777
135. Saravanan A, Karishma S, Jeevanantham S, Jeyasri S, Kiruthika AR, Kumar PS, Yaashikaa PR (2020) Optimization and modeling of reactive yellow adsorption by surface modified Delonix regia seed: Study of nonlinear isotherm and kinetic parameters. Surfaces and Interfaces 20:100520. https://doi.org/10.1016/j.surfin.2020.100520
136. Shadi AMH, Kamaruddin MA, Niza NM, Emmanuel MI, Ismail N, Hossain S (2021) Effective removal of organic and inorganic pollutants from stabilized sanitary landfill leachate using a combined Fe_2O_3 nanoparticles/electroflotation process. J Water Process Eng 40:101988. https://doi.org/10.1016/J.JWPE.2021.101988
137. Shahadat M, Azha SF, Ismail S, Shaikh ZA, Wazed SA (2018) Treatment of industrial dyes using chitosan-supported nanocomposite adsorbents. In: The impact and prospects of green chemistry for textile technology (Issue 2009). Elsevier Ltd. https://doi.org/10.1016/B978-0-08-102491-1.00016-2
138. Sharma B, Tiwari S, Bisht N, Tewari L (2021) Eco-friendly bioprocess using agar plug immobilized Penicillium crustosum PWWS-6 biomass for treatment of wastewater contaminated with toxic Congo red dye for use in agriculture. Indus Crops Products 170:113755. https://doi.org/10.1016/J.INDCROP.2021.113755
139. Shin J, Bae S, Chon K (2021) Fenton oxidation of synthetic food dyes by Fe-embedded coffee biochar catalysts prepared at different pyrolysis temperatures: a mechanism study. Chem Eng J 421:129943. https://doi.org/10.1016/J.CEJ.2021.129943
140. Shrestha R, Ban S, Devkota S, Sharma S, Joshi R, Tiwari AP, Kim HY, Joshi MK (2021) Technological trends in heavy metals removal from industrial wastewater: a review. J Environ Chem Eng 9(4):105688. https://doi.org/10.1016/J.JECE.2021.105688
141. Sillanpää M, Shestakova M (2017) Electrochemical water treatment methods. In: Electrochemical water treatment methods: fundamentals, methods and full scale applications. https://doi.org/10.1016/B978-0-12-811462-9.00002-5
142. Singh NB, Nagpal G, Agrawal S, Rachna (2018) Water purification by using adsorbents: a review. Environ Technol Innov 11:187–240. https://doi.org/10.1016/j.eti.2018.05.006
143. Singh R (2020) Recycling of agricultural waste for wastewater treatment. In: Encyclopedia of renewable and sustainable materials. Elsevier Ltd. https://doi.org/10.1016/b978-0-12-803581-8.11444-4
144. Skaf DW, Punzi VL, Rolle JT, Cullen E (2021) Impact of Moringa Oleifera extraction conditions on zeta potential and coagulation effectiveness. J Environ Chem Eng 9(1):104687. https://doi.org/10.1016/J.JECE.2020.104687

145. Soldatkina L, Zavrichko M (2019) Equilibrium, kinetic, and thermodynamic studies of anionic dyes adsorption on corn stalks modified by cetylpyridinium bromide. Colloids and Interfaces 3(1). https://doi.org/10.3390/colloids3010004
146. Soliman NK, Moustafa AF (2020) Industrial solid waste for heavy metals adsorption features and challenges; a review. J Market Res 9(5):10235–10253. https://doi.org/10.1016/j.jmrt.2020.07.045
147. Somasekhara Reddy MC, Sivaramakrishna L, Varada Reddy A (2012) The use of an agricultural waste material, Jujuba seeds for the removal of anionic dye (Congo red) from aqueous medium. J Hazard Mater 203–204:118–127. https://doi.org/10.1016/j.jhazmat.2011.11.083
148. Srinivasan P, John Bosco A, Kalaivizhi R, Arockia Selvi J, Sivakumar P (2018) Adsorption isotherm and kinetic study of direct orange 102 dyes on TNJ activated carbon. Mater Today: Proceed 34:389–394. https://doi.org/10.1016/j.matpr.2020.02.198
149. Stavrinou A, Aggelopoulos CA, Tsakiroglou CD (2018) Exploring the adsorption mechanisms of cationic and anionic dyes onto agricultural waste peels of banana, cucumber and potato: adsorption kinetics and equilibrium isotherms as a tool. J Environ Chem Eng 6(6):6958–6970. https://doi.org/10.1016/j.jece.2018.10.063
150. Sulyman M, Kucinska-Lipka J, Sienkiewicz M, Gierak A (2021) Development, characterization and evaluation of composite adsorbent for the adsorption of crystal violet from aqueous solution: isotherm, kinetics, and thermodynamic studies. Arabian J Chem 14(5). https://doi.org/10.1016/j.arabjc.2021.103115
151. Svetozarević M, Šekuljica N, Knežević-Jugović Z, Mijin D (2021) Agricultural waste as a source of peroxidase for wastewater treatment: Insight in kinetics and process parameters optimization for anthraquinone dye removal. Environ Technol Innov 21:101289. https://doi.org/10.1016/J.ETI.2020.101289
152. Talaiekhozani A, Reza Mosayebi M, Fulazzaky MA, Eskandari Z, Sanayee R (2020) Combination of TiO_2 microreactor and electroflotation for organic pollutant removal from textile dyeing industry wastewater. Alex Eng J 59(2):549–563. https://doi.org/10.1016/j.aej.2020.01.052
153. Taoufik N, Boumya W, Achak M, Sillanpää M, Barka N (2021) Comparative overview of advanced oxidation processes and biological approaches for the removal pharmaceuticals. J Environ Managem 288. https://doi.org/10.1016/j.jenvman.2021.112404
154. Thulasi A, Kannaiyan S, Kumar K (2021) Diverse processes for the treatment of textile wastewater. The Future of Effluent Treatment Plants 661–682. https://doi.org/10.1016/B978-0-12-822956-9.00032-5
155. Tianzhi W, Weijie W, Hongying H, Khu ST (2021) Effect of coagulation on bio-treatment of textile wastewater: quantitative evaluation and application. J Cleaner Prod 312:127798. https://doi.org/10.1016/J.JCLEPRO.2021.127798
156. Torres NH, Souza BS, Ferreira LFR, Lima ÁS, dos Santos GN, Cavalcanti EB (2019) Real textile effluents treatment using coagulation/flocculation followed by electrochemical oxidation process and ecotoxicological assessment. Chemosphere 236:124309. https://doi.org/10.1016/J.CHEMOSPHERE.2019.07.040
157. Toumi K-H, Bergaoui M, Khalfaoui M, Benguerba Y, Erto A, Dotto GL, Amrane A, Nacef S, Ernst B (2018) Computational study of acid blue 80 dye adsorption on low cost agricultural Algerian olive cake waste: statistical mechanics and molecular dynamic simulations. J Mol Liq 271:40–50. https://doi.org/10.1016/j.molliq.2018.08.115
158. Toumi KH, Benguerba Y, Erto A, Dotto GL, Khalfaoui M, Tiar C, Nacef S, Amrane A (2018) Molecular modeling of cationic dyes adsorption on agricultural Algerian olive cake waste. J Mol Liq 264:127–133. https://doi.org/10.1016/J.MOLLIQ.2018.05.045
159. Ugbe FA, Anebi PO, Ikudayisi VA (2018) Biosorption of an anionic dye, eosin yellow onto pineapple peels: isotherm and thermodynamic study. Int Annals Sci 4(1):14–19. https://doi.org/10.21467/ias.4.1.14-19
160. Varjani S, Rakholiya P, Shindhal T, Shah AV, Ngo HH (2021) Trends in dye industry effluent treatment and recovery of value added products. J Water Process Eng 39

161. Vieira WT, De Farias MB, Spaolonzi MP, Da Silva MGC, Vieira MGA (2021) Latest advanced oxidative processes applied for the removal of endocrine disruptors from aqueous media—a critical report. J Environ Chem Eng 9(4):105748. https://doi.org/10.1016/J.JECE.2021.105748
162. Vigneshwaran S, Sirajudheen P, Karthikeyan P, Meenakshi S (2021) Fabrication of sulfur-doped biochar derived from tapioca peel waste with superior adsorption performance for the removal of Malachite green and Rhodamine B dyes. Surfaces and Interfaces 23:100920. https://doi.org/10.1016/j.surfin.2020.100920
163. Wang W, Huang G, An C, Xin X, Zhang Y, Liu X (2017) Transport behaviors of anionic azo dyes at interface between surfactant-modified flax shives and aqueous solution: Synchrotron infrared and adsorption studies. Appl Surf Sci 405:119–128. https://doi.org/10.1016/j.apsusc.2017.01.311
164. Wekoye JN, Wanyonyi WC, Wangila PT, Tonui MK (2020) Kinetic and equilibrium studies of Congo red dye adsorption on cabbage waste powder. Environ Chem Ecotoxicol 2:24–31. https://doi.org/10.1016/j.enceco.2020.01.004
165. Wong S, Ghafar NA, Ngadi N, Razmi FA, Inuwa IM, Mat R, Amin NAS (2020) Effective removal of anionic textile dyes using adsorbent synthesized from coffee waste. Sci Rep 10(1):1–13. https://doi.org/10.1038/s41598-020-60021-6
166. Yadav A, Bagotia N, Sharma AK, Kumar S (2021) Advances in decontamination of wastewater using biomass-basedcomposites: a critical review. Sci Total Environ 784. https://doi.org/10.1016/j.scitotenv.2021.147108
167. Yadav S, Yadav A, Bagotia N, Sharma AK, Kumar S (2021) Adsorptive potential of modified plant-based adsorbents for sequestration of dyes and heavy metals from wastewater—a review. J Water Process Eng 42. https://doi.org/10.1016/j.jwpe.2021.102148
168. Yaseen DA, Scholz M (2019) Textile dye wastewater characteristics and constituents of synthetic effluents: a critical review. Int J Environ Sci Technol 16(2). Springer, Berlin, Heidelberg. https://doi.org/10.1007/s13762-018-2130-z
169. Yu H, Wang T, Dai W, Li X, Hu X, Ma N (2015) Single and bicomponent anionic dyes adsorption equilibrium studies on magnolia-leaf-based porous carbons. RSC Adv 5(79):63970–63977. https://doi.org/10.1039/c5ra11568j
170. Zainith S, Ferreira LFR, Saratale GD, Mulla SI, Bharagava RN (2021) Membrane-based hybrid processes in industrial waste effluent treatment. Membrane-Based Hybrid Processes for Wastewater Treatment 205–226.https://doi.org/10.1016/B978-0-12-823804-2.00008-2
171. Zhang J, Lu W, Li H, Zhan S, Wang X, Ma C, Qiu Z (2021) Polyethyleneimine-impregnated alkali treated waste bamboo powder for effective dye removal. Water Sci Technol 83(5):1183–1197. https://doi.org/10.2166/wst.2021.041
172. Zhao C, Zhou J, Yan Y, Yang L, Xing G, Li H, Wu P, Wang M, Zheng H (2021) Application of coagulation/flocculation in oily wastewater treatment: a review. Sci Total Environ 765:142795. https://doi.org/10.1016/J.SCITOTENV.2020.142795
173. Zhou Y, Lu J, Zhou Y, Liu Y (2019) Recent advances for dyes removal using novel adsorbents: a review. Environ Pollut 252:352–365. https://doi.org/10.1016/j.envpol.2019.05.072

Application of Composite Polymeric Membranes in Textile Wastewater

Leila D. Fiorentin-Ferrari, Maria Carolina S. Gomes, Fernanda L. Seixas, Veronice Slusarski-Santana, Monica L. Fiorese, Poliana R. Frare, Larissa C. Grigoleto, and Ana C. Spaciari

Abstract Currently, large amounts of colored effluent are generated by textile industries. Dye wastewater is a typical kind of bio refractory organic waste with complex composition. In addition to the dyes, large amounts of salts are present, which makes the treatment of these effluents even more difficult. Membrane separation processes are an efficient alternative for the textile wastewater treatment. Due to its low cost, among other advantages, the application of polymeric membranes is highlighted in this chapter. Alternatively, there is the possibility of incorporating different materials into the membrane, which can improve membrane performance, favoring the retention of compounds present in the effluent to be treated. Thus, we review (1) the general aspects related to polymeric membranes, as well as (2) several methods of modification of polymeric membranes. In addition, a section will address the (3) composite polymeric membranes applied to the treatment of textile wastewater.

L. D. Fiorentin-Ferrari (✉) · V. Slusarski-Santana · P. R. Frare
Department of Chemical Engineering, State University of West Parana, P.O. Box 520, Toledo, PR 85903-000, Brazil
e-mail: leila.ferrari@unioeste.br

V. Slusarski-Santana
e-mail: veronice.santana@unioeste.br

M. C. S. Gomes · F. L. Seixas · L. C. Grigoleto · A. C. Spaciari
Program of Post-Graduation in Chemical Engineering, Federal University of Technology—Paraná, P.O. Box 635, Apucarana, PR 86812-460, Brazil
e-mail: mariagomes@utfpr.edu.br

F. L. Seixas
e-mail: fernandaseixas@utfpr.edu.br

L. C. Grigoleto
e-mail: larissagrigoleto@alunos.utfpr.edu.br

A. C. Spaciari
e-mail: anaspaciari@alunos.utfpr.edu.br

M. L. Fiorese
Program of Post-Graduation in Chemical Engineering, State University of West Parana, P.O. Box 520, Toledo, PR 85903-000, Brazil
e-mail: monica.fiorese@unioeste.br

S. S. Muthu and A. Khadir (eds.), *Textile Wastewater Treatment*, Sustainable Textiles: Production, Processing, Manufacturing & Chemistry,
https://doi.org/10.1007/978-981-19-2852-9_8

Finally, a (4) case study will present the production of a polymeric membrane functionalized with sisal fiber and its application in dye removal. Both the adsorption process and the membrane separation process have been applied to the removal of dyes from effluents. The vegetal fiber of sisal, a civil construction waste, could be incorporated during the fabrication of the membrane promoting the yours adsorptive characteristic. The composite membrane, when applied in the treatment of laboratory wastewater containing dyes, showed high permeate flux (152 kg h^{-1} m^{-2}), selectivity in relation to color (color rejection: 63%), low fouling and full recovery of the permeate flux.

Keywords Ultrafiltration · Composite polymeric membrane · Adsorption · Functionalization · Modification · Natural sisal fiber · Lignocellulosic waste · Membrane fouling · Textile effluents · Wastewater treatment

1 Introduction

In textile industry processes, substantial amounts of water, mineral salts, heavy metals, emulsifiers, and synthetic dyes are used to obtain high quality fabrics. Dyes are organic micropollutants classified according to their water solubility [83]. They are made up of a chromophore group responsible for the coloration and another auxochrome group that helps in the fixation of the chromophore group to the textile fiber [12, 89]. It has been estimated that more than 10–15% of the total dyestuff is released into the environment during their synthesis and dyeing process [60, 83]. As a consequence, there is the generation of wastewater which contains complex composition, high organism concentration, potential toxicity, heavily colored and poor biodegradability [51, 74].

With this, it is necessary to seek efficient treatment alternatives for these effluents. Currently, many of the world's textile manufacturers are equipped with their own wastewater treatment plants, which usually combine an aerobic biological process and a physicochemical process. Nevertheless, most of these traditional methods were found inadequate due to the large variability of composition of textile wastewater [48, 50, 68, 69, 78].

There are several techniques that can be used to remove these contaminants, such as photocatalysis [99], coagulation/flocculation, ozonation, adsorption on activated carbon [56, 94], electrochemical [52], and electrocoagulation [105]. These technologies are promising, but they all still suffer limitations which require further research and development [50]. Membrane separation processes are an efficient alternative to manage the amount of effluents improperly discharged into water due to the use of dyes in industrial processes [39].

Membrane technology is recommended in the treatment of industrial effluents containing dyes due to its high performance and operational versatility in combination with other processes [62, 78]. Furthermore, the possibility of functionalizing membranes with the addition of organic, inorganic, composite, or nanocomposite materials allows the combination of different techniques in a single separation, concentration, and/or purification process [22, 95].

The modification of membranes by incorporating materials provides characteristics such as increased selectivity and decreased fouling [82]. Among the various materials used for incorporation into membranes, lignocellulosic materials have the potential to provide adsorptive characteristics to membranes. Thus, some lignocellulosic residues can be used as additives in polymeric matrices, enabling an improvement in membrane separation performance, as well as a green process by reusing these waste materials [19, 100].

Both the adsorption process and the membrane separation process have been applied to the removal of dyes from effluents, so that the production of adsorptive polymeric membranes has a great potential in the treatment of textile effluent, with the advantage of carrying out both processes in a single step [104].

2 Polymeric Membrane

The development of the separation process by synthetic membranes and your industrial consolidation occurred quickly and efficiently in recent decades. This behavior occurred due to the fact that synthetic membranes can be used for separation, concentration, and purification of molecules and fine particles in a single process step [9, 24, 61].

In relation to structural form, the membranes are classified as isotropic or anisotropic as shown in Fig. 1. Porous isotropic membranes have a uniform distribution of interconnected pores and diameter ranging from 0.01–10 μm. Anisotropic

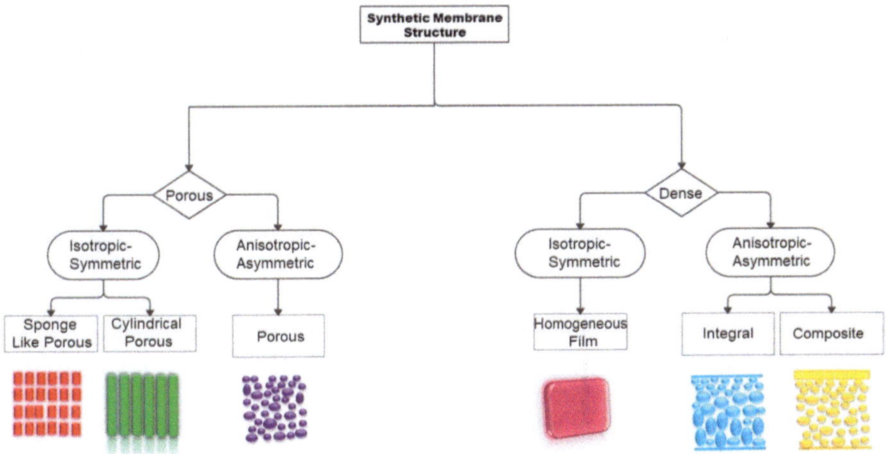

Fig. 1 Schematic representation of the general structure of synthetic membranes showing the classification in terms of morphology

Table 1 Characteristics of the main polymers used in the production filtration membranes

Polymer	Characteristics
Cellulose acetate	high water permeability and mechanical strength pH range 4–7 hydrophilicity poor chlorine resistance
Polysulphone and polyethersulphone	good strength pH range 1–14 hydrophobicity superior chlorine resistance wide temperature tolerance
Polyamide	high temperature resistance (>50 °C) pH range 4–10 poor chlorine resistance
Polyvinylidene fluoride	resistance hydrocarbons and oxidizing compounds pH range 3–10 temperature resistance up to 50 °C

membranes are characterized by the formation of a thin layer on the surface that limits the permeation rate as a porous sublayer. Due to the high performance, the anisotropic membranes are more commercially employed [24, 61].

The structural form of the membranes presented in Fig. 1 is a function of the type of material used and the method of the production membrane. The membranes can be fabricated with organic materials, preferably polymeric, inorganic such as metals, ceramic, and composite materials [61]. In this chapter, the polymeric membrane will be highlighted due to their lower cost compared to synthetics and their wide industrial use Table 1 shows the main polymers used in the manufacture of industrially used membranes and their main structural characteristics [64, 73, 85, 90].

Generally, it is desired that polymeric membranes have high hydrophilicity, chemical, physical and thermal stability, and the presence preferably of functional groups such as -OH, -NH, -SH, -COOH [58]. The characteristics are obtained from the type of polymer or polymer mixture used in production of the membranes. In the case of a mixture of polymers, the physicochemical, mechanical, thermal, and rheological properties of the material must be evaluated to obtain good solution miscibility. Another point that needs to be evaluated in the blend is the morphology and interfacial bonding of the phases involving the size and geometry of polymer chains, repeating units and the effect of the arrangement of molecules [14, 90]. Thus, when polymers are mixed it's necessary to evaluate the individual properties of each material as well as the microstructure formed in the polymer mixture.

Polymeric membranes can be fabricated using the techniques shown in Fig. 2. In the sintering technique, fine particles of powder or suspension are made into polymeric films with pore sizes in the range of 0.1–10 μm using pressing, casting, extrusion, and sintering below the melting point of the materials. In stretching, a

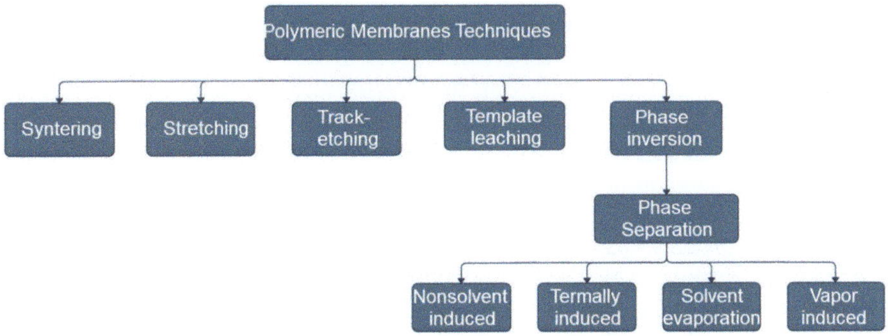

Fig. 2 Schematic representation of polymeric membrane production methods

mixture consisting of polymer is extruded and pores in the range of 0.1–3 μm are formed on the surface during stretching of the membrane. In track-etching, a thin polymeric film or sheet is irradiated with energetic ions followed by chemical corrosion carried out in an alkaline or acidic medium and porous membranes with pore sizes in the range of 0.02 10 μm are obtained. In the case of the template leaching method, one of the components present in the polymeric film with a pore size of 0.005 μm is leached through immersion in a solvent [22, 61].

The phase inversion technique is widely used due to the possibility of obtaining anisotropic and isotropic membranes, mainly in the micro and ultrafiltration range. Besides it can be used in the preparation of various membrane configurations such as flat sheet, hollow fiber, multi hole hollow fiber, and tubular polymeric [1, 9, 61]. In phase inversion, a polymer or polymer blend is added to a solvent forming a homogeneous solution. In contact with non-solvent, the solvent non-solvent interaction will promote polymeric destabilization making the polymer insoluble causing its precipitation [57].

Inversion from liquid polymeric to solid phase in Fig. 2 can be taken by solvent induction method of Loeb and Sourirajan, in which the polymeric solution is spread over a surface and placed in contact with a bath of non-solvent [40]. Asymmetric and symmetric porous membranes are formed with pore sizes in the range of 0.1–20 μm. The thermal induction method is used when it is not possible to dissolve polymers in solvents. In this case, it chooses to use a sequence of high and low temperature variations until obtaining the polymeric membrane. The solvent evaporation method consists of spreading the polymer solution that has been dissolved in one or more solvents with different volatilities on a surface. From that, the controlled evaporation of the polymeric film is carried out until the solid membrane is obtained. In the induced steam technique, the non-solvent is a vapor that precipitates the polymer to the solid phase [1, 32, 61].

Membranes are classified into microfiltration, ultrafiltration, nanofiltration, and reverse osmosis [24, 61]. The micro membranes (100 a 1000 nm) and ultrafiltration (5 a 100 nm) are considered porous membranes while the reverse osmosis (0.1 a 1 nm) is dense. The nanofiltration membrane (1–5 nm) is considered dense; however, it has larger pores than reverse osmosis. Microfiltration membrane ones are closer to classic filtration and use the lowest pressures (0.1–2 bar), while the reverse osmosis uses higher pressures (10–100 bar). In this classification range, membranes retain from microorganisms, macromolecules, proteins to inorganic ions [9, 64, 80].

In micro and ultrafiltration membranes, the pore size mechanism is responsible for the separation of substances [24]. In reverse osmosis membranes, the separation occurs by diffusion across the membrane and the performance of the membrane is determined by the constituent material of the membrane, ionic charge, and interactions between permeated chemical species and the membrane, since separation is a function of the solubility of molecules within the membrane [61]. So, microfiltration and ultrafiltration porous membranes separate particles by pore size difference, while dense nanofiltration and reverse osmosis membranes separate particles by diffusion and solubility [86].

The membrane filtration process can be carried out in continuous modules using conventional dead-end filtration and cross flow. These modules can be connected in series or in parallel, depending on the performance you want to achieve during the process [25, 61]. The type of membrane and method of filtration used directly influence the permeate flow. For filtrations using only distilled water without the presence of contaminating solutes, the permeate flow usually has a constant kinetic. However, in some cases, even using distilled water, for example, an initial decline in flow as a function of time is observed. The decline of the flow is associated with mechanical instability of the membrane and possible swelling of the polymer matrix due to water permeation.

However, the permeate flux with solutions or suspensions has a declining behavior similar to that presented by membranes with mechanical instability when permeating liquids without contaminants or solutes Fig. 3. In the dead-end filtration, the liquid feed is forced through the membrane with the aid of pressure. The particles will be retained by the membrane and an increase in concentration will occur near the surface where the rejected particles end up accumulating. As a result, there is a gradual decline in flow as a function of time. In this sense, in industrial operations it

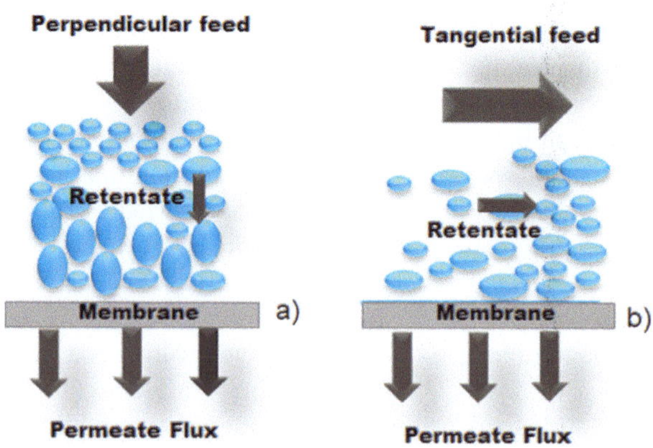

Fig. 3 Comparative scheme of flow modes in membrane filtration processes **a** dead-end; **b** cross flow

is preferable to cross-flow filtration which due to the flow parallel to the surface of the feed membrane, there is less deposition of rejected particles and, consequently, less drop in the permeate flow [7, 9, 25].

Regardless of the type of operation used for filtration dead-end or cross flow, when solutions or suspensions in the retention range from micro to nanofiltration are permeated, there will always be a higher concentration of solute near the membrane surface with a decline in permeate flux, regardless of the diffusive movement of the particles into the solution. This increase in solute with additional resistance to mass transfer across the membrane is called concentration polarization. To reduce the concentration polarization, it's customary to vary the flow conditions of the feed. For the cross flow, when the higher the velocity the smaller the thickness of deposited solids. However, care must be taken with selectivity, as it's often inversely proportional to the Reynolds number [23, 61].

Even with the increase in feed speed, it is common for the permeate flux to continue to decrease after stabilization of the concentration polarization [61]. In this case, another phenomenon starts to interfere in the process known as fouling, which can be reversible or irreversible. In general, fouling can start with adsorption of contaminating particles, followed by complete clogging, internal, or random of the membrane pores and also the deposition of the material forming filter cake [37, 49]. Generally, [35] models are used to demonstrate fouling mechanisms that occur during the permeation of contaminants [55, 77].

In order for polymeric membranes to be industrially viable, it is necessary that after their use, it is possible to carry out cleaning and reuse [15]. In this sense, for micro and ultrafiltration membranes a hydraulic backwash system has been used, which consists of the passage of water with the membrane inverted inside the experimental module. Combined with the backwash cleaning it is customary to carry out periodic cleaning with acid or alkaline solutions for a certain period of time [16, 92].

Thus, it is evident that the choice of the type of polymer, combined with the morphological characteristics that are obtained as a function of the membrane fabrication technique, as well as the type of filtration used, is essential to have membranes with high selectivity and permeate flux and with the longest possible operating time without considerable fouling intervention. Besides, the membranes need to have mechanical, chemical, and thermal resistance to withstand the cleaning processes through several reuse cycles.

So that the membranes can be reused in the process, maintaining the desired performance in filtration of liquid solutions, such as in the case of industries that generate effluents, one option is to combine the membrane separation process with other techniques such as coagulation/flocculation, adsorption, advanced oxidative processes and biological treatment. This joining of processes can be carried out in the form of sequential combination or hybrid. In the hybrid combination, the polymeric membrane is structurally modified during the application of the preparation techniques. In this case, different particulate materials can be added, such as adsorbents, metal oxides and compounds, carbon nanotubes and nanocomposites. Thus, a hybrid membrane has the characteristic of associating two processes in a single equipment, reducing costs and operating time.

3 Methods of Modification of Polymeric Membranes

In membrane separation processes, the main challenge is to achieve high selectivity by operating with a high and constant permeate flux. As already mentioned, the decline in flux over time is caused by fouling, which occurs due to the accumulation of particles that are retained on the surface of the membrane, as well as by adsorption in the pores. The occurrence of fouling provides lower productivity and requires the use of aggressive cleaning methods that cause a reduced membrane life [10].

One of the main strategies to minimize fouling, as well as to improve their mechanical, chemical, and thermal properties, is the modification of membranes. A widely used method for modification is the incorporation of materials into the membrane, which can improve membrane performance, favoring the retention of compounds present in the effluent to be treated and increasing membrane selectivity [36, 67]. Thus, the incorporation of materials aims to provide new functionalities for the polymeric membrane, such as hydrophilicity, roughness, porosity, surface charge, surface functional groups, mechanical resistance, and biological activities surface [8, 31].

The modification or functionalization of membranes can occur by incorporating the material into the polymer solution, followed by membrane synthesis, or the material can be incorporated after the production of the membrane, on its surface or pores [82]. The functionalization of the membrane by incorporating the material into the polymer solution provides greater homogeneity and a large number of immobilized molecules grafted onto the polymer chain. However, this method can interfere with the performance of the membrane by changing its porosity. Functionalization on the membrane surface, on the other hand, offers the greatest advantage of enabling the immobilization of large molecules, such as other polymers with a lower molecular mass, macroheterocyclic compounds, or even proteins or enzymes [82].

These membrane modifications can be classified into physical modifications (coating, blending, and using composite membranes) and chemical modifications (functionalization of polymers, plasma treatment, and graft polymerization) [8, 63].

Coating is a method wherein the coating material forms a thin layer that non-covalently adheres to the membrane surface. The thin film composite can be formed by physical adsorption, coating with a monolayer using Langmuir–Blodgett or other technique, plasma polymerization, casting of a polymer solution on the membrane surface and spin coating deposition [63, 70, 71].

Blending is a process in which two (or more) polymers are physically mixed to obtain specific properties. Some works report that the mixture of two or more polymers with distinctive physical, chemical, and separation characteristics can provide membranes with desired characteristics such as hydrophilicity, selectivity, and permeate flux. Furthermore, blending polymers can improve thermal and mechanical properties in polymeric membranes [11, 13, 70, 72, 87].

Composite membranes can also be obtained by adding materials with physicochemical characteristics different from the polymer and which are incorporated in the polymer matrix [63]. Additive plays an important role in the formation of membrane structure by enhancing pore formation, providing adsorption capacity and improving anti-fouling performance by increasing hydrophilicity and electrostatic repulsion characteristics. In addition, the incorporation of materials can also provide an increase in membrane mechanical strength [28, 59, 70, 79].

For chemical modification, the membrane material is treated with modifying agents to introduce various functional groups on the membrane surface. Grafting is a method wherein monomers are covalently bonded onto the membrane and may be performed via chemical, photochemical, enzymatic, using high-energy radiation and plasma. The choice for a specific grafting technique depends on the chemical structure of the membrane and the desired characteristics after surface modification [63, 96, 102].

A combination of these modification methods can also be employed to obtain specific characteristics in polymeric membranes. Considering the membrane modification alternatives, the use of polymer blends and additives can provide the development of membranes with high permeability and selectivity. These prepared membranes are called mixed matrix membranes that usually contain nanoparticles in the continuous polymer phase. The development of these mixed matrix membranes can increase the performance of polymeric membranes by adding appropriate porous or non-porous fillers, such as metal – organic frameworks (MOFs), zeolites, and carbon nanotubes or other organic and inorganic fillers to the polymer matrix [3, 8 2, 87]

Thus, in order to increase the removal of pollutants, several nano- or micro-scale porous adsorbents can be used to obtain membranes with adsorptive characteristics [21, 84, 101, 109].

Nanoparticles can be used to modify and improve hydrophilicity and porosity, providing anti-fouling characteristics. The incorporation of nanoparticles such as zeolite, silica, and titanium dioxide nanoparticles modify the structure of the membrane network and provide a direct pathway for water transport and thereby increasing water permeability and anti-fouling properties [82, 93]. Carbon nanotubes have been used in various applications due to their specific characteristics such as mechanical strength, high specific area, thermal and electrical conductivity, and optical properties [97].

Carbon-based nanomaterials have also been used to increase the adsorptive properties of membranes [2, 65, 91, 97]. Other carbon-based materials, such as graphene and graphene oxide, have also been applied to reduce membrane fouling tendency. The properties of graphene, such as high surface area and excellent mechanical strength, indicate that it can be used as a material for the production of membranes with specific characteristics [17, 27].

Considering the needs and perspectives of synthesize membranes in a completely eco-friendly way, where raw materials, membrane preparation, post-treatment, and other involved procedures are all green, the search for alternative materials that can be incorporated into the polymer matrix is relevant [41].

Cellulose nanofibrils have attracted much attention in water filtration membranes as a substitute of commonly incorporated materials due to distinctive advantages, such as good mechanical properties, high specific surface area, thermal stability, and biodegradability. Lignin–cellulose nanofibrils have been employed to improve the hydrophilicity and mechanical properties of ultrafiltration membranes by blending with polyethersulfone in the casting solution [20, 106, 107].

The incorporation of bamboo-based lignocellulose nanofibrils in the production of cellulose acetate ultrafiltration membranes using phase inversion was evaluated by Yang et al. [107]. The incorporation of lignocellulose nanofibrils into the polymeric matrix, in addition to providing anti-fouling characteristics, also endowed antibacterial activity to composite membranes without any chemical modification, which is beneficial to prolong the service life of membranes.

Considering the perspectives of incorporation of lignocellulosic materials in polymeric membranes, it is interesting to note that several studies present the application of lignocellulosic waste as adsorbents in the treatment of effluents for the removal of dyes [19, 45, 66, 100]. These low cost residues can be an alternative to the additives used in the production of adsorptive membranes.

These low cost agricultural wastes can be an alternative to the traditional nanoadsorbents used in the production of adsorptive membranes. Ghazali et al. [29] investigate the use of pineapple peel waste as an additive in the production of polyether block amide nanocomposite membrane to be applied in gas separation. It was reported that the incorporation of this nanoadsorbent changed the membrane structure and chemical properties, thus providing an increase in membrane selectivity.

The scarcity of studies on the use of waste as additives in polymer membranes to obtain specific characteristics indicates the prospect of further work in this area. In the case study presented in this chapter, the use of sisal fiber for the production of membranes will be addressed. Furthermore, research using membranes incorporated with waste for the treatment of textile wastewater is also scarce. The next section covers the application of composite polymeric membranes in the treatment of this type of effluent.

4 Composite Polymeric Membranes Applied to the Treatment of Textile Wastewater

Fabric dyeing processes use large amounts of dyes and other chemical compounds, consequently, a huge volume of effluent containing contaminants is produced [18]. Dye wastewater is a typical kind of bio refractory organic waste with complex composition, high organism concentration, and poor biodegradability [74]. In addition to dyes, there are some inorganic salts such as NaCl and Na_2SO_4 in textile effluents that can make the treatment of these effluents even more difficult [109]. Thus, it is extremely important to develop efficient treatment for textile effluents.

There are several techniques that can be used to remove these contaminants, such as photocatalysis, electrochemical, and electrocoagulation [52, 99, 105]. Among these treatment methods, membrane separation has some advantages, such as: high selectivity, low energy consumption, no additional waste generation, and ease application [39, 103]. However, the major disadvantage of this technique is the low water permeability with the high-energy cost [108].

Thus, according to Song et al. [88], in order to efficiently separate or reuse salt-containing dyes from aqueous solutions, it is necessary for the surface functionalized composite membranes endowing not merely improved surface hydrophilicity, smoothness, charge consistent with the target dyes, and multivalent ions, but also regularly increased the effective porosity in the skin-layer internal structure.

In this sense, several studies evaluate the preparation of loose nanofiltration membrane, which is usually organic inorganic hybrid membrane. The loose nanofiltration membrane is prepared by doping inorganic nano materials into organic membranes. Loose nanofiltration membranes are characterized by the presence of large water transport channels. These structures are larger than conventional nanofiltration membranes and such channels enable the permeation of monovalent ions [46].

Kang et al [46] developed graphene oxide composite loose nanofiltration polyethersulfone membranes with low negatively charged surface by one-step co-deposition. The authors noted that the incorporated tannic acid/Ni complex increased the flux of water due to the increased of the distance between layers of the graphene oxide nanosheets. Compared to the graphene oxide coated membrane, the permeation of graphene oxide-tannic acid-nickel (20) (10 mM TA and 20 mM Ni^{2+}) increased significantly (43.4 Lmh/bar to 71.7 Lmh/bar). In addition, high dye rejections were achieved (92.9% for methyl blue rejection, 98.8% for congo red rejection).

Song et al. [88] evaluated the manufacture of phosphorylated chitosan nanofiltration membrane via surface functionalization with graphene oxide nanosheets through the formation of covalent bonds. The authors concluded that in optimized condition, the 10 mg L^{-1} graphene oxide phosphorylated chitosan membrane exhibited better anti-fouling properties, reaching an excellent flow decline ratio of 8.67% and an especially irreversible fouling rate of just 0.17% for Black Direct 38. The membranes showed negative surface charge, high hydrophilicity, and roughness. Furthermore, it showed good separation ability of anionic dye and salts.

Another promising line of research applied to the treatment of textile effluents is that of adsorbent membranes. The adsorbent has high porosity and specific surface, consequently the adsorbent membranes have a high dye removal capacity. Xu et al. [104] produced nanofibrous polyethersulfone membranes by electrospinning (single step) from the blend of amphiphilic copolymer of acrylic acid and methyl methacrylate. The membranes showed adsorption capacity of 2257.88 mg g^{-1} for the methylene blue dye. Moreover, the nanofiber membrane exhibited excellent recyclability (81.45% after 5 cycles), due to chemical and physical stability and high filtration-purification efficiency (above 99%, in a high flow of 100 mL min^{-1}).

An ecologically correct and sustainable alternative for the treatment of effluents with dyes is the immobilization of enzymes in functionalized membranes. Jun et al. [43] immobilized the Jicama peroxidase in a functionalized multi-walled carbon nanotube membrane Buckypaper/polyvinyl alcohol membrane and used it for the degradation of the methylene blue dye. The immobilized Jicama peroxidase membrane showed a maximum adsorption capacity of 90 mg g^{-1}. In addition, it can also be recycled and reused up to eight times, maintaining 64.3% of its original activity, which can reduce the costs of using this type of treatment and enabling its implementation in textile wastewater treatment plants.

Januário et al. [39] evaluated the functionalization of polyethersulfone microfiltration membranes with titanium dioxide, graphene oxide, and sulfuric acid by the layer-by-layer self-assembly method. The excellent separation for dyes was found to twilight yellow (69.98%) due to size exclusion; to safranin orange (100%) due to electrostatic repulsion; and to red bordeaux (93.35%) due to adsorptive processes. In addition, the proposed modification improved the anti-fouling characteristics of the membrane, the results of which were 65.13% and 95.21% for the modified and pure membrane, respectively.

There are several works that promote the incorporation of metal–organic structure (MOFs) in membranes [30, 42, 47]. These membranes have been efficient for the treatment of textile effluents. Gnanasekaran et al. [30] evaluated the production of nanofiltration membranes by incorporating metal–organic framework MIL-100 (Fe) in chitosan films using the casting technique. The composite membrane showed an increase in water flux from 5 to 52 L/m^{2}h, in addition, it also showed 99% of rejection capacity for methylene blue and methyl orange dyes. The chitosan/MIL-100 (Fe) membrane exhibited an improved pure water flux from 5 to 52 L/m^{2}h as well as 99% rejection efficiency for cationic methylene blue and anionic methyl orange.

The process of separation of dyes by composite polymeric membranes proves to be very efficient. There are several materials (metal–organic framework, activated carbon, graphene oxide) that can be incorporated into membranes through different methods, as shown in Table 2.

As already presented in the previous section, in general, the incorporation of such related materials into the polymer matrix improved surface hydrophilicity and increased the effective porosity of the membrane structure, thereby increasing water permeability and anti-fouling properties. Moreover, it is observed that the incorporation methods used increased the physical stability of the membranes allowing the membranes to be recycled and reused. Fact that can reduce the costs of using this type of treatment and enabling its implementation in textile wastewater treatment plants.

Furthermore, the incorporation of traditional adsorbent composites enables the removal of large amounts of dyes. It was possible to observe from this review that, in general, the membranes presented rejection capacity values, for different dyes, above 99%. Fact that demonstrates the high separation efficiency obtained by this type of material. Also, there is a wide variety of agro-industrial residues with high adsorption capacity that can be evaluated for incorporation into polymeric membranes.

However, as already mentioned, research related to the application of these composites materials in the treatment of textile wastewater is scarce. Thus, the next section addresses a case study that uses polymeric membranes incorporated with sisal fiber for the treatment of textile effluent.

Table 2 Comparative review of the application of composite membranes for dye removal

Membrane material	Composite	Methods	Dyes rejection (%)	Reference
Polyethersulfone	Sulfuric acid, titanium dioxide and graphene oxide	Layer-by-layer self-assembly method via electrostatic interaction through a pressure-assisted filtration system	Twilight yellow: 68.9 Safranin orange: 100 Red bordeaux: 93.4	Januário et al. [39]
Polyethersulfone	Nano-SiO_2	Phase inversion	Reactive Black 5: 97.9 Reactive Green 19: 99	Zhang et al. [109]
Phosphorylated chitosan	Graphene oxide nanosheets	Surface coating and chemical cross-linking	Direct black 38: 99.7	Song et al. [88]
Polyethersulfone and GO	Tannic acid and nickel	One-step co-deposition	Methyl Blue: 92.9 Congo Red: 98.8	Kang et al. [46]
Polyvinyl chloride mixed matrix containing multi-walled carbon nanotubes	Polypyrrole	Phase inversion	Reactive Blue 50: 95	Haghighat and Vatanpour [34]
Polysulfone	Tannic acid	Phase inversion	Reactive black: 99 Reactive Orange 16: 90	Ibrahim et al. [38]
Polyethersulfone	Acrylic acid and methyl methacrylate	One-step electrospinning	Methylene blue: 99	Xu et al. [104]
Chitosan	Metal–organic framework	Phase inversion	Methylene blue: 99 Methyl orange: 99	Gnanasekaran et al. [30]
Polysulfone	Activated Carbon	Layer-by-layer	Methylene blue: 99.29 Crystal violet: 99.54 Rhodamine B: 99.04	Goswami et al. [33]

5 Case Study

Agro-industrial lignocellulosic materials such as sugarcane bagasse, corn straw, grain husks, cellulosic residues from paper and wood production, herbaceous biomass such as hay and sisal fiber [4] can be used in the functionalization of polymeric membranes. Lignocellulosic biomass, another name given to these materials, is basically composed of lignin, hemicellulose, and cellulose. Due to degradation to smaller molar masses, lignin is extracted from agro-industrial materials and used in various chemical processes [44, 54].

In the membrane separation process, it is customary to pre-treatment the biomass with the application of physical and chemical techniques to extract lignin, hemicellulose, and cellulose and then use them individually in the fabrication and/or functionalization of membranes [4, 5, 53, 75, 76]. However, no studies were found that use lignocellulosic biomass in its entirety and in natural form, without pre-treatment before being used in the fabrication and/or functionalization of membranes. In this sense, the plant sisal fiber (Agave sisalana) is a light material of low density, which has excellent rigidity and resistance, is biodegradable and is highly available. Fiber is currently being used extensively in the synthesis of composites to replace synthetic fibers that are not biodegradable and cause serious environmental problems after their use [81, 98].

The aim of this case study is to demonstrate the application of the concepts discussed in this chapter in relation to the functionalization of polymeric membranes with materials in order to obtain greater efficiency in the treatment of colored wastewater. Thus, in this case study is reported the production of a polymeric membrane functionalized with Natural Sisal Fiber (NSF). The membrane is applied in the treatment of laboratory wastewater that contains a dye mixture.

The laboratory wastewater was collected in the Pollution Control Laboratory of State University of West Parana—Department Chemical Engineering. All the textile effluent used in other academic works is discarded in a gallon. Thus, was collected a volume the effluent of this discard gallon to carry out the experimental tests in this Case Study. The laboratory wastewater contained a mixture of acid, basic and reactive dyes, with predominance of the dye Reactive Blue 198 (RB198).

Laboratory wastewater was characterized in terms of pH, electrical conductivity, color, and Chemical Oxygen Demand (COD—[6]). The effluent contains in higher concentration the dye RB198 and the presence of other dyes in the UV–Vis range from 400 to 800 nm, in addition to acidic solutions used in washing membranes and mainly in adsorption tests. For comparison purposes, a synthetic solution of 20 mg L^{-1} was prepared with the dye RB198 since this dye was found in higher concentration in the effluent.

Natural sisal fiber (NSF) was collected from civil construction waste where it is used for the application of gypsum. The NSF was washed under running water and left to dry in the sun during the summer (30–35 °C). The polymeric membrane was manufactured using the technique of phase inversion in a non-solvent bath. The polymer dispersion was prepared from a mixture of 18% polyethersulfone (PES), 1% sodium chloride (NaCl), 5% natural sisal fiber (NSF) (%wt) in N'N—dimethylformamide solvent. The polymeric film was spread over a polyester support and left in a bath of distilled water (23 °C) for 12 h until the phase inversion occurred as described in Fig. 4. In this Case Study the membrane will be called 18%PES/1%NaCl/5%NSF.

The 18%PES/1%NaCl/5%NSF membrane was characterized in relation to water absorption, porosity, hydrophilicity, and zero charge point (pH_{pzc}) [26]. The compaction and filtration tests of the synthetic solution and waste effluent were carried out in a dead-end filtration module, with power supplied by a nitrogen

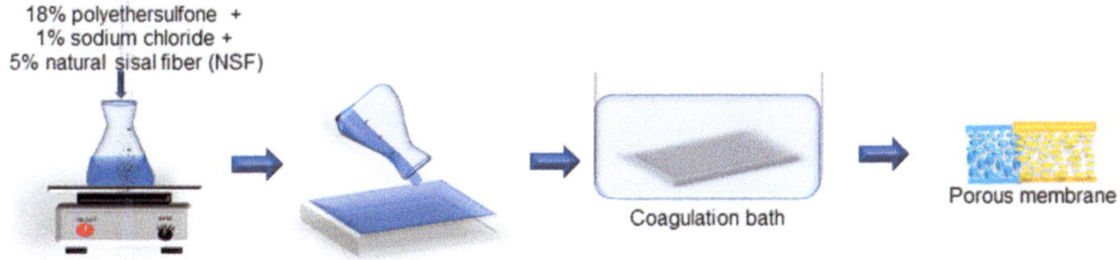

Fig. 4 Schematic representation of the steps in the production of 18%PES/1%NaCl/5%NSF polymeric membrane by the phase inversion method

cylinder. The membrane with a useful area of 2.53 cm^2 was inserted on a support inside the module and submitted to the tests. All tests were carried out with a pressure of 2 bar until stabilization of the permeate flow. After filtration of the solution and the wastewater, cleaning was performed with a backwash for 15 min at a pressure of 1 bar, and the reestablishment of the flow was verified [26].

The characterizations showed that the 18%PES/1%NaCl/5%NSF membrane has a porosity of 48%, estimated from the water absorption (104%), in addition to a hydrophilic characteristic with a contact angle of less than 90° as shown in Fig. 5.

The results show that even using a polymer with a hydrophobic character as is the case of polyethersulfone, the high water absorption presented implies that the Loeb and Souriraj methodology was effective for the junction of polyethersulfone/sodium chloride/natural sisal fiber to solvent. Another factor that helped to obtain the porous and hydrophilic membrane was the addition of sodium chloride in the solution and its migration to the membrane surface during contact with distilled water in the phase inversion bath [26].

Regarding the surface charge, it was observed that the membrane has a neutral characteristic, presenting a pH_{pzc} of 6.5 as seen in Fig. 6a. When $pH_{solution}$ lower than pH_{pzc}, the surface is positively charged; and if $pH_{solution}$ is higher than pH_{pzc}, the surface is positively charged. Thus, it is believed that the 18%PES/1%NaCl/5%NSF membrane presents a favorable characteristic for the retention of anionic dyes. In Fig. 6b, there is the structure of organic dye Reactive Blue 198 belongs to the class of water-soluble anionic dyes that are used in the dyeing of cotton and wool, in which the –OH reactive group helps in the fixation of the cellulose fibers while –OH, –SH, and $–NH_2$ in wool fixation [83]. It is observed the presence of different functional groups in the structure of the RB198 dye, especially sulfonated groups which are negatively charged. Thus, the negative charges present in the molecular structure of

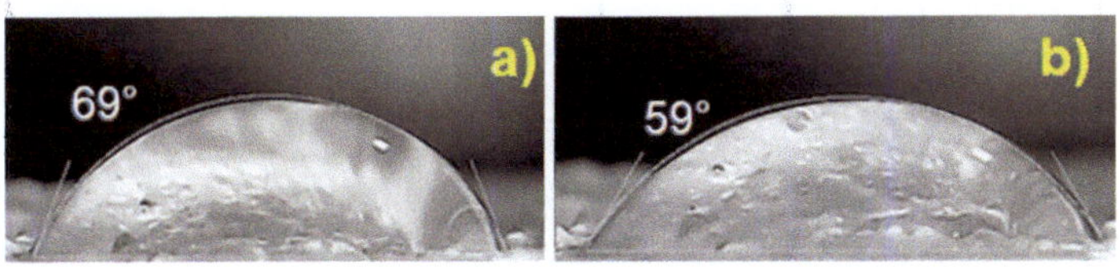

Fig. 5 Contact angle profile or water droplet on the membrane surface after **a** 5 s; **b** 5 min

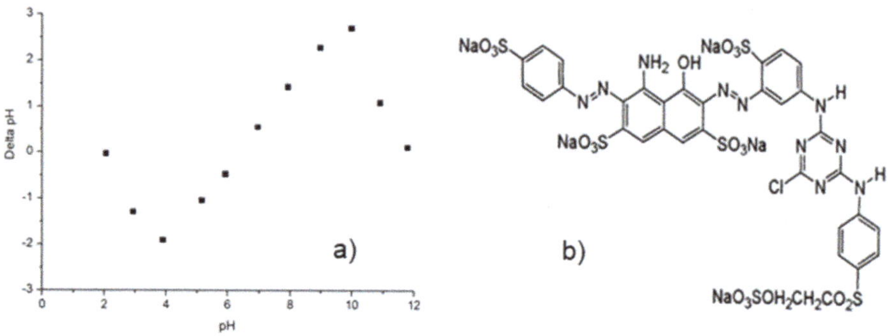

Fig. 6 Membrane surface charge variation and comparison with the dye structure in order to verify the interaction capacity **a** pH_{pzc} of the 18%PES/1%NaCl/5%NSF (%wt) membrane surface; **b** structure of the RB198 dye

the RB198 dye are easily interacting with the 18%PES/1%NaCl/5%NSF membrane protonated groups.

Tests show a high flux decline within the first 80 min of filtration primarily for water and dye solution as shown in Fig. 7. For laboratory wastewater the flow decay was smaller over time. After 100 min of filtration, the three filtrations evaluated reached a constant permeate flux. The decline in permeate flux with the dye solution (58%) was lower than laboratory wastewater (96%) when compared with flow of

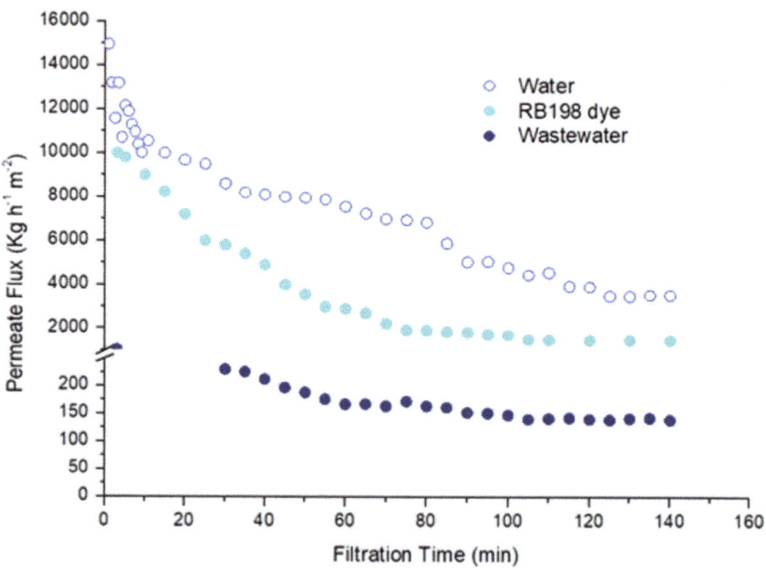

Fig. 7 Permeate flux over filtration time through 18%PES/1%NaCl/5%NSF membrane by using dead-end filtration mode under 2 bar

Table 3 Characterization of the feed and permeate solutions in the filtration of synthetic dye solution RB198 and laboratory wastewater

	pH	Conductivity (μS cm^{-1})	Color rejection (%)	COD** (mgO$_2$ L^{-1})
RB198 dye solution	7.0 ± 0.1	26.5 ± 0.2	–	–
Treated RB198 dye solution permeate	6.8 ± 0.1	25.3 ± 0.2	20*	–
Laboratory wastewater	1.6 ± 0.1	13.52 ± 0.2	–	460
Treated Laboratory wastewater permeate	1.5 ± 0.1	13.65 ± 0.2	63*	400

* 600 nm wavelength
** Chemical Oxygen Demand

distilled water. The flux decline is accordance with the concepts involved in the polarization of concentration followed by the development of fouling [49].

The particles and macromolecules present in the laboratory wastewater are deposited on the surface of the 18%PES/1%NaCl/5%NSF membrane and even the diffuse movement of molecules into the wastewater is not sufficient to prevent the decrease of the flux of the wastewater, 152 kg h^{-1} m^{-2}. However, even with flux reduction with laboratory wastewater, after cleaning with backwash the membrane had its flux reestablished to that obtained during water filtration, 3500 kg^{-1} m^{-2}, thus demonstrating that the fouling occurred in a reversible way.

To complete this study, Table 3 results obtained from the characterization of the feed and permeate solutions in the filtration of synthetic dye solution RB198 and laboratory wastewater.

Initially, it is observed in Table 3 that the pH and electrical conductivity were kept constant during filtration. The synthetic solution maintained a neutral pH while for the effluent the pH was acidic. The acidity of the laboratory wastewater is attributed to the mixture of acid solutions used in the backwash of membranes and in adsorption processes. Regarding discoloration, the membrane rejected 63% of the color of the wastewater at the characteristic wavelength of the RB198 dye. However, when observing the UV–vis spectrum, Fig. 8, there is also a rejection of pollutants, approximately 80% in the range from 400 to 600 nm. The result demonstrates that the membrane also rejects other dyes and macromolecules present in the laboratory wastewater visible within the UV–vis range.

Regarding the type of rejection that occurred for the two contaminants, the results indicate that in the synthetic solution the color rejection occurred predominantly due to the exclusion between the membrane pore size and the molecular size of the RB198 dye molecule. However, for the disposal wastewater, in addition to the exclusion by pore size difference, there was also the adsorption of the dye on the natural sisal fiber. The rejection mechanism is supported by observing the pH$_{pzc}$ of 6.8 determined for the 18%PES/1%NaCl/5%NSF membrane. In this case, the surface of the membrane favors the adsorption of acidic compounds, such as those present in the laboratory

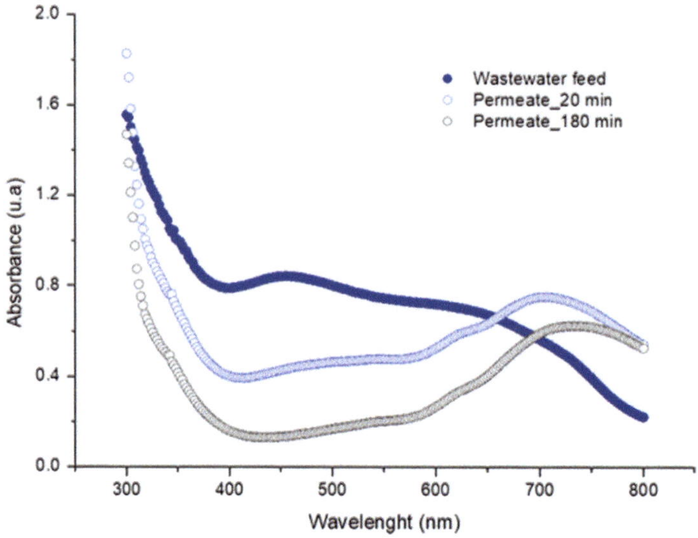

Fig. 8 UV–Vis absorption spectrum of laboratory wastewater permeate in the 18%PES/1%NaCl/5%NSF membrane

wastewater, which make it have a pH of 6 (Table 3). So in the RB198 dye solution, the adsorption process was unfavorable due to its neutral character (pH 7.0). In relation to COD, it is observed that the laboratory wastewater has a low initial organic load and as the manufactured membrane has high porosity and high permeate flow with effluent, COD rejection was small, about 13%.

Thus, this case study demonstrated the natural fiber sisal, a civil construction waste, can be incorporated during the fabrication of the 18%PES/1%NaCl/5%NSF membrane promoting the yours adsorptive characteristic. The adsorptive membrane showed high permeate flux, selectivity in relation to color and low fouling, with full recovery of the permeate flux. The Case Study demonstrates that structural modification of polymeric membranes with biomass lignocellulosic in nature is possible and feasible, because in a single continuous process it was possible to associate membrane separation and adsorption. In addition, the insertion of the natural sisal fiber made it possible to obtain a membrane with characteristics closer to biodegradability concepts.

6 Conclusions

The generation of effluents in the textile industry is a concern, since these effluents contain organic and inorganic substances that can cause environmental problems if improperly disposed. Several methodologies have been developed and applied in the treatment of these effluents, including membrane technology and the use of adsorbent materials for the removal of dyes presented.

The production of polymeric membranes modified by incorporating materials is an alternative to provide the specific characteristics of these materials to membranes, such as the use of adsorbent materials, and improve the separation efficiency. The incorporation of adsorbent composites enables the removal of large amounts of dyes. The most used adsorbent compounds are metal–organic framework, activated carbon, and graphene oxide. However, there is a wide variety of agro-industrial residues with high adsorption capacity that can be evaluated for incorporation into polymeric membranes. Furthermore, the development of research evaluating the application of such materials in the treatment of real textile effluents is of great importance.

The case study which was reported demonstrated the technical feasibility of producing an adsorptive polymeric membrane by the phase inversion method with incorporation of a lignocellulosic residue, sisal fiber. This composite membrane showed selectivity to dyes present in the laboratory wastewater that was treated and provided a reduced fouling in the membrane. The incorporation of materials in polymeric membranes can provide specific characteristics and better filtration performance, in addition to enabling the use of waste materials, thus minimizing environmental impact.

Acknowledgements The authors would like to thank the State University of West Parana (UNIOESTE—Toledo) and Federal University of Technology—Parana (UTFPR—Apucarana) for the support and availability of the laboratory structure to carry out the experiments.

References

1. Abdallah H (2017) A review on catalytic membranes production and applications. Bull Chem React Eng Catal 12:136–156. https://doi.org/10.9767/BCREC.12.2.462.136-156
2. Amaral P, Partlan E, Li M, Lapolli F, Mefford OT, Karanfil T (2016) Ladner DA (2016) Superfine powdered activated carbon (S-PAC) coatings on microfiltration membranes: effects of milling time on contaminant removal and flux. Water Res 1(100):429–438. https://doi.org/10.1016/j.watres.2016.05.034
3. Amid M, Nabian N, Delavar M (2020) Fabrication of polycarbonate ultrafiltration mixed matrix membranes including modified halloysite nanotubes and graphene oxide nanosheets for olive oil/water emulsion separation. Sep Purif Technol 251:117332. https://doi.org/10.1016/J.SEPPUR.2020.117332
4. Amusa AA, Ahmad AL, Adewole JK (2020a) Study on lignin-free lignocellulosic biomass and PSF-PEG membrane compatibility. BioResources 16:1063–1075. https://doi.org/10.15376/BIORES.16.1.1063-1075
5. Amusa AA, Ahmad AL, Adewole JK (2020) Mechanism and compatibility of pretreated lignocellulosic biomass and polymeric mixed matrix membranes: a review. Membranes (Basel) 10:1–29. https://doi.org/10.3390/MEMBRANES10120370
6. APHA (2005) American Public Health Association. Standard methods for examination of water and watwater, 21st ed. Washington
7. Aptel, P. (1993). Membrane pressure driven processes in water treatment. In: João G Crespo, Karl W Boddeker (V. 272). Membrane processes in separation and purification. Springer-Science+Business Media, B. Portugal, pp 263–281

8. Ayyavoo J, Nguyen TPN, Jun BM, Kim IC, Kwon YN (2016) Protection of polymeric membranes with antifouling surfacing via surface modifications. Colloids Surfaces A Physicochem Eng Asp 506:190–201. https://doi.org/10.1016/J.COLSURFA.2016.06.026
9. Baker RW (2004) Membrane technology and applications. John Wiley & Sons, England
10. Bengani-Lutz P, Zaf RD, Culfaz-Emecen PZ, Asatekin A (2017) Extremely fouling resistant zwitterionic copolymer membranes with ~1 nm pore size for treating municipal, oily and textile wastewater streams. J Memb Sci 543:184–194. https://doi.org/10.1016/J.MEMSCI.2017.08.058
11. Benkhaya S, Achiou B, Ouammou M, Bennazha J, Younssi SA, M'rabet S, El HA (2019) Preparation of low-cost composite membrane made of polysulfone/polyetherimide ultrafiltration layer and ceramic pozzolan support for dyes removal. Mater Today Commun 19:212–219. https://doi.org/10.1016/J.MTCOMM.2019.02.002
12. Benkhaya S, M' rabet S, El Harfi A (2020a) A review on classifications, recent synthesis and applications of textile dyes. Inorg Chem Commun 115:107891. https://doi.org/10.1016/J.INOCHE.2020.107891
13. Benkhaya S, M'rabet S, Hsissou R, El HA (2020) Synthesis of new low-cost organic ultrafiltration membrane made from Polysulfone/Polyetherimide blends and its application for soluble azoic dyes removal. J Mater Res Technol 9:4763–4772. https://doi.org/10.1016/J.JMRT.2020.02.102
14. Braun D, Cherdron H, Rehahn M, Ritter H, Voit B (2013) Polymer synthesis: theory and practice: fundamentals, methods, experiments. Springer Heidelberg, New York
15. Chang H, Liang H, Qu F, Liu B, Yu H, Du X, Li G, Snyder SA (2017) Hydraulic backwashing for low-pressure membranes in drinking water treatment: a review. J Memb Sci 540:362–380. https://doi.org/10.1016/J.MEMSCI.2017.06.077
16. Chang H, Liang H, Qu F, Ma J, Ren N, Li G (2016) Towards a better hydraulic cleaning strategy for ultrafiltration membrane fouling by humic acid: effect of backwash water composition. J Environ Sci 43:177–186. https://doi.org/10.1016/J.JES.2015.09.005
17. Cohen-Tanugi D, Grossman JC (2012) Water desalination across nanoporous graphene. Nano Lett 12:3602–3608. https://doi.org/10.1021/NL3012853
18. Cusioli LF, Quesada HB, Baptista ATA, Gomes RG, Bergamasco R (2019) Soybean hulls as a low-cost biosorbent for removal of methylene blue contaminant. Environ Prog Sustain Energy 39:e13328. https://doi.org/10.1002/EP.13328
19. Degermenci GD, Değermenci N, Ayvaoğlu V, Durmaz E, Çakır D, Akan E (2019) Adsorption of reactive dyes on lignocellulosic waste; characterization, equilibrium, kinetic and thermodynamic studies. J Clean Prod 225:1220–1229. https://doi.org/10.1016/j.jclepro.2019.03.260
20. Ding Z, Zhong L, Wang X, Zhang L (2016) Effect of lignin–cellulose nanofibrils on the hydrophilicity and mechanical properties of polyethersulfone ultrafiltration membranes 28:1192–1200. https://doi.org/10.1177/0954008315621611
21. Ebadi R, Maghsoudi H, Babaluo AA (2021) Fabrication and characterization of Pebax-1657 mixed matrix membrane loaded with Si-CHA zeolite for CO2 separation from CH4. J Nat Gas Sci Eng 90:103947. https://doi.org/10.1016/J.JNGSE.2021.103947
22. Esfahani MR, Aktij SA, Dabaghian Z, Firouzjaei MD, Rahimpour A, Eke J, Escobar IC, Abolhassani M, Greenlee LF, Esfahani AR, Sadmani A, Koutahzadeh N (2019) Nanocomposite membranes for water separation and purification: fabrication, modification, and applications. Sep Purif Technol 213:465–499. https://doi.org/10.1016/J.SEPPUR.2018.12.050
23. Field R (2017) Concentration polarization, fouling, and its mitigation. In: Field R, Bekassy-Molnar E, Linizki F, Vatai G (eds) Engineering aspects of membrane separation and application in food processing. CRC Press, Boca Raton, pp 41–66
24. Field R, Molnar-Bekassy E, Lipnizki F, Vatai G (2017) Engineering aspects of membrane separation and application in food processing. CRC Press. Taylor & Francis Group, New York
25. Field R, Lipnizki F (2017). Membrane separation processes and overview. In: Field R, Bekassy-Molnar E, Linizki F, Vatai G (eds) Engineering aspects of membrane separation and application in food processing. CRC Press, Boca Raton, pp 4–40

26. Fiorentin-Ferrari LD, Celant KM, Gonçalves BC, Teixeira SM, Slusarski-Santana V, Módenes AN (2021) Fabrication and characterization of polysulfone and polyethersulfone membranes applied in the treatment of fish skin tanning effluent. J Clean Prod 294:126127. https://doi.org/10.1016/J.JCLEPRO.2021.126127
27. Gahlot S, Sharma PP, Gupta H, Kulshrestha V, Jha PK (2014) Preparation of graphene oxide nano-composite ion-exchange membranes for desalination application. RSC Adv 4:24662–24670
28. Ghaemi N, Zereshki S, Heidari S (2017) Removal of lead ions from water using PES-based nanocomposite membrane incorporated with polyaniline modified GO nanoparticles: performance optimization by central composite design. Process Saf Environ Prot 111:475–490. https://doi.org/10.1016/J.PSEP.2017.08.011
29. Ghazali AA, Abd Rahman S, Abu Samah R (2021) Preparation and characterisation of pineapple peel waste as nanoadsorbent incorporated into Pebax 1657 nanocomposite membrane for CO2/CH4 separation. Mater Today Proc 41:88–95. https://doi.org/10.1016/J.MATPR.2020.11.1012
30. Gnanasekaran G, Sudhakaran MSP, Kulmatova D et al (2021) Efficient removal of anionic, cationic textile dyes and salt mixture using a novel CS/MIL-100 (Fe) based nanofiltration membrane. Chemosphere 284:131244. https://doi.org/10.1016/J.CHEMOSPHERE.2021.131244
31. Goh PS, Ismail AF (2020) Chemically functionalized polyamide thin film composite membranes: the art of chemistry. Desalination 495:114655. https://doi.org/10.1016/J.DESAL.2020.114655
32. Gohil JM, Choudhury RR (2019) Introduction to nanostructured and nano-enhanced polymeric membranes: preparation, function, and application for water purification. Nanoscale Mater Water Purif 25–57. https://doi.org/10.1016/B978-0-12-813926-4.00038-0
33. Goswami R, Gogoi M, Borah A, Sarmah H, Ingole PG, Hazarika S (2021) Functionalized activated carbon and carbon nanotube hybrid membrane with enhanced antifouling activity for removal of cationic dyes from aqueous solution. Environ Nanotechnol Monit Manag 16:100492. https://doi.org/10.1016/J.ENMM.2021.100492
34. Haghighat N, Vatanpour V (2020) Fouling decline and retention increase of polyvinyl chloride nanofiltration membranes blended by polypyrrole functionalized multiwalled carbon nanotubes. Mater Today Commun 23:100851. https://doi.org/10.1016/J.MTCOMM.2019.100851
35. Hermia J (1982) Constant pressure blocking filtration laws—application to power law NON-Newtonian fluids. Trans Inst Chem Eng pp 183–187
36. Hosseini SS, Bringas E, Tan NR, Ortiz I, Ghahramani M, Shahmirzadi MAA (2016) Recent progress in development of high performance polymeric membranes and materials for metal plating wastewater treatment: a review. J Water Process Eng 9:78–110. https://doi.org/10.1016/J.JWPE.2015.11.005
37. Huang B, Gu H, Xiao K, Qu F, Yu H, Wei C (2020) Fouling mechanisms analysis via combined fouling models for surface water ultrafiltration process. Membranes (Basel) 10:1–12. https://doi.org/10.3390/MEMBRANES10070149
38. Ibrahim GPS, Isloor AM, Moslehyani A, Ismail AF (2017) Bio-inspired, fouling resistant, tannic acid functionalized halloysite nanotube reinforced polysulfone loose nanofiltration hollow fiber membranes for efficient dye and salt separation. J Water Process Eng 20:138–148. https://doi.org/10.1016/J.JWPE.2017.09.015
39. Januário D, Freitas E, Beluci NCL, Vidovix TB, Vieira MF, Bergamasco R, Vieira AMS (2020) Functionalization of membrane surface by layer-by-layer self-assembly method for dyes removal. Process Saf Environ Prot 134:140–148. https://doi.org/10.1016/j.psep.2019.11.030
40. Jian LY (2016) Asymmetric membrane. Drioli, E, Giorno, L (2016) Encyclopedia of membranes. Springer, Heidelberg, Verlag Berlin, pp 126–127
41. Jiang S, Ladewig BP (2020) Green synthesis of polymeric membranes: recent advances and future prospects. Curr Opin Green Sustain Chem 21:1–8. https://doi.org/10.1016/J.COGSC.2019.07.002

42. Johari NA, Yusof N, Lau WJ, Abdullah N, Salleh WNW, Jaafar J, Aziz F, Ismail AF (2021) Polyethersulfone ultrafiltration membrane incorporated with ferric-based metal-organic framework for textile wastewater treatment. Sep Purif Technol 270:118819. https://doi.org/10.1016/J.SEPPUR.2021.118819
43. Jun LY, Karri RR, Mubarak NM, Yon LS, Bing CH, Khalid M, Jagadish P, Abdullah EC (2020) Modelling of methylene blue adsorption using peroxidase immobilized functionalized Buckypaper/polyvinyl alcohol membrane via ant colony optimization. Environ Pollut 259:113940. https://doi.org/10.1016/J.ENVPOL.2020.113940
44. Kadla JF, Kubo S, Venditti RA, Gilbert RD, Compere AL, Griffith W (2002) Lignin-based carbon fibers for composite fiber applications. Carbon N Y 40:2913–2920. https://doi.org/10.1016/S0008-6223(02)00248-8
45. Kallel F, Bouaziz F, Chaari F, Belghith L, Ghorbel R, Chaabouni SE (2016) Interactive effect of garlic straw on the sorption and desorption of Direct Red 80 from aqueous solution. Process Saf Environ Prot 102:30–43. https://doi.org/10.1016/J.PSEP.2016.02.012
46. Kang X, Cheng Y, Wen Y, Qi J, Li X (2020) Bio-inspired co-deposited preparation of GO composite loose nanofiltration membrane for dye contaminated wastewater sustainable treatment. J Hazard Mater 400. https://doi.org/10.1016/J.JHAZMAT.2020.123121
47. Karmakar S, Roy D, De S (2021) Multicomponent transport model-based scaling up of long-term adsorptive filtration of MOF incorporated mixed matrix hollow fiber membrane: treatment of textile effluent. Chem Eng J 403:125103. https://doi.org/10.1016/J.CEJ.2020.125103
48. Kansedo J, Katheresan V, Lau SY (2018) Efficiency of various recent wastewater dye removal methods: a review. J Environ Chem Eng 6:4676–4697. https://doi.org/10.1016/J.JECE.2018.06.060
49. Koh LLA, Ashokkumar M, Kentish SE (2013) Membrane fouling, cleaning and disinfection. In: Tamime AY (ed) Membrane processing. Dairy and Beverage Applications. Blackwell Publishing, West Sussex, pp 73–102
50. Lau WJ, Ismail AF (2009) Polymeric nanofiltration membranes for textile dye wastewater treatment: preparation, performance evaluation, transport modelling, and fouling control—a review. Desalination 245:321–348. https://doi.org/10.1016/J.DESAL.2007.12.058
51. Lavanya C, Dhankar R, Chhikara S, Sheoran S (2014) Degradation of toxic dyes: a review. J Current Microbiol 3:189–199
52. Lin SH, Peng CF (1994) Treatment of textile wastewater by electrochemical method. Water Res 28:277–282. https://doi.org/10.1016/0043-1354(94)90264-X
53. Lopatina A, Anugwom I, Esmaeili M, Puro L, Virtanen T, Manttari M, Killioinen M (2020) Preparation of cellulose-rich membranes from wood: effect of wood pretreatment process on membrane performance. Cellulose 27:9505–9523. https://doi.org/10.1007/S10570-020-03430-0
54. Lora JH, Glasser WG (2002) Recent industrial applications of lignin: a sustainable alternative to nonrenewable materials. J Polym Environ 10:39–48. https://doi.org/10.1023/A:1021070006895
55. Ma B, Ding Y, Li W, Hu C, Yang M, Liu H, Qu J (2018) Ultrafiltration membrane fouling induced by humic acid with typical inorganic salts. Chemosphere 197:793–802. https://doi.org/10.1016/J.CHEMOSPHERE.2018.01.037
56. Marmagne O, Coste C (1996) Color removal from textile plant effluents. American dyestuff reporter, vol 85
57. Matsuura T (1993) Synthetic membranes and membrane separation processes. CRC Press, ISBN, p 9780367449612
58. Militano F, Giorno L (2016) Affinity membranes. In: Drioli E, Giorno L (eds) Encyclopedia of membranes. Springer Verlag Berlin Heidelberg, pp 15–19
59. Mondal S, Kumar Majumder S (2020) Fabrication of the polysulfone-based composite ultrafiltration membranes for the adsorptive removal of heavy metal ions from their contaminated aqueous solutions. Chem Eng J 401:126036. https://doi.org/10.1016/J.CEJ.2020.126036

60. Moussavi G, Mahmoudi M (2009) Removal of azo and anthraquinone reactive dyes from industrial wastewaters using MgO nanoparticles. J Hazard Mater 168:806–812. https://doi.org/10.1016/J.JHAZMAT.2009.02.097
61. Mulder M (1996) Basic principles of membrane technology. Kluwer academic publishers, London
62. Mulder M (1993) The use of membranes processes in environmental problems. In: Crespo JG, Boddeker KW (eds) (V. 272). Membrane processes in separation and purification. Springer-Science+Business Media, B. Portugal, pp 229–262
63. Nady N, Franssen MCR, Zuilhof H, Eldin MSM, Boom R, Schroën K (2011) Modification methods for poly(arylsulfone) membranes: a mini-review focusing on surface modification. Desalination 275:1–9. https://doi.org/10.1016/j.desal.2011.03.010
64. Nasir A, Masood F, Yasin T, Hameed A (2019) Progress in polymeric nanocomposite membranes for wastewater treatment: preparation, properties and applications. J Ind Eng Chem 79:29–40. https://doi.org/10.1016/J.JIEC.2019.06.052
65. Nechifor G, Voicu SI, Nechifor AC, Garea S (2009) Nanostructured hybrid membrane polysulfone-carbon nanotubes for hemodialysis. Desalination 241:342–348. https://doi.org/10.1016/J.DESAL.2007.11.089
66. Noreen S, Bhatti HN, Nausheen S, Sadaf S, Ashfaq M (2013) Batch and fixed bed adsorption study for the removal of Drimarine Black CL-B dye from aqueous solution using a lignocellulosic waste: a cost affective adsorbent. Ind Crops Prod 50:568–579. https://doi.org/10.1016/J.INDCROP.2013.07.065
67. Ong CS, Goh PS, Lau WJ, Misdan N, Ismail AF (2016) Nanomaterials for biofouling and scaling mitigation of thin film composite membrane: a review. Desalination 393:2–15. https://doi.org/10.1016/J.DESAL.2016.01.007
68. Panda GC, Das SK, Guha AK (2009) Jute stick powder as a potential biomass for the removal of congo red and rhodamine B from their aqueous solution. J Hazard Mater 164:374–379. https://doi.org/10.1016/J.JHAZMAT.2008.08.015
69. Pandey A, Singh P, Iyengar L (2007) Bacterial decolorization and degradation of azo dyes. Int Biodeterior Biodegradation 59:73–84. https://doi.org/10.1016/J.IBIOD.2006.08.006
70. Qalyoubi L, Al-Othman A, Al-Asheh S (2021) Recent progress and challenges of adsorptive membranes for the removal of pollutants from wastewater. Part II: environmental applications. Case Stud Chem Environ Eng 3:100102. https://doi.org/10.1016/j.cscee.2021.100102
71. Qing W, Liu F, Yao H, Sun S, Chen C, Zhang W (2020) Functional catalytic membrane development: a review of catalyst coating techniques. Adv Colloid Interface Sci 282:102207. https://doi.org/10.1016/J.CIS.2020.102207
72. Rahimpour A, Madaeni SS (2007) Polyethersulfone (PES)/cellulose acetate phthalate (CAP) blend ultrafiltration membranes: preparation, morphology, performance and antifouling properties. J Memb Sci 305:299–312. https://doi.org/10.1016/J.MEMSCI.2007.08.030
73. Ravve A (2012) Principles of polymer chemistry, 3rd edn, pp 1–801. https://doi.org/10.1007/978-1-4614-2212-9
74. Reddy PVL, Kim KH (2015) A review of photochemical approaches for the treatment of a wide range of pesticides. J Hazard Mater 285:325–335. https://doi.org/10.1016/J.JHAZMAT.2014.11.036
75. Rodrigues Filho G, Da Cruz SF, Pasquini D, Cerqueira DA, de Prado VS, de Assunção RMN (2000) Water flux through cellulose triacetate films produced from heterogeneous acetylation of sugar cane bagasse. J Memb Sci 177:225–231. https://doi.org/10.1016/S0376-7388(00)00469-5
76. Rodrigues Filho G, Ribeiro SD, Meireles C da S, Silva LG da, Ruggiero R, Ferreira MF, Cerqueira DA, Assunção RMN de, Zeni M, Polleto P (2011) Release of doxycycline through cellulose acetate symmetric and asymmetric membranes produced from recycled agroindustrial residue: sugarcane bagasse. Ind Crops Prod 33:566–571. https://doi.org/10.1016/J.INDCROP.2010.10.037
77. Saha S, Das C (2015) Analysis of fouling characteristics and flux decline during humic acids batch ultrafiltration. https://doi.org/10.4172/2157-7048.1000252

78. Samsami S, Mohamadizaniani M, Sarrafzadeh MH, Rene ER, Firoozbahr M (2020) Recent advances in the treatment of dye-containing wastewater from textile industries: overview and perspectives. Process Saf Environ Prot 143:138–163. https://doi.org/10.1016/J.PSEP.2020.05.034
79. Sarkar S, Chakraborty S (2021) Nanocomposite polymeric membrane a new trend of water and wastewater treatment: a short review. Groundw Sustain Dev 12:100533. https://doi.org/10.1016/J.GSD.2020.100533
80. Selatile MK, Ray SS, Ojijo V, Sadiku R (2018) Recent developments in polymeric electrospun nanofibrous membranes for seawater desalination. RSC Adv 8:37915–37938. https://doi.org/10.1039/C8RA07489E
81. Senthilkumar K, Saba N, Rajini N, Chandrasekar M, Jawaid M, Siengchin S, Alotman OY (2018) Mechanical properties evaluation of sisal fibre reinforced polymer composites: a review. Constr Build Mater 174:713–729. https://doi.org/10.1016/J.CONBUILDMAT.2018.04.143
82. Serbanescu OS, Voicu SI, Thakur VK (2020) Polysulfone functionalized membranes: properties and challenges. Mater Today Chem 17:100302. https://doi.org/10.1016/J.MTCHEM.2020.100302
83. Sharma S, Sarmah J, Soni V (2021) Classification and impact of synthetic textile dyes on Aquatic Flora: a review. Reg Stud Mar Sci 45:101802. https://doi.org/10.1016/J.RSMA.2021.101802
84. Sherugar P, Naik NS, Padaki M, Nayak V, Gangadharan A, Nadig AR, Déon S (2021) Fabrication of zinc doped aluminium oxide/polysulfone mixed matrix membranes for enhanced antifouling property and heavy metal removal. Chemosphere 275:130024. https://doi.org/10.1016/J.CHEMOSPHERE.2021.130024
85. Smith K (2013a) Commercial membrane technology. In: Tamime AY (ed) Membrane processing. Dairy and Beverage Applications. Blackwell Publishing, West Sussex, pp 52–71
86. Smith K (2013b) Development of membrane process. In: Tamime AY (ed) Membrane processing. Dairy and Beverage Applications. Blackwell Publishing, West Sussex, pp 1–15
87. Soleimany A, Hosseini SS, Gallucci F (2017) Recent progress in developments of membrane materials and modification techniques for high performance helium separation and recovery: a review. Chem Eng Process Process Intensif 122:296–318. https://doi.org/10.1016/J.CEP.2017.06.001
88. Song Y, Sun Y, Chen M, Huang P, Li T, Zhang X, Jiang K (2020) Efficient removal and fouling-resistant of anionic dyes by nanofiltration membrane with phosphorylated chitosan modified graphene oxide nanosheets incorporated selective layer. J Water Process Eng 34:101086. https://doi.org/10.1016/J.JWPE.2019.101086
89. Srinivasan A, Viraraghavan T (2010) Decolorization of dye wastewaters by biosorbents: a review. J Environ Manage 91:1915–1929. https://doi.org/10.1016/J.JENVMAN.2010.05.003
90. Subramanian MN (2017) Polymer blends and composites-Chemistry and technology. Scrivener publishing LLC, USA
91. Tagliavini M, Schäfer AI (2018) Removal of steroid micropollutants by polymer-based spherical activated carbon (PBSAC) assisted membrane filtration. J Hazard Mater 353:514–521. https://doi.org/10.1016/J.JHAZMAT.2018.03.032
92. Tang X, Guo T, Chang H, Yue X, Wang J, Yu H, Xie B, Zhu X, Li G, Liang H (2021) Membrane fouling alleviation by chemically enhanced backwashing in treating algae-containing surface water: from bench-scale to full-scale application. Engineering. https://doi.org/10.1016/J.ENG.2021.01.013
93. Tiraferri A, Kang Y, Giannelis EP, Elimelech M (2012) Superhydrophilic thin-film composite forward osmosis membranes for organic fouling control: fouling behavior and antifouling mechanisms. Environ Sci Technol 46:11135–11144. https://doi.org/10.1021/ES302861
94. Torres NH, Souza BS, Ferreira LFR, Lima AS, Dos Santos GN (2019) Real textile effluents treatment using coagulation/flocculation followed by electrochemical oxidation process and ecotoxicological assessment. Chemosphere 236. https://doi.org/10.1016/J.CHEMOSPHERE.2019.07.040

95. Thuyavan Y, Arthanareeswaran G, Ismail AF, Goh PS, Shankar MV, Reddy NL (2020) Treatment of synthetic textile dye effluent using hybrid adsorptive ultrafiltration mixed matrix membranes. Chem Eng Res Des 159:92–104. https://doi.org/10.1016/J.CHERD.2020.04.005
96. Upadhyaya L, Qian X, Ranil Wickramasinghe S (2018) Chemical modification of membrane surface—overview. Curr Opin Chem Eng 20:13–18. https://doi.org/10.1016/J.COCHE.2018.01.002
97. Vatanpour V, Mousavi Khadem SS, Masteri-Farahani M, Mosleh N, Ganjali MR, Badiei A, Pourbashir E, Mashhadzadeh AH, Munir MT, Mahmodi G, Zarrintaj P, Ramsey JD, Kim SJ, Saeb MR (2020) Anti-fouling and permeable polyvinyl chloride nanofiltration membranes embedded by hydrophilic graphene quantum dots for dye wastewater treatment. J Water Process Eng 38:101652. https://doi.org/10.1016/J.JWPE.2020.101652
98. Vigneshwaran S, Sundarakannan R, John KM, Johnson RDJ, Prasath KA, Ajith S, Arumugaprabu V, Uthayakumar M (2020) Recent advancement in the natural fiber polymer composites: a comprehensive review. J Clean Prod 277:124109. https://doi.org/10.1016/J.JCLEPRO.2020.124109
99. Waghmode TR, Kurade MB, Sapkal RT, Bhosale CH, Jeon BH, Govindwar SP (2019) Sequential photocatalysis and biological treatment for the enhanced degradation of the persistent azo dye methyl red. J Hazard Mater 371:115–122. https://doi.org/10.1016/J.JHAZMAT.2019.03.004
100. Wong S, Tumari HH, Ngadi N, Mohamed NB, Hassan O, Mat R, Amin NAS (2019) Adsorption of anionic dyes on spent tea leaves modified with polyethyleneimine (PEI-STL). J Clean Prod 206:394–406. https://doi.org/10.1016/J.JCLEPRO.2018.09.201
101. Wu G, Fan Q, Sun W, Yu Z, Jia Z, Ma J (2021) Regulatable pervaporation performance of Zn-MOFs/polydimethylsiloxane mixed matrix pervaporation membranes. Chinese J Chem Eng. https://doi.org/10.1016/J.CJCHE.2021.02.011
102. Xu W, Chen Q, Ge Q (2017) Recent advances in forward osmosis (FO) membrane: chemical modifications on membranes for FO processes. Desalination 419:101–116. https://doi.org/10.1016/J.DESAL.2017.06.007
103. Xu Y, Li Z, Su K, Fan T, Cao L (2018) Mussel-inspired modification of PPS membrane to separate and remove the dyes from the wastewater. Chem Eng J 341:371–382. https://doi.org/10.1016/J.CEJ.2018.02.048
104. Xu Y, Bao J, Zhang X, Li W, Xie Y, Sun S, Zhao W, Zhao C (2019) Functionalized polyethersulfone nanofibrous membranes with ultra-high adsorption capacity for organic dyes by one-step electrospinning. J Colloid Interface Sci 533:526–538. https://doi.org/10.1016/J.JCIS.2018.08.072
105. Ya V, Martin N, Choo KH, Chou YH, Lee SJ, Le NC, Li CW (2019) High-pressure electrocoagulation system with periodic air replenishment for efficient dye wastewater treatment: reaction dynamics and cost evaluation. J Clean Prod 213:1127–1134. https://doi.org/10.1016/J.JCLEPRO.2018.12.249
106. Yang S, Wang T, Tang R, Yan Q, Tian W, Zhang L (2020) Enhanced permeability, mechanical and antibacterial properties of cellulose acetate ultrafiltration membranes incorporated with lignocellulose nanofibrils. Int J Biol Macromol 151:159–167. https://doi.org/10.1016/J.IJBIOMAC.2020.02.124
107. Yang X, Reid MS, Olsén P, Berglund LA (2020) Eco-Friendly cellulose nanofibrils designed by nature: effects from preserving Native State. ACS Nano 14:724–735. https://doi.org/10.1021/ACSNANO.9B07659
108. Yao L, Zhang L, Wang R, Chou S, Dong Z (2016) A new integrated approach for dye removal from wastewater by polyoxometalates functionalized membranes. J Hazard Mater 301:462–470. https://doi.org/10.1016/J.JHAZMAT.2015.09.027
109. Zhang Y, Song Q, Liang X, Wang J, Jiang Y, Liu J (2020) High-flux, high-selectivity loose nanofiltration membrane mixed with zwitterionic functionalized silica for dye/salt separation. Appl Surf Sci 515:146005. https://doi.org/10.1016/j.apsusc.2020.146005

Chitosan and Chitosan-Based Nanocomposite Membranes in the Removal of Synthetic Dye from Effluent Wastewaters

Suseela Lanka, Sarvasai Nikhilesh Gantala, and Santhi Latha Pandrangi

Abstract The important groups of corrosive organic compounds are synthetic dyes that are often found as environmental contaminants due to their widespread application in industries. There are nearly more than one lakh dyes available commercially. Contamination is commonly caused by these substances that are mainly toxic, mutagenic. In recent times the public has observed the presence of colored effluents from the industry as causes of dangerous pollution. Contamination of water by the effluent dyes is a major environmental issue that should be addressed by proper treatment measures before the discharge of wastewater into the environment. The artificial origin and the aromatic structure of dyes make them resistant to many of the treatment procedures. The process is used in combustion (physical, chemical, and biological) to get the desired quality of water. In addition, there is a requirement to develop new removal strategies and decolorization methods that are more efficient, eco-friendly and have wide application in industries. Present treatment strategies are mainly focused on the use of natural materials as efficient adsorbents for dye removal. The most popular biosorbent is the cross linked chitosan-based hydrogels compared to other proposed materials. The objective of numerous fundamental studies is the polymeric matrices that can be reliable biosorbents for environmental applications.

Keywords Synthetic dyes · Carcinogenic · Effluent wastewater · Biosorption

---

S. L. Pandrangi (✉)
Department of Biochemistry and Bioinformatics, GITAM School of Science, Onco-Stem Cell Research Laboratory, GITAM Deemed to be University, Visakhapatnam, India
e-mail: dpandran@gitam.edu

S. Lanka
Department of Biosciences and Biotechnology, Krishna University, Machilipatnam, India

S. N. Gantala
Department of Biotechnology, Institute of Technology, GITAM Deemed to be University, Visakhapatnam, India

S. S. Muthu and A. Khadir (eds.), *Textile Wastewater Treatment*, Sustainable Textiles: Production, Processing, Manufacturing & Chemistry,
https://doi.org/10.1007/978-981-19-2852-9_9

1 Introduction

Effluent released from the textile industries into the water is one of the leading causes for water pollution. Among the various stages of the textile industry, the most pollutant stage is the dyeing of cloth materials. The wastes obtained from textile dyeing industries contain partially used or unused organic substances, strong color, and high chemical oxygen demand and biological oxygen demand (Samad et al. 2015). For the removal of dyes and other colored contaminants from wastewaters, several chemical, physical, physicochemical, and biological methods are currently in use. The most popular physicochemical method among them has been the removal of dyes using adsorption. It has been reported that many different types of adsorbents made of polymers are effective in removing colors from aqueous effluents (Samad et al. 2015). Polymers are made up of macromolecules that are abundant in nature and serve a variety of purposes in daily life. Biobased polymers (chitin, chitosan, keratin, silk fibroin, starch, etc.) and synthetic polymers (chitin, chitosan, keratin, silk fibroin, starch, etc.) are the two primary types of polymers (polypropylene, terephthalate, polyethylene, etc.). Biobased polymers are non-toxic, cost-effective, biodegradable, and eco-friendly [1]. Chitosan a "Hydrocolloid" is a biopolymer with a positive charge owing to its highly reactive amino groups. Coming to its chemical structure, chitosan is a linear polysaccharide made of N-acetyl D-glucosamine, and D-glucosamine units [2] (Fig. 1).

In food and paint applications, chitosan has its attention for having its versatile characteristics like degradability, pH-sensitivity, weak acid solubility, film-forming property, biocompatibility, low-cost, non-antigenic, non-toxic nature. Besides these, chitosan has its natural origin and has applications in the preparation of biodegradable films, coatings, blends, composites, nanocomposites, etc. [2].

1.1 Textile Industries-Effluent Wastewaters-Synthetic Dyes—Environmental Issues

A serious environmental and public health issue concerned with water pollution is the use and release of synthetic dyes. Synthetic dyes are one of the important

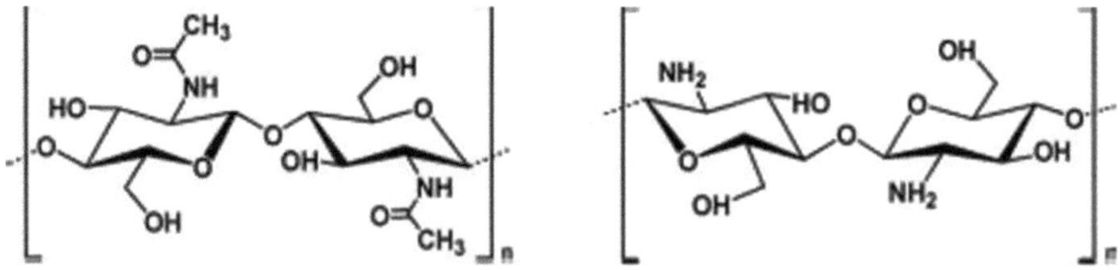

Fig. 1 Structures of chitin and chitosan [2]

classes of recalcitrant organic compounds that are widely available in nature owing to their widespread industrial application [3]. Industries like chemicals, gin and paper, textiles, paint, leather industry, clothing, packaging, food, pharmacy consume more amounts of water. Textile industries (54%) are the main contributors of the available dye water in nature across the globe. The high solubility of the dyes makes them the most common contaminants that can often be traced back to their industrial effluents [2]. Many industries use RhodamineBandCongo red as a common dye that must be removed from the wastewater due to its carcinogenic and mutagenic effects. Similarly, Cr (VI) has no biodegradable property and is highly toxic and reflects a major threat to human health that is derived from several industries like electroplating, textile, storage batteries, and leather tanning [12]. Textile dyes coupled with various other industrial pollutants form extremely toxic and potentially poisonous environmental contaminants that are in turn fatal to animals if discharged into aquatic systems without following proper treatment measures [1]. Textile industries produce nearly seven lakh tons of synthetic dyes all over the world every year and in India, and it comes to about 80,000 tons [2]. Public view toward the water quality is strongly influenced by color generating an increased number of complaints and concerns [2]. The removal of these dye impurities by introducing solid substances called adsorbents or biosorbents is a simple, useful, and effective process. The adsorbent used can be a mineral, organic, or biological [2] in origin. The most preferable adsorbent for industrial application is the carbon but the high cost restricted its use. Since 30 years, numerous methods have been studied for the production of effective, low-cost, and efficient biosorbents to eliminate dyes from contaminated water bodies [2]. Figure 2 shows the effects of dye effluents on the environment, plants, human beings, and aquatic animals.

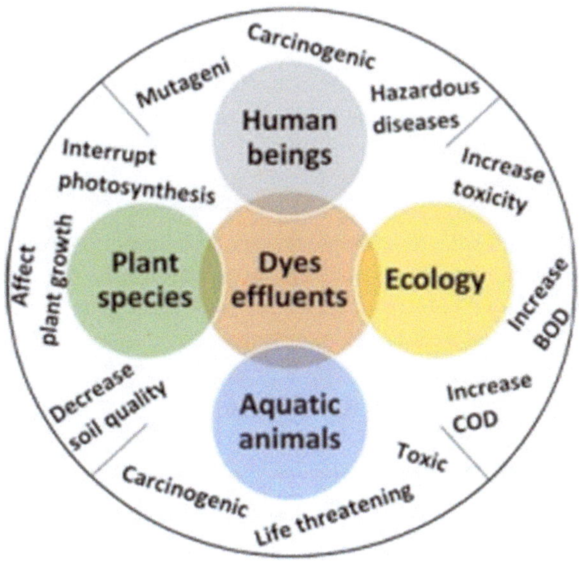

Fig. 2 The effects on the environment and the living beings by the dye effluents [1]

2 Role of Chitosan and Chitosan-Based Nanocomposite Membranes

In general, traditional wastewater treatments are composed of mechanical, physical, chemical, and biological processes and operating operations that remove soluble contaminants from insoluble particles and wastewater and have legal inactivation. Among the various machining measures quoted to remove dye, it is common in the industrial field and is commonly adopted in the industrial sector for economic reasons. Efficient, industrialized and environmentally friendly new removal strategies and decoloring methods. Many methods, liquid grades, and adsorption processing are selectable procedures, liquid grades, and adsorption processing in the therapeutic facility, and the adsorption process for removing the dye proposed by the second (final) step, the liquid grade, and the adsorption process are selecting procedures and can be used to remove various types of color materials.

Activated carbon is widely used on an industrial scale as well as for dye removal. Used to adsorb pollutants from wastewater as well as drinking water sources such as rivers, lakes, and reservoirs. Although activated carbon is the best material, its high material and processing cost limits its widespread use. To overcome this, many approaches have been explored to develop inexpensive and effective new materials such as chitosan-based materials [3]. Used as biosorbents over the past 20 years, these materials have been primarily used for metal chelation and dye removal in water and wastewater treatment. In fact, chitosan has a very high affinity for many types of dyes including metals and metalloids as well as direct acids, 2444 mediators and reactive dyes. In a comprehensive review, biosorbent for chitosan emerged as a promising alternative to conventional sorbents used for bleaching, metal chelation or extraction, and organic matter removal. Nutraceuticals and cosmetics, water and wastewater treatment sectors are the next market for chitosan development [3]. Chitosan is a low-cost, renewable, non-toxic, and biodegradable resource, with unique properties such as hydrophilicity. It is the best alternative and eco-friendly agent for its characteristics such as the presence of polyelectrolytes at acidic pH, high reactivity, coagulation, aggregation and adsorption properties, reactive hydroxyls, and most amine groups in the macromolecular chain. The use of wastewater treatment is justified as two different important benefits. First, the excellent binding of the contaminants and excellent selectivity, and the second actually chitosan is a strong affinity to interact with the contaminated material present in the trace level [3].The materials can be used in solid form to eliminate pollutants from wastewater dissolved in water and adsorption process or liquid state, i.e., dissolved in acid media. Earborbar filtration). Among these procedures, the bio fusion for chitosan hydrogels is one of the most common methods of dye removal [3].

2.1 Coastal Areas-Sea Food Waste—Crustacean Shell Waste—Environmental Impact

Shells of crabs, shrimps, lobsters, and krill (crustaceans) are the sources of chitosan that are produced in millions of tons every year due to widespread consumption by the public as protein-rich seafood worldwide. Crustaceans' shells and the other parts that are not edible for consumption form one-half of their total body mass serve as a consistent source of chitin and chitosan. After cellulose, chitin is the second largest natural polysaccharide on earth, but its insolubility in many of the regularly used solvents makes it not fit for fabrication of products or as a food commodity. The deacetylation of chitin obtained from crustaceans waste results in the production of chitosan, the value-added product that had several industrial applications such as food, agriculture, and biomedicine, etc. The United States Food and Drug Administration has recommended chitosan as a generally recognized as safe material for use in various applications [2].

2.2 Value-Added Products from Crustacean Shell Waste (Chitin and Chitosan Extraction) and Their Applications

Generally, chitin is extracted from the exoskeleton of crustaceans by acid treatment followed by alkali treatment to separate calcium carbonate and protein and the process mainly involves three steps—demineralization, deproteinization, and decolorization. The demineralization step involves the removal of calcium carbonate and calcium chloride, which are vital inorganic compounds in the crustacean outer frame [2]. Dilute hydrochloric acid is usually used for the digestion reactions, and then followed by filtration, washing, and drying. In the removal of mineral components, the emission of carbon dioxide is an important indication [2]. In step two, the deproteinization is done using an alkaline solution, i.e., with the dilute sodium hydroxide (NaOH). The filtrate so obtained is then filtered, washed, and dried. The final step involves the purification process in which color pigments like astaxanthin and β-carotene are removed in a process called depigmentation, and the process is done using different organic and inorganic solvents such as sodium hypochlorite, acetone, and hydrogen peroxide. For obtaining pure chitin, the traditional process for oxidation of chitin is done at high temperatures using a solution of concentrated sodium or potassium hydroxide. The acetyl groups are removed from the polymer chain of chitin leading to the formation of the amine group [2].

2.3 Production of Chitosan-Based Nanocomposite Membranes

In the past studies, many have reported the chemical modification of chitosan either by coupling with small molecules or grafting with polymers, for the better utilization of inherent properties of chitosan. Recently, chitosan-based nanocomposites have emerged, containing polymer, and nanoparticles [2]. The dispersed nanomaterials present in the chitosan polymer matrix will improve the physical, mechanical, and thermal stability of chitosan; however, it also endogenizes the compound with its inherent qualities, like the high surface area or the variable physicochemical properties [2].

Chitosan and Metal/Metal Oxide

Novel metals like silver, an effective antimicrobial agent, can be synthesized on the pot in a chitosan matrix due to its metal ion chelation capability of chitosan. Owing to the antimicrobial properties of the silver nanoparticles, the chitosan-silver coatings on metal, glass, and wood surfaces have been studied. Silica nanoparticles in chitosan are observed to be selectively absorbing the rare earth elements, while zinc oxide nanoparticles are reported to produce photo-catalytically reactive oxygen species, leading to high anti-diatom and antibacterial activities under sunlight [2].

Chitosan and Carbon Materials

CNT, graphene, graphene oxide are the carbon materials that are well studied. Introducing the graphene oxide into chitosan enhances the mechanical properties of chitosan. In addition, it also enhances the antimicrobial and pollutant removal abilities. Chitosan is being applied to promote practical applications of CNTs by improving their solubility and by reducing toxicity as it is a biocompatible polymer [2].

Chitosan and Polymer Mixture or Copolymer

Chitosan is usually combined with different types of polymers to form a mixture or a co-polymer that will uplift the properties for specific applications. Polymers having activated carboxyl groups can be covalently bonded with chitosan because of the presence of chemically reactive primary amine groups in chitosan. Carbodiimide crosslinking reagents are used to form an amide bond between amine and carboxyl groups [2].

Chitosan and Clay Composites

Clay is a fine-grained clay that contains metal oxides or hydroxides as well as traces of organic stuff. Clay is widely employed in numerous applications such as water purification, odor absorber, and lubricant in the construction industry due to its particle size and outstanding colloidal characteristics. Because of the structural features of chitosan, nanotubes, and platelets may be employed to introduce the active ingredient and increase the creation of anticorrosive coating barriers. Because of the interactions between chitosan and clay, composites are frequently formed through adsorption, gelation, or intercalation. Clay in chitosan-clay nanocomposites has distinct features, such as particular species absorption or hemostatic capabilities while chitosan can provide a higher loading or a network of chitosan connected as a cross or clay scaffolding [2].

2.4 Application of Chitosan and Chitosan-Based Nanocomposite Membranes in Synthetic Dye Removal from Textile Industry Effluents

The application of synthetic dyes in industries is increasing day by day, especially in the textile industry, and the release of which into the water without proper treatment results in serious water pollution. The textile industry produces over 7×10^5 tons of synthetic dyes annually worldwide. Most synthetic dyes are released into ecosystems without proper treatment, causing global environmental problems. Removing the dye from the water is difficult because of its inert properties and presence at low concentrations. Chitosan is an inexpensive biopolymer with applications in water purification due to the greater number of its reactive amine and hydroxyl groups [2]. The electrostatic interaction between negatively charged dye ions and protonated amino groups of chitosan produces the acidic dye adsorption chitosan and modified chitosan. The dye removal from wastewater using chitosan nanocomposites involves the use of several methods such as physical adsorption, ion exchange, hydrogen bonds, hydrophobic attractions, and chemical bonding. In the metal ion uptake, chitosan-based composite fibers and pellets of zinc photocatalysts supported with metallic and bimetallic nanoparticles are being synthesized. The chitosan-based composite fibers are used as dip-catalysts in the reduction of nitrophenols and azo dyes like methyl orange and congo red [2] (Table 1).

3 Conclusions

The most available chemical compounds that are organic and vital biomass resources on the planet are cellulose and the chitin polysaccharides that is synthesized by plants and lower animals, respectively, and chitosan is reactive and useful because of its NH_2 groups at the C2 atom and hydroxyl groups at C3 and C6 positions. The textile industry produces enormous amounts of dye effluent with extremely and potentially toxic properties along with a wide range of other industrial pollutants, which have been linked to environmental problems and fatal diseases in animals [1]. Plant growth gets inhibited by the dyes affecting the food chain, resulting in bioaccumulation and the repulsion can cause mutagenicity, toxicity, and carcinogenicity. Chitosan-based adsorbents have been in increased use for dye adsorption due to biocompatibility,

Table 1 Chitosan-based nanocomposites for the removal of synthetic dyes from textile wastewater

S. No	Chitosan-based Nanocomposites	Synthetic dye removed from Textile Industry effluents	References
1	CS/MoO_3/TiO2	Methyl orange	Magesan et al. [11]
2	CS/AgNPs	Methyl orange	Nithya et al. [4]
3	CS/lignin/titania	Brilliant black	Masilompane et al. [6]
4	CS/Bio-silica 1	Acid red 88	Darvishi et al. [7]
5	CS/SnO2 intercalated polyaniline (PANI)	Methylene blue reactive yellow 15 20	Karpuraranjith et al. [10]
6	CS/SnO2 9	Methyl orange, Rhodamine—B	Gupta et al. [8]
7	CS/Polyaniline/CdS 1	Reactive blue-19	Rasoulifard et al. [5]
8	CS, Zirconium Tungstate	Reactive blue-21, Reactive CS/Zirconium red-141, Rhodamine- 6G and Binary mixture; RB + RR, RR + RH, RB + RH	Vanamudan et al. [9]

duality. It has low degradability, low toxicity, and convenient chemical modification, which shows high adsorption capacity in neutral solutions. Dye adsorption using chitosan-based nanostructures have been studied and still more detailed studies on many other dyes or organic contaminants should be made. There is a need for experimentation to identify the potential of nano chitosan for dye adsorption.

4 Future Directions

It is difficult to differentiate between composites due to the various materials used in the formation of chitosan composites. Some of the factors such as various experimental conditions, scarcity in the data, and differences in the given data are also associated with difficulties in making comparisons. Few highlights are made in the current paper that may help for future research. The cost factor is prioritized because low production costs combined with high removal efficiencies are truly acceptable. It primarily decides whether the adsorbent chosen is suitable for use on a wide scale. Chitosan-based materials are cost-effective because they are simple to make and use inexpensive chemical reagents. Lab-scale studies should not be the sole focus of research. When contrasted to laboratory solutions, industrial effluents contain several contaminants. It is better to design a multi-purpose adsorbent capable of effectively removing a variety of contaminants. Though the multi-purpose adsorbent complicates the adsorbent's characterization procedure, it really represents a more realistic approach that suits the design for multiple concurrent industrial applications.

References

1. Qamar SA, Ashiq M, Jahangeer M, Riasat A, Bilal M (2020) Chitosan-based hybrid materials as adsorbents for textile dyes–a review. /https://doi.org/10.1016/j.cscee.2020.100021
2. Kumar S, Ye F, Dobretsoy S, Dutta J (2019) Chitosan nanocomposite coatings for food, paints, and water treatment applications. Appl Sci 9:2409. https://doi.org/10.3390/app9122409
3. Crini G, Torri G, Lichtfouse E, Kyzas GZ, Wilson LD, Morin-Crini N (2019) Dye removal by biosorption using cross-linked chitosan-based hydrogels. 17:1645–1666
4. Nithya A, JeevaKumari HL, Rokesh K, Ruckmani K, Jeganathan K, Jothivenkatachalam K (2015) A versatile effect of chitosan-silver nanocomposite for surface plasmonic photocatalytic and antibacterial activity. 153:412–422
5. Rasoulifard MH, Seyed Dorraji MS, Amani-Ghadim AR, Keshavarz-babaeinezhad N (2016) Visible-light photocatalytic activity of chitosan/polyaniline/CdS nanocomposite: kinetic studies and artificial neural network modeling. Appl Catal A Gen 514:60–70
6. Masilompane TM, Chaukura N, Mishra SB, Mishra AK (2018) Chitosan-lignin-titania nanocomposites for the removal of brilliant black dye from aqueous solution. 120:1659–1666
7. Darvishi Cheshmeh Soltani R, Khataee AR, Safari M, Joo SW (2013) Preparation of bio silica/chitosan nanocomposite for adsorption of a textile dye in aqueous solutions. 85:383–391
8. Gupta VK, Saravanan R, Agarwal S, Gracia F, Khan MM, Qin J, Mangalaraja RV (2017) Degradation of azo dyes under different wavelengths of UV light with chitosan-SnO2 nanocomposites. 232:423–430
9. Vanamudan A, Sadhu M, Pamidimukkala PS (2018) Nanostructured zirconium tungstate and its bionanocomposite with chitosan: wet peroxide photocatalytic degradation of dyes. 85:74–82
10. Karpuraranjith M, Thambidurai S (2016) Biotemplate-SnO_2 particles intercalated PANI matrix: enhanced photo catalytic activity for degradation of MB and RY-15 dye. 133:108–118
11. Magesan P, Sanuja S, Umapathy MJ (2015) Novel hybrid chitosan blended MoO_3–TiO_2 nanocomposite film: evaluation of its solar light photocatalytic and antibacterial activities. 5:42506–42515
12. Gu F, Geng J, Li M, Chang J, Cui Y (2019) Synthesis of chitosan—lignosulfonate composite as an adsorbent for dyes and metal ions removal from wastewater. 21421–21430

Application of Bio-Based and Inorganic Composite Materials to Remove Heavy Metals from Textile Effluent

Rutto Toroitich Evans, Seth Apollo, Leonard Genson Murithi, and Njogu M. Kimani

Abstract Textiles are among the basic needs for human. The manufacture of textile in various industries involves various operations. The large volumes of water used in the textile processing are later generated as wastewater, containing various dyes, heavy metals and other contaminants. This water pollutes soil and water sources if disposed without proper treatment. This chapter reviews the treatment of textile effluent to eliminate heavy metals and puts an emphasis on the adsorption process as a cheap and efficient alternative. Adsorbents made of inorganic and organic/bio-based materials as well as their composites are discussed. Bacteria, algae, fungi, chitin and, chitosan, cellulose and lignocellulosic materials are discussed as biosorbent. Inorganic adsorbents such as clay and zeolites are also reviewed. The interest of this chapter is on the use of composites of inorganic bio-based materials to eliminate heavy metals from textile wastewater by adsorption. Clay–cellulose, cellulose–zeolites and clay–chitosan composites are discussed as adsorbents. Combination of biosorbent with inorganic adsorbent is interesting as the important property of recyclability of a good adsorbent is achieved.

Keywords Textile wastewater · Heavy metals · Adsorption · Inorganic bio-based adsorbent

1 Introduction

Textiles are among the basic needs for human beings. Textile processing industries, through diverse production operations, convert fibres into yarn. This yarn is finally transformed into fabrics or allied commodities by dyeing. The use of sizing materials, auxiliary chemicals and numerous dyes during the wetting process leads to generation of wastes. The wastewater adversely affects the environment, therefore making it

R. T. Evans · S. Apollo (✉) · L. G. Murithi · N. M. Kimani
School of Pure and Applied Sciences, Department of Physical Sciences, University of Embu, P.O Box 6, Embu 60100, Kenya
e-mail: sethapollo@gmail.com

S. S. Muthu and A. Khadir (eds.), *Textile Wastewater Treatment*, Sustainable Textiles: Production, Processing, Manufacturing & Chemistry,
https://doi.org/10.1007/978-981-19-2852-9_10

necessary for the wastewater from wet process operations to be completely treated prior to being discharge to the ecosystem [42].

Large volumes of water are used in textile industries which in turn generate highly polluted wastewater whose composition is creating a concern to wastewater treatment plants. Self-degradation of pollutants in wastewater has been compromised by the presence of toxic metallic ions, high colour, and dyes in the textile effluents [25]. Textile effluent contains toxicants, organics, inhibitory compounds, colour, surfactants, salts, chlorinated compounds, among other pollutants [15]. Every textile industry is unique with respect to chemicals used, technology and type of production.

The concentration of various contaminants in textile wastewaters, therefore, varies depending on the practices taken to manage the wastewater and their dilution after production. Textile processing applies various chemicals which depend on the type of the inputs and also the end products [43]. Some common main processes in textile industries include dyeing, singeing, bleaching, desizing, printing and finishing. Mineral oils and suspended solids are present in textile wastewater. Phenols are produced in the wet finishing process, while bleaching by use of solvents produces halogenated organics [11].

Creation of more textile industries leads to increase in environmental risk due to increase in wastes containing heavy metals. Heavy metals like manganese, cadmium, zinc chromium, copper, lead and many others originate from dye processing during textile production [27]. Pollution of water by cations of these toxic elements makes it very toxic. Accumulation of heavy metals may also occur in the environment through the food chain. This accumulation is harmful for mankind. Pollution of water assets by ions of heavy metals is a danger to public health and the environment because these metal ions bioaccumulate and are non-degradable [52]. High quantities of water are estimated to be used in textile industries than any other industry in the globe and almost all of its effluent are polluted to a high extent. It therefore causes an environmental concern [30]. Metallic elements are required for various dyes to impart colour. Chromium, lead, zinc, nickel, cobalt, manganese and copper are the commonly applied heavy metals which are each linked to various textile dyes. The discharge of wastes containing heavy metal ions into soils and water bodies is a potential risk to human well-being and also the environment [53]. Humans when exposed to dyes have had complications such as lung and skin irritation, nausea and headaches.

Many treatment techniques extensively investigated in the past for removing heavy metal ions from solutions. They include adsorption on activated carbon, membrane technologies, activated sludge process and ion exchange [57]. The techniques have various merits and demerits unique to each method. Various methods employed in the past to treat industrial effluent have been successful but not much efficient. Reverse osmosis requires highly trained personnel for its application. Some of the methods are very expensive and produce waste sludge which are difficult to manage. Adsorbing metal ions by use of modified carbons, silica gel, and commercial alumina sand blended materials has also been studied [48, 69]. Adsorption is an efficient method of reducing the mass per unit volume of dyes dissolved in the effluent therefore removing colour. Organic and inorganic compounds can be removed efficiently from an effluent by adsorption.

1.1 Textile Wastewater Characteristics

Due to various operations in the industries that produce textiles, a wide range of dyestuffs and chemicals are utilized. It is not all of these components that end up in the final product, but some are found in the wastewater which should be treated before it is finally discharged [59]. Textile effluents contain a mixture of both inorganic and organic chemicals of different characteristics and amounts. Among the characteristics of textile effluents include high chemical oxygen demand, total solids both dissolved and suspended solid, biochemical oxygen demand and low values of dissolved oxygen with strong colour [42].

Industrial effluent, generated by operations intended to produce paper and, food processing factories, slaughterhouses, activities to mine and process coal can add significant colour to the stream in which the wastes are discharged. True colour results when chemicals such as dyes, minerals and organics dissolve in water. It is necessary to eliminate colour in order to make water fit for industrial and domestic use. Coloured industrial effluent needs to be treated to remove colour prior to discharging it into streams as even a small amount of colour is seen as water contaminant, which leads to aesthetic and health problems to the consumers. Although colourless water is perceived to be clean, it may not be necessarily safe for human consumption [31]. pH is the measure of acid–base equilibrium attained by various compounds when dissolved. CO_2, CO_3 and HCO_3 equilibrium system controls pH in most natural waters by an equilibrium system between CO_3, HCO_3 and CO_2. The pH of composite textile wastewater mostly ranges from 7.0 to 9.0 [7].

The measure of oxygen consumed to oxidatively break the organic material contents of a sample in acidic conditions, using strong oxidizing agents, is referred to as the chemical oxygen demand [38]. Chemical oxygen demand is measured in milligrams per litre indicating the quantity of oxygen consumed in a litre of solution. The rate of pollution of an effluent or water varies proportionally with the chemical oxygen demand of the effluent or water [65]. Textile wastewater is characterized by high chemical oxygen demand ranging from 150 to 12,000 mg per litre. These values are much higher than the permissible levels of below 60 mg per litre as per the Iranian standards [15].

The quantity of oxygen used by bacteria to stabilize the organic matter content of a sample when temperature and time conditions are controlled is called the biochemical oxygen demand [35]. Biochemical oxygen demand measures the organic pollution of water which can undergo biological degradation. Biochemical oxygen demand is measured in milligrams of oxygen per litre of solution or sample. High biochemical oxygen demand in textile wastewater results from the presence of various toxic chemicals and synthetic dyes which contains sulphur, nitrates, alkalis, acids, surfactant-dispersing agents and hydrogen peroxide used in the production process [62]. The biochemical oxygen demand of composite textile wastewater ranges from 80 to 6000 mg of per litre [31]

Toxic heavy metals are among the major contaminants of concern in textile wastewater. Heavy metals present a danger to human beings when present in the textile industry. Higher concentrations of these heavy metals in textile industrial effluent have negatively affected our environment [71]. Most common heavy metals in textile wastewater include lead, chromium, copper and cadmium. Copper and chromium originate from dye processing in the textile industry [27].

1.2 Heavy Metals Occurrence in Textile Effluent

An element whose atomic weight ranges between 200.6 and 63.5 and whose specific density divided by density of water is greater above 5.0 is termed as a heavy metal [22]. These groups of elements sometimes occur naturally in the environment. They are toxic and have a high densities even when their concentrations are extremely low [6].

Effluents from textile treatment industries contain a variety of contaminants, among them, heavy metals. The key contaminants in the dyeing process include numerous chemicals like salts, heavy metals and sulphides, which are added as enhancers for dye adsorption onto the fibres [42].

Heavy metals, specifically, lead, cadmium, copper and chromium, are commonly used in textile dyes for production of colour pigments. Naturally, heavy metals can exist within the structures of textile or they can permeate into fibres of textile during processing, via agents used for protection during storage. Heavy metals, after being channelled to the environment, become highly toxic because they can bioaccumulate in the bodies of living organisms through food chain. Natural water bodies, aquatic life and also soil are not an exemption [13].

1.3 Environmental Impacts of Heavy Metals

Heavy metals enter the ecosystem, that is soil, water and, air, through anthropogenic activities and through natural processes. Due to their prevalence in the world and their toxic effects, heavy metals like Cd, Al, Sb, Fe, Cr, Ag, Be, Pb, Zn, Ni, Se, As and Hg have been classified as primary pollutants and are considered as major contaminants [28]. Almost all heavy metals or metalloids are potential toxicants depending on the duration of exposure and the dose.

Heavy metals can originate anthropogenic or natural sources. The geological or natural sources include volcanic eruptions and weathering of rocks bearing the metals. However, in the recent past, high rate of urbanization and industrial growth has increased the share of heavy metallic ions coming as a result of anthropogenic sources. Industrial activities and agricultural and activities and mining are anthropogenic operations that introduce heavy metals to our environment [9].

1.4 Occurrence of Heavy Metals in Water

Water, being a universal solvent, dissolves various environmental pollutants, which may be organic or inorganic chemical contaminants. Effluents from textile processing contain various dyes and complex compounded pollutants of heavy metals. These high molecular weight compounds are not readily biodegradable. Industrial waste generated directly from industries contains heavy metals and needs to be treated for discharge into water sources to reduce pollution [40].

1.5 Pollution of Soil

Pollution of soil can be intentional or unintentional. Intentional pollution of soil arises from animal manures, wastewater irrigation, pesticides, fertilizers, mine ore waste, spillage of petroleum distillates, sewage sludge, leaded paint coal, combustion residues, and dumping of wastes. Using untreated sewage for irrigation has led to increase in heavy metal concentration in our agricultural lands. Crops which are later consumed by human beings uptake these heavy metals [46]. Non-intentional soil pollution brought results from overflow of contaminated water from rivers and seas and at times when trucks transporting toxic chemicals are involved in accidents. The heavy metals in their natural forms or when modified are non-biodegradable, a property that enables them to persist in soils and therefore ruins the ecosystem [26].

Heavy metals as well affect the ability of organic contaminants to biodegrade, making their degradation slow. This doubles the effect of environmental pollution. Presence of heavy metals in soil causes harm to the entire biosphere. This is not only a potential hazard the plant that absorbs it but also to the animals, birds and human beings that consumes whole or part of the plant. Heavy metals alter the quality of the soil in question like the natural chemistry, how porous the soil is, colour and pH therefore influencing soils quality and subsequently causing water contamination [18].

The carcinogenicity of heavy metals can be classified on Table 1.

2 Heavy Metals Removal from Aquatic Environment

Since they are harmful to the aquatic environment and the ecosystem at large, the fact is that they do not naturally degrade but accumulate day by day. Heavy metals

Table 1 Classification of the carcinogenicity pf heavy metals

Group	Carcinogenicity level on human beings	Proof/evidence	Classification of heavy metal
I	Carcinogenic	Sufficient evidence in humanity	Aluminium compounds Nickel and nickel compounds Compounds of chromium (VI) cadmium
IIA	Potentially carcinogenic	Little evidence in humans, plenty in animal poisoning	Inorganic lead compounds
II B	Potentially carcinogenic	Little evidence in humans, no significant evidence in animals	Cobalt, vanadium pentoxide metallic nickel and alloys, molybdenum trioxide
III	Potential not classified	No sufficient proof in humans and animals	Mercury and inorganic mercury compounds Copper Selenium and selenium compounds Arsenic organic arsenic compounds that cannot be metabolized by human

Gupta et al. [26]; Muchuweti et al. [46].

removal from effluents from industries and aquatic environment is becoming the interest of many researchers. Particular protocols control the disposal of pollutants in wastewaters. High enforcement removal method is needed to avoid some properties that inhibit the removal of wastewater [22]. Each technique has its own shortcomings. Some of the available technologies for sequestering heavy metals from effluents include supercritical fluid extraction, ion exchange, adsorption, precipitation, filtration, electrodialysis [51], the electrochemical process, microbial system, membrane bioreactors and advanced oxidation [4]. Methods mentioned above are widely classified into biological, chemical or physical sections with regard to the nature of pollutants the treatment methods efficiently remove when used [22].

Physical methods including adsorption, irradiation, ion exchange and filtration are among the mostly applied methods treatment of effluent from textile industries. The low cost incurred when applying physical methods as well as the high removal efficiency has led to their wide application [25]. Generally, bioremediation is achieved by biosorption, biotransformation, biomineralization using microorganisms and bioaccumulation processes. Biosorption, among them, is promising as the technique plays a central role to remove heavy metals. Biosorption applies a physical–chemical pathways of uptake and is a reversible process [44].

Biological ways of treating water to eliminate heavy metals and sediments are categorized into biosorption, a process that uses dead biomass and their metabolites and constituents as mediators to bind metals via physico-chemical adsorption and bioaccumulation, where living microorganisms are used to bind metals within their cells [36].

Chemical treatment methods include chemical oxidation and coagulation/flocculation. Coagulation and flocculation practices using coagulants including lime [Ca $(OH)_2$], alum or aluminium sulphate [$Al_2(SO_4)_3.18H_2O$] and ironic salts, like ironic sulphate [$Fe_2(SO_4)_3.7H_2O$] or ironic chloride [$FeCl_3$.7H2O], are efficient methods of chemically treating wastewater. They are commonly applied to treat textile effluent as their management is easy [42].

2.1 Application of Adsorption in Textile Wastewater

Different conventional methods, for example coagulation, chemical precipitation, solvent extraction, electrochemical removal, membrane filtration and ion exchange, can be used to treat aquatic media lowering the concentrations of heavy metals. Despite their success, these methods have some drawbacks for example, consumption of energy, incomplete removal, being sensitive to operating conditions, low efficiency and generation of toxic sludge making costly the disposal of wastes generated [19]. Many approaches aimed at reducing the amount of effluents and treating the effluents have been proposed, in order to overcome these shortcomings [30]. The majority of the methods are adsorption-based processes, because it is cost effective and easy to operate apart from having a great impact on transport of heavy metals, their toxicity as well as availability in biological systems [24, 60].

Adsorption occurs by accumulation of an adsorbate (solute) on the surface of a solid material forming an atomic or molecular film. There are various adsorbents from different materials which can be used with or without modification to eliminate toxic heavy metal cations from wastewater. Adsorbents can either be naturally inorganic or organic. The most commonly used adsorbents are zeolites, biomaterials, clay minerals, activated carbons and industrial solid waste [19]. The adsorption process is a heterogeneous process with an initial rapid adsorption rate followed by a slower rate. This was particularly noticed for the Fe^{3+} and Zn^{2+} cations adsorption on zeolite (Fig. 1b, d). In the initial stages, the adsorption sites are available, the cations interact easily with the sites, and hence, a higher rate of adsorption was observed as illustrated in Fig. 1 adopted from [55].

2.1.1 Biosorbents

The use of biological materials to accumulate and remove heavy metal ions species, particulates, or compounds from a solution is termed as biosorption. These biosorbents, such as fungi, algae and bacteria, possess unique properties to sequester metal ions and therefore are applicable in reduction of the concentrations of ions of heavy metals in aqueous media to ppb from ppm levels. Biosorbents are effective

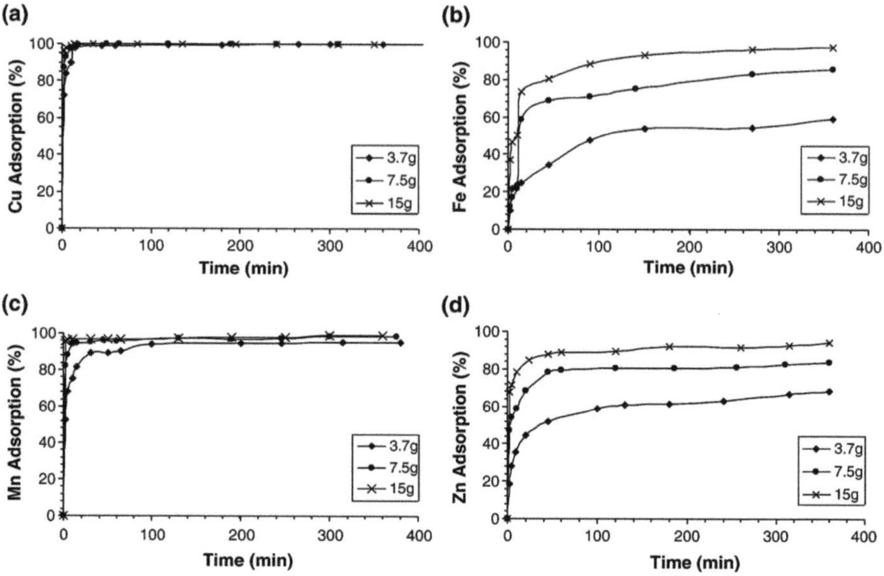

Fig. 1 Kinetics of heavy metal ion adsorption using zeolite (grain size: 1–3 mm, 100 ml of solution, pH 3.5)

in elimination of dissolved ions from dilute complex aqueous media quickly and efficiently, thence are ideal materials treat high volumes of lowly concentrated complex wastewaters [68].

2.1.2 Bacteria Biosorbent

Out of the various examples of biosorbents that is, yeasts, bacteria, fungi, weeds, and agricultural wastes, bacteria are predicted to remove heavy metals from industrial effluents, with high effectiveness and reliability. The small size of bacteria and large surface area enables it to be utilized as biosorbents. Bacteria are able to multiply under specified conditions and can persist and adapt to a change in conditions of the environment making good adsorbents. Bacteria may be specific to one element or able to adsorb many elements depending on the species [68].

Table 2 outlines examples of bacteria utilized in heavy metal removal from water.

Fungal biosorbents.

Fungal cells, both dead and living, have a remarkable a potential to uptake toxic heavy metals. As an example, removal of lead (II) ions was studied using *B. cinerea* and the findings of the study confirmed that heated inactivated B. *cinerea* is a cheap biosorbent that can selectively, easily and effectively remove lead (II) ions from aqueous solutions [5].

Table 2 Outlines examples of bacteria utilized in heavy metal removal from water

Bacteria type(name)	Heavy metal removed	Adsorption capacity(mg/g)
Gloecapsa gelatinosa	Lead	256.41
Streptomyces rimosus	Lead	135
Spirulina sp	Cadmium	99.5
Streptomyces pimprina	Cadmium	30
Pseudomonas putida	Copper	96.9
Staphylococcus xylosus	Chromium (VI)	143

Gautam et al. [23]

Cellulose biosorbents.

Research in the last decade has focused on the utilization of cellulosic waste materials as adsorbents. Cellulose is extracted from agricultural waste materials like banana petioles, coconut husk and cassava bagasse. Although various natural fibres have been extensively investigated, an exploration is needed on the use of rice husk as a cheap natural source of cellulose [34]. Accumulation of heavy metals is associated with the presence of functional groups in the cellulose structure which include hydroxyl (phenol) or carboxylic, aromatic rings, carboxylic phenol and ester [64].

Chitosan and chitin.

Chitin and chitosan are emerging biopolymers used for biosorption of dyes and heavy metals. Chitosan and chitin are abundant, biodegradable and renewable resources. Many of the studies on chitosan and chitin have shown that biosorbents based on chitosan are competent materials which highly attract many types of dyes. These materials are flexible and capable of being used as sorbents in differing forms, for example bead, gels types, fibres or flakes [11].

Lignocellulosic materials.

The interest of current researchers is to do feasibility studies on the use of lignocellulosic materials as cheap alternatives for the elimination of heavy metal ions out of an industrial effluent by adsorbing them. The lignocellulosic substances can be agro-industrial wastes and by-products or natural substances.

Agricultural materials remove heavy metals from water by biosorption. In biosorption, metal ions are attracted and bound on the adsorption sites present in the materials by complex processes including chemisorption. Biosorption processes in general can significantly reduce total treatment costs incurred in comparison with the conventional methods [2].

Similar types of chemical components and functional groups are found in lignocellulosic by-products, but the amounts vary. They are responsible in adsorbing heavy

Table 3 Some common lignocellulosic materials and their respective chemical compositions

Type	% Cellulose	% Lignin	% Hemicellulose	% Ash	% Silica
Straws of rice	26–35	9–15	20–30	15–20	9–15
Straws of wheat	31–36	15–21	25–32	4.5–9	4–7
Barley straw	32–36	14–16	25–29	6–8	3–5
Sugarcane bagasse	31–43	20–25	26–34	1.5–5	Below 4
Grass	30–40	10–25	35–50	5–15	–
Corncob	35–45	5–15	35–45	1–2	Below 1

Abdolali et al. [2]; Cagnon et al. [20]

metal ions. The following table shows some common lignocellulosic materials and their respective chemical compositions (Table 3).

Lignin, lipids, hemicellulose, proteins, water, simple sugars, starch and hydrocarbons, are among the basic components of the lignocellulosic waste materials, contain various functional groups which facilitate metal complexation leading to removal of heavy metals. Some examples of the agricultural wastes used as adsorbents include rice bran, rice husk [32], wheat bran, barks of some trees, wheat husk, groundnut shells, waste tea leaves, coconut shells, sugarcane bagasse, black gram husk, hazelnut shells, cotton seed hulls, maize corn cob, peels of, apple, bananas and, oranges, sugar beet pulp, water hyacinth and coffee beans sunflower stalks [63].

Agricultural waste products including groundnuts, rice and black gram husk, grape stalks and olive stone wastes, peanut and walnut shells, chitosan, lemongrass and orange peels have been tested as adsorbents for lead [8]. Results from certain examinations have demonstrated an adsorption limit in the scope of between and 263.0 mg/g for coconut waste and chitosan at 8.3 mg/g [47].

2.2 *Inorganic Adsorbents*

2.2.1 Clay

Clays are hydrated aluminosilicates whose structure is characterized by tetrahedral sheets of silicates with linkages to octahedral aluminate sheets [33]. Some major clay minerals include kaolin, smectite and montmorillonite, bentonite clay and chlorite. Kaolin is basic and has different minerals of clay like kaolinite and dickite, which develops after decomposition. Kaolinite group is characterized by the basic structural formula $Al_2Si_2O_5(OH)_4$ but major of its portion is filled by Al_2O_3, H_2O SiO_2,while the minor portion is magnesium, potassium and iron [21]. The high specific surface area, small particle size and complex structure with pores of clay and clay minerals allow it to strongly interact chemically and physically with dissolved heavy metal ions.

Clay minerals can adsorb heavy metal ions from water but are less effective. Montmorillonite is more effective towards cations while kaolinite is more effective towards anions. Clay as an adsorbent has good potential because clay is cheap; however, clay exists in aqueous solutions either as colloids or dispersed particles. This makes the removal of clay from filters difficult [67].

Adsorption occurs at the clay surface and is associated with the chemistry of charges found on the surface. Clay minerals adsorb metal ions by various mechanisms such as physical adsorption and micro-precipitation involving ion exchange at permanent charged sites chemical adsorption [37]. Acid-activated clay is an effective sorbent for lead removal from industrial effluents [58]. Clay removes lead heavy metal by the precipitation of lead carbonate. This is a phenomenon which would be explained by the low solubility product constant ($Ksp = 7.4 \times 10^{-14}$) and precipitation pH (pH 5.3) of $PbCO_3$ at 25 °C [61]. Original clay samples demonstrated substantial removal amounts of metals when compared with treated forms. The removal efficiency was discovered to depend upon the physicochemical characteristics of the clay sample and the metal removed. In single element system, the calculated Langmuir capacities ranged from 131.58 to 32.89 mg/g for removal of lead and 27.40 to 12.97 mg/g for removal of copper removal by various clay samples from Tunisia [61].

2.2.2 Zeolites

They are a huge group of glasslike aluminosilicates that are hydrated and have remarkable features such as the ability to exchange anions and cations [45]. The Si-Al zeolites include faujasites and mordenite [39]. Natural zeolites, clinoptilolite, have been utilized as inorganic matrices as their surface area is large and is microporous, in addition to their high mechanical and thermal stability [41]. Natural zeolites hybrids functionalized with amidoxime exhibit an excellent and high limit of adsorption for lead (II) ions from solutions [56]. This could be due to the accessibility of more sorption sites resulting from the increased area of the surface [50].

Zeolite is an effective adsorbent. It is able to adsorb ions of heavy metals from wastewater solutions. The general formula of zeolites is $(M^+, M^{2+})\,O.\,Al_2\,O_3.\,gSiO_2.\,zH_2\,O$. The number of moles of silicon and water is indicated by g and z, respectively. M^{2+} is normally potassium ion or sodium ion, and M^{2+} is calcium ion, iron ion or magnesium ion. At rare instances, strontium, lithium and barium ions may take the place M^+ or M^{2+}. Iron (III) is assumed to substitute into position of the tetrahedral framework. Heavy metals cations like that of chromium, lead, cadmium and copper can substitute into the structure of zeolites replacing the M^+ cations [13]. It is noted that the extraction efficiency of mordenite for all heavy metal ions is observed higher than stilbite and heulandite. The obtained results clearly showed that mordenite could be useful to reduce Co^{2+} concentration from wastewater. These zeolites were less efficient in removal of Mn^{2+}ions [49]. Figure 1 shows the procedure of preparing mordenites zeolites adopted from [54] (Fig. 2).

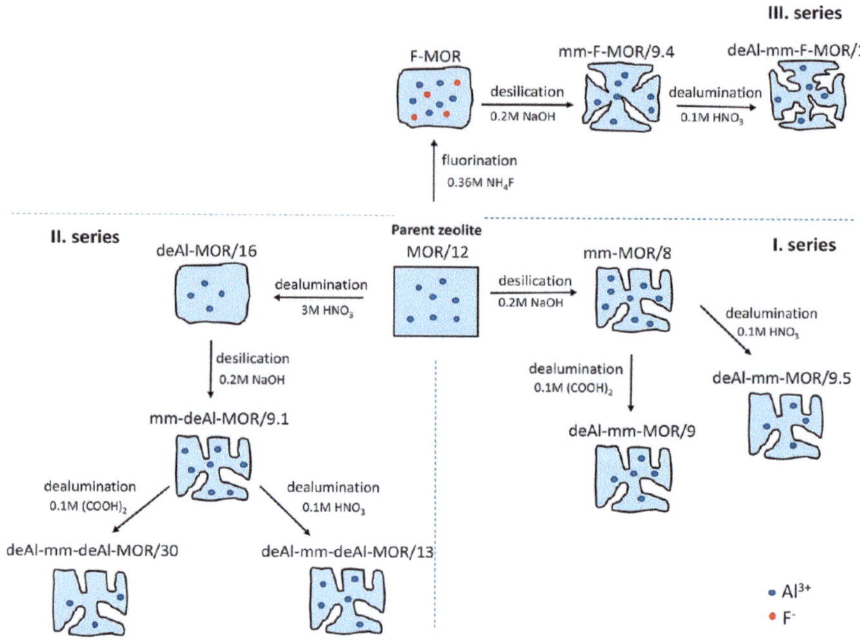

Fig. 2 Scheme of the preparation of micromesoporous mordenite zeolites using postsynthesis alkaline−acid (series I), acid−alkaline−acid (series II), and fluorination−alkaline−acid (series III) leaching procedures

2.3 *Composites of Biosorbents and Inorganic Adsorbent*

2.3.1 Clay–Cellulose Composite

There are different forms in which clay exists, but the chemical structure of clay mineral is formed largely of hydrous octahedral aluminates and tetrahedral silicates crystals. Clays form complexes with crystals of other minerals which are porous and whose specific surface area is high [1]. Cellulose, as a biopolymer, is made of recurrent monosaccharides chains. The length of the chain of cellulose or the level of polymerization attained varies depending on the source/origin of the cellulose and also the technique used in its synthesis. Cellulose is suitable for nanocomposite preparation due to chirality and high specific surface area. Cellulose is excellently reactive due to hydroxyl groups. These properties make cellulose suitable for exploitation in water treatment applications or it can be combined with other materials such as clay to make composites, which are very useful adsorbents [3]. Modification of clay can increase its adsorption capacity. Clay–cellulose biopolymer composite could remove chromium (VI) ions from industrial wastewater at efficiencies as high as 99.5%, using about 0.5–0.6 g of the adsorbent and the column method [66]. Islam et al. [32] observed a gradual increase in adsorption capacity as the cellulose content in composites increased. This could be attributed to better interaction of clay and crystalline cellulose in composites where cellulose percentage is high. The best adsorption capacity was observed in the case of the composite with the highest percentage of cellulose (2.37 mg/g). It was better than crystalline cellulose (0.39 mg/g) or raw clay (0.67 mg/g). The adsorption capacity increased almost about five to six times [32].

2.3.2 Cellulose–Zeolites Composites

Zeolites are highly porous aluminosilicates, cheaper than clay and able to remove trace amounts of heavy metal contaminants from wastewater through adsorption [11]. Natural clinoptilolite zeolites are a good sorbent for heavy metal ions from industrial effluent leading to an effluent with no genotoxic or toxic potential [70]. Cellulose is a natural polymer together with its derivatives and has low chemical and mechanical resistance. The ability of faujasite–cellulose composite membranes to remove divalent ions of heavy metals (calcium, magnesium, lead and copper) was studied, and the composite was excellent with efficiencies above 90% at different pH and concentrations [12]. The theoretical maximum adsorption capacity of graphene oxide/cellulose composite adsorbent for Ce (III) uptake was studied and was observed to be 109.1 mg/g. The graphene oxide/cellulose composite adsorbent has a high adsorption capacity in a wide pH range of 3–8. This suggested that graphene oxide/cellulose composite adsorbent could be used in some extreme conditions because it is highly resistant to acidic pH [29].

2.3.3 Clay–Chitosan Composite

Chitosan is extracted from chitin and also from the exoskeletons of some insects and also crustacean fungi. Chitosan is inexpensive because chitin is readily available, and after cellulose, it follows as the most abundant polymer [17]. There are many ways of modifying chitosan, and most of these ways increase the cost of production, although this is accompanied by an increased number of adsorption sites. Oppositely active hydrophobicity and hydrophilicity of clay and chitosan are the main barriers of combining clay with chitosan. An increase in temperature increases desorption rate for most chitosan–clay composites. Specific derivatives of chitosan can be used in place of chitosan to improve the functionality of this composite at elevated temperatures. Many chitosan derivatives have upgraded the property of recyclability to a substantial level during composite fabrication with clay composite, an important requirement for a good adsorbent fabrication and wastewater treatment [10, 72]. In case of clay modification, many chemicals have been used to extensively increase hydrophobicity to become compatible with hydrophobic chitosan and to increase interlayer distance to have strong interaction with chitosan. However, the interlayer space usually depends on the type of clay mineral and the extent of fabrication techniques used. Besides, acid treatment of the clay provides a scope to purify raw clay by eliminating unwanted impurities or by activating the clay surface [16]. A study established that chitosan/clay is able to adsorb hexavalent chromium from aqueous solutions at low concentrations. The low cost of manufacture of the matrix, the good uptake of chromium (vi) at acidic pH (17.28 mg/g) and the low desorption of the adsorbed metal ions from the film in water suggested that the combination of clay and chitosan can be tested in industrial settings [14].

3 Conclusion

Over the years, pollution of the environment due to the release of textile wastewater has been an issue of concern. Textile wastewater releases an effluent containing heavy metals, sometimes without treatment or with inefficient treatment. Water quality is compromised by the presence of heavy metals. Effective treatment is therefore necessary for any textile wastewater before it is discharged. It is therefore a requirement to create an efficient, effective and cost-effective technique of treating industrial effluent. Adsorption method has been given attention by many scholars in the past. Activated carbons, which are expensive adsorbents, can be replaced by low-cost adsorbents including agricultural waste, from maize cobs, sugarcane bagasse and rice husks. Inorganic materials such as zeolites and clay are also good adsorbents. It is however interesting to combine biosorbents such as algae, fungi, chitosan and cellulose with inorganic clays and zeolites. This leads to an adsorbent with increased adsorption sites and good potential to be recycled. This recyclability is an important property of any good adsorbent.

References

1. Abd El-Aziz ME, Kamal KH, Ali KA, Abdel-Aziz MS, Kamel S (2018) Biodegradable grafting cellulose/clay composites for metal ions removal. Int J Biol Macromol 118:2256–2264. https://doi.org/10.1016/j.ijbiomac.2018.07.105
2. Abdolali A, Guo WS, Ngo HH, Chen SS, Nguyen NC, Tung KL (2014) Typical lignocellulosic wastes and by-products for biosorption process in water and wastewater treatment: a critical review. Biores Technol 160:57–66. https://doi.org/10.1016/j.biortech.2013.12.037
3. Abu-Danso E, Peräniemi S, Leiviskä T, Kim TY, Tripathi KM, Bhatnagar A (2020) Synthesis of clay-cellulose biocomposite for the removal of toxic metal ions from aqueous medium. J Hazard Mater 381(June 2019):120871. https://doi.org/10.1016/j.jhazmat.2019.120871
4. Ahmed JK, Ahmaruzzaman M (2016) Journal of water process engineering a review on potential usage of industrial waste materials for binding heavy metal ions from aqueous solutions. J Water Process Eng 10:39–47. https://doi.org/10.1016/j.jwpe.2016.01.014
5. Akar T, Tunali S, Kiran I (2005) Botrytis cinerea as a new fungal biosorbent for removal of Pb(II) from aqueous solutions. Biochem Eng J 25(3):227–235. https://doi.org/10.1016/j.bej.2005.05.006
6. Akwasi A, Clifford F, Daniel A, Abdul, A (2012) Occurrence of toxic heavy metals (Hg, Pb and Cd) in fish on Ghanaian markets. 48(Mar):9669–9671
7. Al-Kdasi A, Idris A, Saed K, Guan CT (2004) Treatment of textile wastewater by advanced oxidation processes– a review. Global NEST J 6(1):222–230

8. Alalwan HA, Kadhom MA, Alminshid AH (2020) Uncorrected Proof agricultural byproducts uncorrected proof. pp 1–15. https://doi.org/10.2166/aqua.2020.133
9. Ali H, Khan E, Ilahi I (2019) Environmental chemistry and ecotoxicology of hazardous heavy metals: environmental persistence, toxicity, and bioaccumulation. J Chem 2019(Cd):1–15. https://doi.org/10.1155/2019/6730305
10. Anirudhan TS, Rijith S (2012) Synthesis and characterization of carboxyl terminated poly(methacrylic acid) grafted chitosan/bentonite composite and its application for the recovery of uranium(VI) from aqueous media. J Environ Radioact 106:8–19. https://doi.org/10.1016/j.jenvrad.2011.10.013
11. Anjum NA, Gill SS, Tuteja N (2017) Enhancing cleanup of environmental pollutants. Enhancing Cleanup Environ Pollutants 2:183–207. https://doi.org/10.1007/978-3-319-55423-5
12. Baghdad K, Hasnaoui AM (2020) Zeolite-cellulose composite membranes: synthesis and applications in metals and bacteria removal. J Environ Chem Eng 8(4):104047. https://doi.org/10.1016/j.jece.2020.104047
13. Bakhtiari AR, Zakaria MP, Yaziz MI, Lajis MNHL, Bi X (2014) Environment Asia. Environ Asia 7(1):104–111. https://doi.org/10.14456/ea.2010.51
14. Batista ACL, Villanueva ER, Amorim RVS, Tavares MT, Campos-Takaki GM (2011) Chromium (VI) ion adsorption features of chitosan film and its chitosan/zeolite conjugate 13X film. Molecules 16(5):3569–3579. https://doi.org/10.3390/molecules16053569
15. Bazrafshan E, Alipour MR, Mahvi AH (2016) Textile wastewater treatment by application of combined chemical coagulation, electrocoagulation, and adsorption processes. Desalin Water Treat 57(20):9203–9215. https://doi.org/10.1080/19443994.2015.1027960
16. Biswas S, Fatema J, Debnath T, Rashid TU (2021) Chitosan-clay composites for wastewater treatment: a state-of-the-art review. ACS ES&T Water 1(5):1055–1085. https://doi.org/10.1021/acsestwater.0c00207
17. Biswas S, Rashid TU, Debnath T, Haque P, Rahman MM (2020) Application of chitosan-clay biocomposite beads for removal of heavy metal and dye from industrial effluent. J Compos Sci 4(1):1–14. https://doi.org/10.3390/jcs4010016
18. Briffa J, Sinagra E, Blundell R (2020) Heavy metal pollution in the environment and their toxicological effects on humans. Heliyon 6(9):e04691. https://doi.org/10.1016/j.heliyon.2020.e04691
19. Burakov AE, Galunin EV, Burakova IV, Kucherova AE, Agarwal S, Tkachev AG, Gupta VK (2018) Ecotoxicology and environmental safety adsorption of heavy metals on conventional and nanostructured materials for wastewater treatment purposes : a review. Ecotoxicol Environ Safety 148(Nov 2017):702–712. https://doi.org/10.1016/j.ecoenv.2017.11.034
20. Cagnon B, Py X, Guillot A, Stoeckli F, Chambat G (2009) Contributions of hemicellulose, cellulose and lignin to the mass and the porous properties of chars and steam activated carbons from various lignocellulosic precursors. Biores Technol 100(1):292–298. https://doi.org/10.1016/j.biortech.2008.06.009
21. Cao L, Li Z, Xiang S, Huang Z, Ruan R, Liu Y (2019) Preparation and characteristics of bentonite–zeolite adsorbent and its application in swine wastewater. Biores Technol 284(April):448–455. https://doi.org/10.1016/j.biortech.2019.03.043
22. Carolin CF, Kumar PS, Saravanan A, Joshiba GJ, Naushad M (2017) Efficient techniques for the removal of toxic heavy metals from aquatic environment: a review. J Environ Chem Eng 5(3):2782–2799. https://doi.org/10.1016/j.jece.2017.05.029
23. Gautam RK, Mudhoo A, Lofrano G, Chattopadhyaya MC (2014) Biomass-derived biosorbents for metal ions sequestration: adsorbent modification and activation methods and adsorbent regeneration. J Environ Chem Eng 2(1):239–259. https://doi.org/10.1016/j.jece.2013.12.019
24. Gonçalves AC, Meneghel AP, Rubio F, Strey L, Dragunski DC, Coelho GF (2012) Applicability of moringa oleifera lam. pie as an adsorbent for removal of heavy metals from waters. Revista Brasileira de Engenharia Agricola e Ambiental 17(1):94–99. https://doi.org/10.1590/S1415-43662013000100013
25. Gosavi VD, Sharma S (2014) Journal of environmental science, computer science and engineering & technology a general review on various treatment methods for textile wastewater. J Environ Sci, Comput Sci Eng Technol 3(1):29–39

26. Gupta N, Khan DK, Santra SC (2012) Heavy metal accumulation in vegetables grown in a long-term wastewater-irrigated agricultural land of tropical India. Environ Monit Assess 184(11):6673–6682. https://doi.org/10.1007/s10661-011-2450-7
27. Hadiyanto AG, Pradana AB, Buchori L, Sri Budiyati C (2014) Biosorption of heavy metal Cu2+ and Cr2+ in textile wastewater by using immobilized algae. Res J Appl Sci, Eng Technol 7(17):3539–3543. https://doi.org/10.19026/rjaset.7.706
28. Handan Dökmeci A (2020) Environmental impacts of heavy metals and their bioremediation. In: Heavy metals—their environmental impacts and mitigation [Working Title] (Issue Dec 2020, pp 1–18). https://doi.org/10.5772/intechopen.95103
29. Hao Y, Cui Y, Peng J, Zhao N, Li S, Zhai M (2019) Preparation of graphene oxide/cellulose composites in ionic liquid for Ce (III) removal. Carbohydr Polymers 208(June 2018):269–275. https://doi.org/10.1016/j.carbpol.2018.12.068
30. Holkar CR, Jadhav AJ, Pinjari DV, Mahamuni NM, Pandit AB (2016) A critical review on textile wastewater treatments: possible approaches. J Environ Manage 182:351–366. https://doi.org/10.1016/j.jenvman.2016.07.090
31. Hussain J, Hussain I, Arif M (2004) Characterization of textile wastewater. J Ind Pollution Control 20(1):1–66. https://doi.org/10.1016/b978-0-12-802326-6.00002-2
32. Islam MM, Khan MN, Biswas S, Rabia Choudhury T, Haque P, U Rashid T, Mizanur Rahman M (2017). Preparation and characterization of bijoypur clay-crystalline cellulose composite for application as an adsorbent. Adv Mater Sci 2(3):1–7. https://doi.org/10.15761/ams.1000126
33. Jiménez-Castañeda ME, Medina DI (2017) Use of surfactant-modified zeolites and clays for the removal of heavy metals from water. Water (Switzerland) 9(4):1–12. https://doi.org/10.3390/w9040235
34. Johar N, Ahmad I, Dufresne A (2012) Extraction, preparation and characterization of cellulose fibres and nanocrystals from rice husk. Ind Crops Prod 37(1):93–99. https://doi.org/10.1016/j.indcrop.2011.12.016
35. Jouanneau S, Recoules L, Durand MJ, Boukabache A, Picot V, Primault Y, Lakel A, Sengelin M, Barillon B, Thouand G (2014) Methods for assessing biochemical oxygen demand (BOD): a review. Water Res 49(1):62–82. https://doi.org/10.1016/j.watres.2013.10.066
36. Kikuchi T, Tanaka S (2012) Biological removal and recovery of toxic heavy metals in water environment biological removal and recovery of toxic. 3389. https://doi.org/10.1080/10643389.2011.651343
37. Krupskaya VV, Zakusin SV, Tyupina EA, Dorzhieva OV, Zhukhlistov AP, Belousov PE, Timofeeva MN (2017) Experimental study of montmorillonite structure and transformation of its properties under treatment with inorganic acid solutions. Minerals 7(4):1–15. https://doi.org/10.3390/min7040049
38. Kumar A, Bisht BS, Joshi VD, Singh AK, Talwar A (2010) Physical, chemical and bacteriological study of water from rivers of Uttarakhand. J Hum Ecol 32(3):169–173. https://doi.org/10.1080/09709274.2010.11906336
39. Kustov L (2020) New horizons in zeolites and zeolite-like materials. Curr Comput-Aided Drug Des 10(8):1–2. https://doi.org/10.3390/cryst10080714
40. Latinwo GK, Jimoda LA, Agarry SE, Adeniran JA (2015) Biosorption of some heavy metals from textile wastewater by green seaweed biomass. Univers J Environ Res Technol 5(4):210–219
41. Lin H, Liu QL, Dong YB, He YH, Wang L (2015) Physicochemical properties and mechanism study of clinoptilolite modified by NaOH. Microporous Mesoporous Mater 218:174–179. https://doi.org/10.1016/j.micromeso.2015.07.017
42. Madhav S, Ahamad A, Singh P, Mishra PK (2018) A review of textile industry: wet processing, environmental impacts, and effluent treatment methods. Environ Qual Manage 27(3):31–41. https://doi.org/10.1002/tqem.21538
43. Mahmood Q, Zheng P, Islam E, Hayat Y, Hassan MJ, Jilani G, Jin RC (2005) Lab scale studies on water hyacinth (Eichhornia). Caspian J Env Sci 3(2):83–88
44. Mary J, Karthik C, Ganesh R, Kumar SS, Prabakar D, Kadirvelu K, Pugazhendhi A (2018) Biological approaches to tackle heavy metal pollution: a survey of literature. J Environ Manage 217:56–70. https://doi.org/10.1016/j.jenvman.2018.03.077

45. Moazeni M, Parastar S, Mahdavi M, Ebrahimi A (2020) Evaluation efficiency of Iranian natural zeolites and synthetic resin to removal of lead ions from aqueous solutions. Appl Water Sci 10(2):1–9. https://doi.org/10.1007/s13201-019-1135-1
46. Muchuweti M, Birkett JW, Chinyanga E, Zvauya R, Scrimshaw MD, Lester JN (2006) Heavy metal content of vegetables irrigated with mixtures of wastewater and sewage sludge in Zimbabwe: implications for human health. Agr Ecosyst Environ 112(1):41–48. https://doi.org/10.1016/j.agee.2005.04.028
47. Nadeem R, Manzoor Q, Iqbal M, Nisar J (2016) Biosorption of Pb(II) onto immobilized and native Mangifera indica waste biomass. J Ind Eng Chem 35:185–194. https://doi.org/10.1016/j.jiec.2015.12.030
48. Nejadshafiee V, Islami MR (2019) Adsorption capacity of heavy metal ions using sultone-modified magnetic activated carbon as a bio-adsorbent. Mater Sci Eng, C 101(March):42–52. https://doi.org/10.1016/j.msec.2019.03.081
49. Obaid SS, Gaikwad DK, Sayyed MI, Al-Rashdi K, Pawar PP (2018) Heavy metal ions removal from waste water by the natural zeolites. Mater Today: Proc 5(9):17930–17934. https://doi.org/10.1016/j.matpr.2018.06.122
50. Olaofe O, Olagboye SA, Akanji PS, Adamolugbe EY, Fowowe OT, Olaniyi AA (2014) Kinetic studies of adsorption of heavy metals on clays. Int J Chem 7(1):48–54. https://doi.org/10.5539/ijc.v7n1p48
51. Osei P, Huang Y, Hua M, Zhang Q, Wu J, Onumah J, Sam-amoah LK, Osei P (2015) Ecotoxicology and environmental safety sorption of heavy metal ions onto carboxylate chitosan derivatives—A. Ecotoxicol Environ Saf 116:113–120. https://doi.org/10.1016/j.ecoenv.2015.01.012
52. Palin D, Rufato KB, Linde GA, Colauto NB, Caetano J, Alberton O, Jesus DA, Dragunski DC (2016) Evaluation of Pb (II) biosorption utilizing sugarcane bagasse colonized by Basidiomycetes. Environ Monit Assess 188(5). https://doi.org/10.1007/s10661-016-5257-8
53. Pang YL, Abdullah AZ (2013) Current status of textile industry wastewater management and research progress in Malaysia: a review. Clean—Soil, Air, Water 41(8):751–764. https://doi.org/10.1002/clen.201000318
54. Pastvova J, Kaucky D, Moravkova J, Rathousky J, Sklenak S, Vorokhta M, Brabec L, Pilar R, Jakubec I, Tabor E, Klein P, Sazama P (2017) Effect of enhanced accessibility of acid sites in micromesoporous mordenite zeolites on hydroisomerization of n-Hexane. ACS Catal 7(9):5781–5795. https://doi.org/10.1021/acscatal.7b01696
55. Process IJM, Motsi T, Rowson NA, Simmons MJH (2009) Adsorption of heavy metals from acid mine drainage by natural zeolite. Int J Miner Process 92(1–2):42–48. https://doi.org/10.1016/j.minpro.2009.02.005
56. Puspitasari T, Kadja GTM, Radiman CL, Darwis D, Mukti RR (2018) Two-step preparation of amidoxime-functionalized natural zeolites hybrids for the removal of Pb2+ ions in aqueous environment. Mater Chem Phys 216:197–205. https://doi.org/10.1016/j.matchemphys.2018.05.083
57. Razi MAM, Gheethi AA, ZA IA (2018) Removal of heavy metals from textile wastewater using sugarcane bagasse activated carbon. Int J Eng Technol 7(4.30):112. https://doi.org/10.14419/ijet.v7i4.30.22066
58. Resmi G, Thampi SG, Chandrakaran S (2012) Removal of lead from wastewater by adsorption using acid-activated clay. 3330:290–297. https://doi.org/10.1080/09593330.2011.572917
59. Sabur MA, Khan AA, Safiullah S (2012) Treatment of textile wastewater by coagulation precipitation method. J Sci Res 4(3):623–633. https://doi.org/10.3329/jsr.v4i3.10777
60. Santhosh C, Velmurugan V, Jacob G, Kwan S, Nirmala A, Bhatnagar A (2016) Role of nanomaterials in water treatment applications: a review. Chem Eng J 306:1116–1137. https://doi.org/10.1016/j.cej.2016.08.053
61. Sdiri A, Higashi T, Chaabouni R, Jamoussi F (2012) Competitive removal of heavy metals from aqueous solutions by montmorillonitic and calcareous clays. Water Air Soil Pollut 223(3):1191–1204. https://doi.org/10.1007/s11270-011-0937-z

62. Sela SK, Hossain AKMN-U, Hussain SZ, Hasan N (2020) Utilization of prawn to reduce the value of BOD and COD of textile wastewater. Cleaner Eng Technol 1(Nov):1–5. https://doi.org/10.1016/j.clet.2020.100021
63. Sud D, Mahajan G, Kaur MP (2008) Agricultural waste material as potential adsorbent for sequestering heavy metal ions from aqueous solutions—a review. 99:6017–6027. https://doi.org/10.1016/j.biortech.2007.11.064
64. Sun M, Hong L (2011) Impacts of the pendant functional groups of cellulose precursor on the generation of pore structures of activated carbons. Carbon 49(7):2173–2180. https://doi.org/10.1016/j.carbon.2011.01.031
65. Tasnim, A. (2015). Estimation of chemical oxygen demand in wastewater using UV-VIS spectroscopy. School of Mechatronic System Engineering, pp 10–18
66. Uddin MK (2017) A review on the adsorption of heavy metals by clay minerals, with special focus on the past decade. Chem Eng J 308:438–462. https://doi.org/10.1016/j.cej.2016.09.029
67. Unuabonah EI, Olu-Owolabi BI, Fasuyi EI, Adebowale KO (2010) Modeling of fixed-bed column studies for the adsorption of cadmium onto novel polymer-clay composite adsorbent. J Hazard Mater 179(1–3):415–423. https://doi.org/10.1016/j.jhazmat.2010.03.020
68. Wang J, Chen C (2009) Biosorbents for heavy metals removal and their future. Biotechnol Adv 27(2):195–226. https://doi.org/10.1016/j.biotechadv.2008.11.002
69. Wawrzkiewicz M, Wiśniewska M, Wołowicz A, Gun'ko VM, Zarko VI (2017) Mixed silica-alumina oxide as sorbent for dyes and metal ions removal from aqueous solutions and wastewaters. Microporous Mesoporous Mater 250:128–147. https://doi.org/10.1016/j.micromeso.2017.05.016
70. Zanin E, Scapinello J, de Oliveira M, Rambo CL, Franscescon F, Freitas L, de Mello JMM, Fiori MA, Oliveira JV, Dal Magro J (2017) Adsorption of heavy metals from wastewater graphic industry using clinoptilolite zeolite as adsorbent. Process Saf Environ Prot 105:194–200. https://doi.org/10.1016/j.psep.2016.11.008
71. Zeiner M, Rezić I, Steffan I (2007) Analytical methods for the determination of heavy metals in the textile industry. Kemija u Industriji/J Chem Chem Eng 56(11):587–595
72. Zhang H, Ma J, Wang F, Chu Y, Yang L, Xia M (2020) Mechanism of carboxymethyl chitosan hybrid montmorillonite and adsorption of Pb(II) and Congo red by CMC-MMT organic-inorganic hybrid composite. Int J Biol Macromol 149:1161–1169. https://doi.org/10.1016/j.ijbiomac.2020.01.201

Potential of Pyrolysis Biochar as an Eco-Friendly Biosorbent for Dye Removal from Industrial Wastewaters

Prakash Binnal, S. Rajashekhara, and Poornima G. Hiremath

Abstract Presently, the removal of dyes from wastewater of industries such as textile, paint, paper and pulp is of major concern. Adsorption has been considered to be one of the best unit operations for separation of various dyes. However, the economics of adsorption is affected by high costs of adsorbents. Recently, biochar, which is one of the products from pyrolysis, is gaining wide attention as an economic, efficient adsorbent for various dyes. Researchers all over the world have contributed to investigations on production of biochar, its modification, characterization, evaluating suitability for dye removal, mechanism of adsorption and optimization of conditions for dye removal process. Several authors have also reported analysis of adsorption isotherms, kinetic studies, thermodynamics of adsorption and also on possibilities of recycling the biochar. In this regard, the present book chapter focuses on a review of various biochar feedstocks used, the type of pyrolysis reactors used for biochar production, the important properties of biochar (elemental and proximate analysis, BET surface area, pore volume, pore diameter), optimum conditions for dyes adsorption, results of adsorption isotherms, kinetics of adsorption, thermodynamics of adsorption and finally, the reduction in activity of biochar when recycled after regeneration.

Keywords Dyes · Biochar · Adsorption · Pyrolysis · Textiles industry

P. Binnal (✉) · S. Rajashekhara · P. G. Hiremath
Department of Chemical Engineering, Siddaganga Institute of Technology, Tumkur 572103, India
e-mail: pb@sit.ac.in

S. Rajashekhara
e-mail: rs@sit.ac.in

P. G. Hiremath
e-mail: pgh@sit.ac.in

S. S. Muthu and A. Khadir (eds.), *Textile Wastewater Treatment*, Sustainable Textiles: Production, Processing, Manufacturing & Chemistry,
https://doi.org/10.1007/978-981-19-2852-9_11

1 Introduction

Dyes are coloured organic materials that are consumed by various industries such as textile, paper, leather, plastic, pharmaceuticals, food and cosmetics. Nearly seven lakh tonnes of different colouring agents derived from nearly one lakh commercially available dyes are manufactured each year [1]. Most often, the dyes are discarded after serving their purpose to water bodies. Due to their toxic nature, the dye effluents have potential to endanger the animal and human lives. As compared to other industries, the textile industries utilize largest quantity of dye (nearly ten thousand tonnes per year worldwide) and also release the highest amount of dye effluent (54%). The share of other industries is: dyeing industries (21%), paper and pulp industries (10%), paint and tannery industries (8%) and the dye manufacturing industries (7%) [2]. Hence, researchers all over the world have developed various technologies for dye removal from effluent wastewaters.

1.1 Classification of Dyes

The dyes can be classified based on their chemical structure and based on application. In general, the dyes can be categorized as cationic, anionic and non-ionic dyes. The anionic dyes can be further classified as direct, acidic and reactive dye. Normally, the dyes utilized in the textile manufacturing industries can also be classified as acidic (acid red 35), azoic (Disperse Yellow), basic (CI Basic Blue 6), direct (Direct Orange 40), disperse (Disperse Blue), mordant (Mordant Black 17), reactive (Reactive Blue 19), solvent, sulphur (Sulphur brilliant green, CI 5357) and vat (Vat Blue 20) [3]. The derivatives of the azo dyes are the major class of dyes used in the industrial applications.

1.2 Methods of Dye Removal

Dye removal technologies can be broadly classified into three main categories: physical methods (adsorption, ion exchange, membrane technologies like reverse osmosis, ultrafiltration, nanofiltration), chemical treatment technologies (advanced oxidation process, electrochemical destruction, Fenton reaction, ozonation, photochemical, ultraviolet irradiation) and biological treatment methodologies such as adsorption by microbial biomass, algae degradation, aerobic-anaerobic combination (conventional method, enzyme degradation, fungal cultures, microbial cultures such as mixed bacterial). The merits and demerits of each technology have been recently discussed in detail by Katheresan et al. [3].

1.3 Dye Removal Through Adsorption

It is well known that dyes cannot be effectively removed by any other conventional wastewater treatment methods and adsorption is the most effective method of removing dyes [4]. It also has benefit of not producing any hazardous residues, and no requirement of pre-treatment. Further, the adsorbent can be recycled several times after regeneration, which improves the economy of process [5]. However, one disadvantage of method is the high cost of adsorbents. This issue has been overcome by developments of economic and efficient adsorbents. Rafatullah et al. [6] reviewed 185 research articles on low-cost adsorbents.

Recently, activated carbon-based adsorbents are being extensively explored and found to be more efficient than other adsorbents [2, 7]. Ding et al. shown that the surface area of activated carbon obtained from treated rice husk was as high as 2516 m^2/g [8].

2 Biochar

Biochar is a carbon-rich product obtained from the thermochemical conversion of biomass. It has porous structure with significant number of functional groups. It contains abundant surface charges and free radicals. Due to these properties, it finds extensive applications in water treatment, catalyst for degradation of plastics, compositing and fermentation. Recently, biochars produced from biomass are gaining wide attention as an efficient adsorbent for the removal of various dyes due to their high surface area, acidity and the presence of minerals and trace metals [9]. An overview of biochar production technologies, properties of biochar and optimum operating conditions for dye removal using biochar are presented in this book chapter.

2.1 Biochar Production Methods

Various methods such as hydrothermal carbonization, hydrothermal liquefaction, slow/fast pyrolysis and gasification are used for biochar production. Among these methods, slow pyrolysis is the promising technique due to high char yield and superior characteristics biochar, etc.

2.1.1 Hydrothermal Carbonization (HTC)

HTC is a thermochemical conversion of biomass into hydrochars in temperature range of 180–250 °C and pressures of 2–10 MPa. The reactions such as dehydration, polymerization occur followed by carbonization reactions. The hydrochars are used in applications such as fuel cells, supercapacitors and as a sorbent. The energy density of hydrochars produced at high temperatures will be high [10].

2.1.2 Hydrothermal Liquefaction (HTL)

In HTL, the wet biomass is heated to high temperatures (250–550 °C) and relatively high pressures of 5–25 MPa. Through thermal depolymerization, wet biomass is converted to bio-oil, biochar and non-condensable gases. Reactions such as hydrolysis, fragmentation, dehydration, aromatization and repolymerization occur to form biochar [11]. Various parameters such as feedstock composition, temperature, pressure and residence time affect the quality of biochar. The lignin gets dissolved in water by hydrolysis leading to formation of phenols which are the precursors for biochar formation. Similarly, cellulose and hemicellulose undergo decomposition, and various reactions occur among pyrolytic intermediates to form char.

2.1.3 Pyrolysis

In pyrolysis, the biomass undergoes thermal/catalytic decomposition to form biochar, bio-oil and non-condensable gases in the temperature range of 200–500 °C. Factors affecting the characteristics of biochar during pyrolysis are feedstock composition, types of pyrolysis (slow, intermediate and fast), temperature, heating rate and pressure. Apart from this, the efficiency of pyrolysis process is also affected by presence of oxygen, nitrogen or vacuum. The other factors influencing quality of biochar are activation technique used (physical/chemical). Among in-situ activation, copyrolysis of biomass with other biomass and with waste plastics has been reported to be an effective method for enhancing the surface area, porosity and stability of biochar. Further, the post-activation methods such as physical, chemical composite modifications have been found to significantly influence the quality of biochar [9].

2.1.4 Gasification

Gasification is a thermochemical process which converts carbonaceous materials into syngas, tars and biochar at elevated temperatures (>550 °C) and inert atmosphere. In general, yield of biochar from gasification will be less than other thermochemical processes due to conversion of some amount elemental carbon to carbon monoxide [12]. However, since the primary aim of gasification is to obtain syngas, the biochar formation is purposely avoided to enhance the energy efficiency. Hence, a wide range of operating conditions can be explored for gasification to produce biochar.

3 Biochar as an Adsorbent for Dyes

Various authors have conducted studies on application of biochar as adsorbent for removal of industrial dyes. Table 1 lists the important properties of these biochars.

3.1 Protocols Followed, Properties and Applications of Biochars Produced from Various Feedstocks

Biochars produced from various feedstocks for their potential use as adsorbents for removal of various dyes have been reported in literature. A summary of protocols followed to produce biochars and properties of biochars is described in Sects. 3.1.1, 3.1.2, 3.1.3, 3.1.4, 3.1.6 and 3.1.7. The details of mechanisms of adsorption of dyes, results obtained for adsorption isotherms, kinetics studies, thermodynamic studies and recyclability of various biochars for adsorption of dyes are described in Sects. 4, 5, 6, 7 and 8, respectively.

3.1.1 Congo Red

Khan et al. [13] produced biochar from rice husk and Cow dung using a SS pyrolysis reactor (Mixed Bed Rotating Pyrolyzer (MBRP)) having dimensions of: ID: 75 mm; length: 1 m (Fig. 1). The nitrogen gas was passed through the reactor at rate of 30 mL/h to maintain the oxygen limited atmosphere. An analysis of elemental composition of rice husk and cow dung biochar showed the presence of Al, Si, P, K, Cl, Mg, Ca, C, O, Si, Ca. The yield of cow dung biochar was higher than rice husk (29.8% v/s 22.8%). The biochars were utilized for removal of congo red dye and the removal efficiencies were 66.8–96.9% for rice husk biochar and 68.9–98.8% for cow dung biochar. Park et al. [14] studied the effect of removal of congo red, methylene Blue, orange G using biochar obtained from switchgrass. The switchgrass biochar was produced at pyrolysis temperature of 600 °C (SB600) and 900 °C (SB900). The mass yields of SB600 and SB900 were reported as 21.8 and 14.8%. The results suggested that the % of carbon in both biochars was similar. However, FTIR and XRD indicated that significant portion of carbon in biochar SB900 was graphitized. The surface areas of SB600 and SB900 were 255.8 and 641.6 m^2/g, while the pore volumes were 0.029 and 0.058 cm^3/g. Thus, higher temperature had a positive influence on quality of biochars.

Also, the pore size decreased at higher temperature (2.162 nm at 600 °C as compared to 3.037 nm at 600 °C). Dai et al. [15] used calcium-rich biochar from crab shell. The C/N ratio of the biochar was 7.22. It was shown that the high adsorption capacity of adsorbent was due to multi-layered structure, low value of C/N ratio, high value of zeta potential and high ID/IG ratio. The zeta potential of biochar reduced from 20 mV at pH 4 to 1.64 mV at pH 6 and was zero at pH 6.7. It increased gradually

Table 1 Properties of biochar produced from various feedstocks

S. No	Source of biochar	Dye	Characteristics of biochar									References
			Biochar yield (wt %)	pH	% C	% H	% N	% O	Surface area (m^2/g)	Pore volume (cc/g)	Pore size (nm)	
1	Rice husk	Congo Red	22.8	6.3	–	–	–	–	–	–	–	[13]
2	Cow dung		29.8	6.9	–	–	–	–	–	–	–	[13]
3	SB600		21.8	10.9	84.8	1.7	–	–	255.8	0.029	3.037	[14]
4	SB900		14.8	10.7	87.5	0.6	–	–	641.6	0.058	21.62	[14]
5	Calcium rich biochar from crab Shell		–	–	–	–	–	–	81.57	0.0861	4.22	[15]
6	Steam activated spent mushroom substrate (SMS)		70.5	12.6	44.0	0.60	0.55	8.31	332	0.29	3.5	[16]
7	Orange peel waste biochar activated by CO_2		44	–	73.4	2.0	0.6	24.0	158.5	0.01	1.9	[17]
8	Orange peel biochar activated by steam		31	–	78.4	2.1	0.5	19	305.1	0.18	1.3	[17]
9	Phoenix dactylifera leaves biochar (PBC) composite	Congo red	–	–	–	–	–	–	1.07	0.0064	21.6	[32]
10	Mg Al composite	Methylene blue	–	–	–	–	–	–	151.259	0.2003	3.5135	[19]
11	Wheat straw			7.8	63.11	1.57	0.85	–	22.5	–	–	[18]

(continued)

Table 1 (continued)

S. No	Source of biochar	Dye	Characteristics of biochar									References
			Biochar yield (wt %)	pH	% C	% H	% N	% O	Surface area (m^2/g)	Pore volume (cc/g)	Pore size (nm)	
12	Eucalyptus		–	–	44.51	6.06	0.09	47.5	0.69	0.0025	14.34	[20]
13	Municipal waste		–	–	70	–	–	17.7	–	–	–	[21]
14	Pumpkin peel		–	–	–	–	–	–	3.6	0.22	–	[22]
15	Hazelnut husk		–	4.5	68.28	2.02	1.01	–	770	–	–	[23]
16	Date palm	Crystal violet	–	6.55	–	–	–	–	640	0.403	3.37	[24]
17	Wakame biochar	Malachite green	–	–	52.58	1.75	0.64	15.29	69.70	0.67	2.32	[25]
18	Opuntia ficus-indica biochar-1		51.28	9	62.40	3.29	1.51	25.45	1.15	0.018	27.46	[26]
19	Opuntia ficus-indica :Activated biochar -2	Malachite green	42.85	9.1	80.25	2.37	1.52	16.0	32.07	0.093	11.5	[26]
20	Opuntia ficus-indica: Activated biochar-3		37.36	9.6	81.0	2.29	1.51	17.0	33.62	0.095	11.3	[26]
21	BHC 300	Rhodamine	–	–	–	–	–	–	7.59	0.04	18.8	[27]
22	BHC-600		–	–	–	–	–	–	353	0.21	2.39	[27]
23	BHC-800		–	–	–	–	–	–	513	0.27	2.09	[27]

(continued)

Table 1 (continued)

S. No	Source of biochar	Dye	Characteristics of biochar									References
			Biochar yield (wt %)	pH	% C	% H	% N	% O	Surface area (m^2/g)	Pore volume (cc/g)	Pore size (nm)	
24	Magnetic corncob biochar	Acridine orange	–	–	–	–	–	–	552.6	–	11.9	[28]
25	Ag-TiO_2 biochar	Methyl Orange	–	–	–	–	–	–	35.212	0.033	3.60	[32]
26	BMHC450-N	Reactive red	–	–	82.8	-	-	15.2	452	–	–	[29]
27	BMHC600-N		–	–	78.7	-	-	19.3	317	–	–	[29]
28	Chitin	Methyl violet	–	–	81.30	2.15	3.45	13.1	275	0.178	1.3	[30]
29	Bovine bones	Basic red dye	62.82	Fixed carbon: 3.45%, Volatile matter:39.34%, Moisture: 7.86%, Ash : 57.21%					90.3	0.271	60	[31]

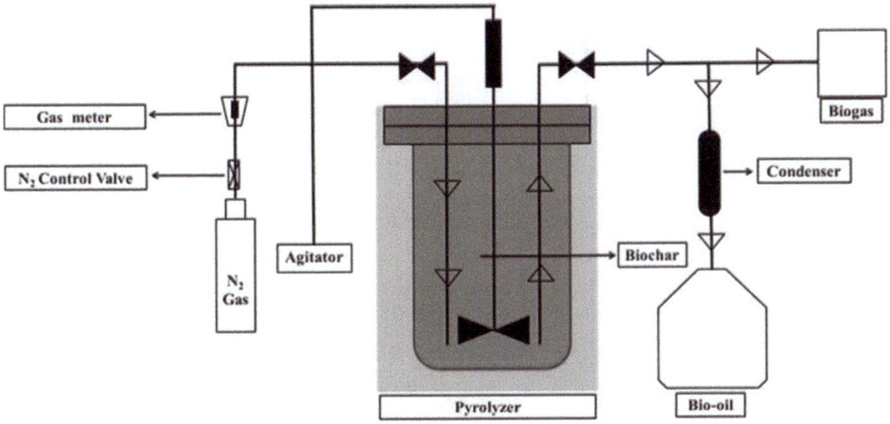

Fig. 1 Systematic flow diagram of Pyrolyzer used by Khan et al. [13] (Reprinted with permission from Elsevier)

decreased to about −4 mV at pH > 9. Due to these properties, exceptionally high congo red dye removal rates were observed during batch experiments (20,317 mg/g). The congo red removal was also studied by Sewu et al. [16], who used spent mushroom substrate (SMS) and its steam activated form for removal of congo red and crystal violet. Dried SMS was transferred into the SS pyrolysis reactor (Height: 1.5 m, Internal Diameter: 2.0 m, Volume: 4.7 m^3) through a strapper and screw conveyor and subsequently pyrolyzed at 450 °C for 4 h (Fig. 2). Proximate analysis showed that fixed carbon and volatiles content of steam activated biochar were higher than inactivated biochar (43.7% and 24.7% vs 32.1% and 20.0%). However,

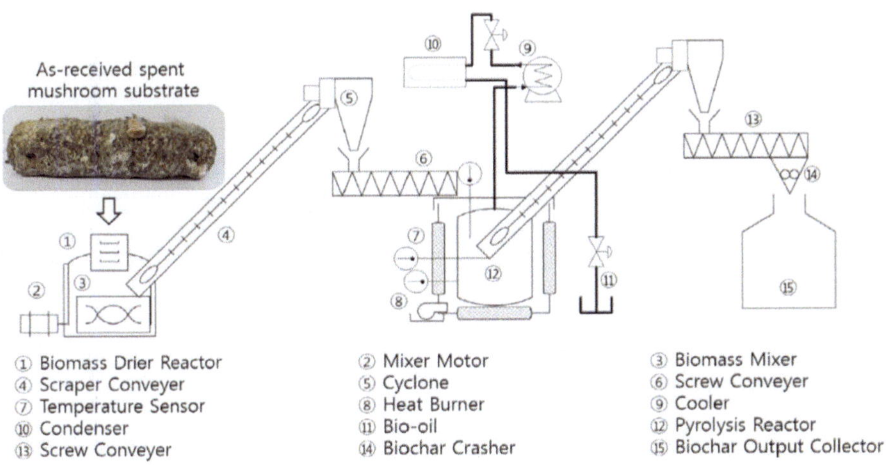

Fig. 2 Schematic diagram of the industrial-scale biochar production setup used in the study of Sewu et al. [16] (Reprinted with permission from Elsevier)

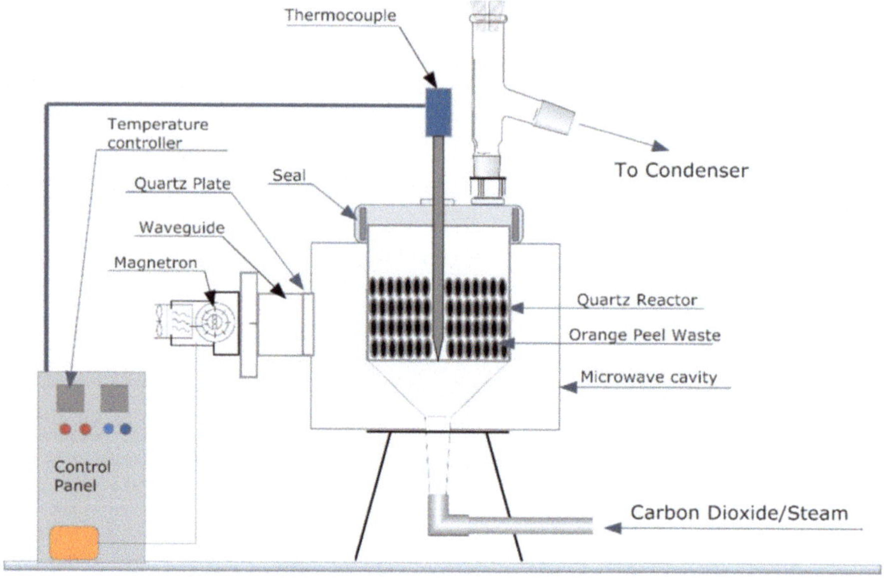

Fig. 3 Single step microwave pyrolysis using CO_2/steam for activation used by Yek et al. [17] (Reprinted with permission from Elsevier)

ash contents were in a reverse order with values of 31.6% and 47.9% for inactivated biochar and steam activated biochar, respectively. The reason was that, during steam activation, the SMS lost most of through water–gas shift reaction.

Yek et al. [17] followed a single-step pyrolysis process which consisted of simultaneous microwave heating and carbon dioxide/steam activation in order to produce orange peel biochar (Fig. 3). Thus, carbonization and activation simultaneously occurred. Carbon dioxide performed a dual role of activation of biochar and purge gas. This approach demonstrated that faster heating rate (120 °C/min), pyrolysis temperature (>800 °C) and short residence time (15 min) promoted higher mass yield of biochar (32–43 wt %), higher amount of fixed carbon (58.7–61.1 wt %), greater BET surface area (158.8–305.2 m^2/g), lower ratio of H/C and O/C (0.3 and 0.2, respectively). Activation with CO_2 produced more micropores than steam (which generated more mesopores). The adsorption efficiency of steam activated biochar was greater than carbon dioxide activated (137 mg/g, 92 mg/g) indicating 87–91% removal of congo red dye.

3.1.2 Methylene Blue

The yields and characteristics of biochars used for removal of methylene blue are shown in Table 2.

Table 2 Optimum conditions of dye removal by various adsorbents

S. No	Source of biochar	Dye	Optimum conditions	Langmuir Isotherms		Freundlich isotherm		Pseudo first order rate constants		Pseudo second order rate constants		References
				Q_e	K_L	N	K_F	$k_1(h^{-1})$	Q_e (mg/g)	k_2 (g/(mg.h))	Q_e (mg/g)	
1	Rice husk	Congo red	pH:7–9, Contact time: 72–84 h, Dosage : 5.0 g/100 mL, Max. removal :96.92% (RH) and 98.84% (CD)	15.81	–	<1	–	–	–	–	–	[13]
2	Cow dung			12.46	–	<1	–	–	–	–	–	[13]
3	SB600		pH: 6, Dosage:2 g/L, Initial dye conc.:2.5–640 mg/L	8	0.04	2.85	1.03	0.13	1.32	0.53	4.75	[14]
4	SB900			22.6	0.05	3.28	3.14	0.18	4.25	0.15	15.72	[14]
5	Crab Shell		pH :4, Contact time: 2 min, Dosage :0.5 g/L, Temperature = 25°C, Max removal: 20317 mg/g	20317.4	0.0061	0.514	643.28	19.51	11785.4	3.9×10^{42}	11785.4	[15]
6	Steam activated spent mushroom substrate (SMS)		pH -6, Dosage: 2 g/L, Temperature: 30 °C Initial dye conc.: 1000 mg/L,	292	0.0103	1.76	1.16	0.0099	247	5.0×10^{-5}	271	[16]

(continued)

Table 2 (continued)

S. No	Source of biochar	Dye	Optimum conditions	Langmuir Isotherms		Freundlich isotherm		Pseudo first order rate constants		Pseudo second order rate constants		References
				Q_e	K_L	N	K_F	$k_1(h^{-1})$	Q_e (mg/g)	k_2 (g/(mg.h))	Q_e (mg/g)	
7	Orange peel waste microwave activated biochar	Congo red	pH:2-3, Dosage: 3 g/L, Max.removal:136 mg/g (Steam activated), 91 mg/g (CO_2 activated)	192.30	0.019	4.76	47.589	0.0004	2.99	0.006	97.1	[17]
8	Phsoenix dactylifera leaves and nZVMn/PBC composite		pH:5.8, Dosage: 500 g/L, Initial dye conc. :10–40 mg/l, Reaction time: 120 min, Max.removal = 77.8 mg/g	117.64	0.033	0.033	117.64	0.076	23.640	0.002	34.9	[34]
9	Wheat straw	Methylene blue	pH :8-9, Dosage:6 g/L, Initial dye conc.:10–150 mg/L	14.16	0.089	2.65	2.64	–	–	–	–	]18]
10	Pine wood		pH: 6.5, Dosage:10 g/L, Initial dye conc.:10–500 mg/L	3.99	0.015	1.22	1.540	–	–	–	–	[20]
11	SB600		pH: 6, Dosage:2 g/L, Initial dye conc.:2.5–640 mg/L	37.6	0.16	4.032	9.94	0.12	9.93	0.05	21.6	[14]

(continued)

Table 2 (continued)

S. No	Source of biochar	Dye	Optimum conditions	Langmuir Isotherms		Freundlich isotherm		Pseudo first order rate constants		Pseudo second order rate constants		References
				Q_e	K_L	N	K_F	$k_1(h^{-1})$	Q_e (mg/g)	k_2 (g/(mg.h))	Q_e (mg/g)	
12	SB900			196.1	1.18	3.05	58.31	0.13	21.69	0.01	95.2	[14]
13	Cotton wood		Dosage:2 g/L, Initial dye conc.:5–1000 mg/L	174	0.014	–	–	–	–	–	–	[30]
14	Municipal waste		pH :5, Time-6 h, Dosage: 5 g/l, Initial conc.100 ppm, Max.removal–7.254 mg/g	365.9	0.0002	1.8746	0.6660	0.1615	6.3878	6.719	0.03	[21]
15	Pumpkin waste		pH:7, Initial dye conc :200 ppm, Max. removal :198.15 $mg.g^{-1}$	185.1	0.3235	5.70125	113.77	81.002	198.15	0.004	163	[22]
16	Eucalyptus	Methylene blue	pH :7, Dosage:0.75 g/L, Initial dye conc.:5–400 mg/L	178.57	0.196	3.337	39.774	0.0531	20.2	0.00638	39.2	[20]
17	Hazzlenut		Dosage: 0.15 g/L, initial dye conc. 250 ppm, Temperature : 25°C, Max.removal = 204 mg/g	204	0.374	5.9	86	0.0003	50	37	52	[23]

(continued)

Table 2 (continued)

S. No	Source of biochar	Dye	Optimum conditions	Langmuir Isotherms		Freundlich isotherm		Pseudo first order rate constants		Pseudo second order rate constants		References
				Q_e	K_L	N	K_F	$k_1(h^{-1})$	Q_e (mg/g)	k_2 (g/(mg.h))	Q_e (mg/g)	
18	Date palm petiole derived biochar	Crystal violet	pH :7, Temperature :30°C Max.removal =27.4 mg/g, 209 mg/g	0.074	226	0.307	45.1	0.574	23.9	0.124	24.1	[24]
19	Calcium rich biochar from Crab Shell	Malachite green	Dosage :0.5 g/L, Contact time :150 min, Temperature :25°C, Max.removal: 12,501.98 mg/g	12501.	0.0906	0.2242	2715.09	0.0302	11688.3	4.08×10^{-6}	12616	[15]
20	Rice husk biochar		pH 6:.2, Initial conc.20 ppm, Time 2 h, Max.removal = 99.98%	37.47	1.26	7.04	4.33	–	–	2.5×10^{-2}	20.01	[36]
21	Opuntia ficus-indica		pH-6, Dosage: 60 mg/100 mL, Temperature: 30°C, contact time :120 min, Max.removal: 1341.38 mg/g	1341.3	0.18	3.47	381.96	0.127	739.06	2.51×10^{-4}	795.3	[26]
22	Activated wakame biochar material	Rhodamine	pH: 2–12, Dosage: 0.06 g/50 mL , Temperature : 20°C, Max.removal = 533.77 mg/g	840.34	0.057	6.938	248.32	0.0195	243	0.00034	444.4	[25]

(continued)

Table 2 (continued)

S. No	Source of biochar	Dye	Optimum conditions	Langmuir Isotherms		Freundlich isotherm		Pseudo first order rate constants		Pseudo second order rate constants		References
				Q_e	K_L	N	K_F	$k_1(h^{-1})$	Q_e (mg/g)	k_2 (g/(mg.h))	Q_e (mg/g)	
16	Eucalyptus	Methylene blue	pH: 7, Dosage: 0.75 g/L, Initial dye conc.: 5–400 mg/L	178.57	0.196	3.337	39.774	0.0531	20.2	0.00638	39.2	[20]
17	Hazzlenut		Dosage: 0.15 g/L, initial dye conc. 250 ppm, Temperature : 25°C, Max.removal = 204 mg/g	204	0.374	5.9	86	0.0003	50	37	52	[23]
18	Date palm petiole derived biochar	Crystal violet	pH: 7, Temperature : 30°C Max.removal = 27.4 mg/g, 209 mg/g	0.074	226	0.307	45.1	0.574	23.9	0.124	24.1	[24]
19	Calcium rich biochar from Crab Shell	Malachite green	Dosage: 0.5 g/L, Contact time: 150 min, Temperature : 25°C, Max.removal: 12,501.98 mg/g	12501.	0.0906	0.2242	2715.09	0.0302	11688.3	4.08×10^{-6}	12616	[15]
20	Rice husk biochar		pH 6:.2, Initial conc.20 ppm, Time 2 h, Max.removal = 99.98%	37.47	1.26	7.04	4.33	–	–	2.5×10^{-2}	20.01	[36]

(continued)

Table 2 (continued)

S. No	Source of biochar	Dye	Optimum conditions	Langmuir Isotherms		Freundlich isotherm		Pseudo first order rate constants		Pseudo second order rate constants		References
				Q_e	K_L	N	K_F	$k_1(h^{-1})$	Q_e (mg/g)	k_2 (g/(mg.h))	Q_e (mg/g)	
21	Opuntia ficus-indica		pH-6, Dosage: 60 mg/100 mL , Temperature: 30°C, contact time :120 min, Max.removal: 1341.38 mg/g	1341.3	0.18	3.47	381.96	0.127	739.06	2.51×10^{-4}	795.3	[26]
22	Activated wakame biochar material	Rhodamine	pH: 2–12, Dosage :0.06 g/50 mL, Temperature : 20°C, Max.removal = 533.77 mg/g	840.34	0.057	6.938	248.32	0.0195	243	0.00034	444.4	[25]
23	Bamboo shoot shells :BC-300	Rhodamine	Temperature: 0 °C, Time :120 min, Max.removal: 27 mg/g	25	0.065	0.385	3.251	0.079	8	0.013	20	[27]

(continued)

Table 2 (continued)

S. No	Source of biochar	Dye	Optimum conditions	Langmuir Isotherms		Freundlich isotherm		Pseudo first order rate constants		Pseudo second order rate constants		References
				Q_e	K_L	N	K_F	$k_1(h^{-1})$	Q_e (mg/g)	k_2 (g/(mg.h))	Q_e (mg/g)	
24	BC-600		Temperature:0 °C, Time : 90 min, Max.removal: 23 mg/g	48	0.026	2.357	3.987	0.085	9.1	0.027	23	[27]
25	BC-800		Temperature: 25 °C, Time : 20 min, Max.removal : 85.8 mg/g	85.8	0.033	2.15	6.315	0.084	37	0.015	37	[27]
26	Caulerpa scalpelliformi	Remazol brilliant blue R	pH-2, Temperature : 30°C, Dosage : 2 g/L, Time : 300 min, Max.removal: 129.00 mg/g	19.207	0.178	2.8410	0.2023	0.0471	0.1553	0.435	0.169	[35]
27	Fruit bunch	Orange G	pH-2, dosage: 1.0 g, contact time: 120 min, contact time: 120 min Max.removal: 32.36 mg/g	32.36	0.4221	1.767	9.713	0.060	25.97	0.064	5.86	[29]

Liu et al. [18] produced biochar from wheat straw using a fluidized bed at 550 °C and applied it for the removal of methylene blue. The yield of biochar was about 30%. The biochar was carbon-rich (Table 1) and had graphitic nature (as illustrated by XRD). p^{H}_{pzc} was 7.8, and hence, pH values of dye solutions were maintained above this value to ensure better adsorption through electrostatic forces of attraction between biochar and the methylene blue cationic ions. Meili et al. [19] prepared LDH-biochar composites of various molar ratios of Mg:Al (2:1, 3:1 and 4:1) by co-precipitation strategy. The presence of Mg-Al-LDH on the surface of the biochar was confirmed through characteristic peaks between 5 and 150° in XRD. With the increasing Mg:Al molar ratio, the BET surface increased, but diameter of pores and volume decreased. Highest adsorption capacity of 406.45 mg/g was obtained at 40 °C. A maximum removal of 95% was observed at pH-12 within 20 min. The results suggested that adsorbent was efficient in methylene blue removal process under alkaline conditions.

Sun et al. [20] produced biochars using eucalyptus saw dust and its modified forms (with citric, tartaric and acetic acids) and used it for removal of methylene blue. It was found that, after modification, the carbon content of citric acid modified biochar, tartaric acid modified biochar and acetic acid modified biochar increased. Further, the ratio (N + O)/C decreased significantly for all three modified biochars, which indicated that the polarity of the biochars was influenced by the organic acids. The average pore width of citric acid modified biochar was greater than other two modified biochars. Basically, micropore walls collapsed by the citric acid and lost microporous structure and exhibited greater adsorption efficiencies than other two biochars for same biochar dosage. The adsorption efficiencies of biochars obtained from citric acid and tartaric acid and acetic acid were 95.2%, 90.7%, and 61.5%, respectively, at adsorbent dosage of 0.5 g/L.

Hoslett et al. [21] produced from municipal waste materials in a pipe reactor. Temperature and process time during pyrolysis was kept at 300 °C for a process time of 12 h. Paper wastes, food wastes, garden wastes and cardboard wastes were used as feedstock. The produced biochar was found to contain functional groups with high amount of oxygen and hence favoured the adsorption of methylene blue and graphitic structures. It was envisaged that π-π interactions existed between functional groups and methylene blue. Raman spectra suggested that the biochar contained aromatic C–C Sp2 hybridized bonds of benzene rings. The highest capacity of methylene blue was 7.1 mg/g at 100 ppm and pH 5. Other sources of biochar used for removal of methylene blue are pumpkin [22] and Hazlenut [23].

3.1.3 Crystal Violet

For the removal of dye crystal violet, Chahinez et al. [24] used biochar produced from date palm petioles at 700 °C. Results indicated that the biochar contained high amount of carbon (86.62%) and had high specific surface area (640 m^2/g). The p^{H}_{PZC} of the biochar was 6.55. Upon increasing the pH from 2 to 12, the quantity of crystal violet adsorbed on biochar increased from 18.8 to 27.4 mg/g.

3.1.4 Rhodamine and Malachite Green

Yao et al. [26] adopted a one-step process involving calcination and activation for the production of wakame biochar. The potential of adsorbent was evaluated on methylene blue, rhodamine B and malachite green. Wakame (Undaria pinnatifida) leaves were dried, powdered and mixed with KOH solution and the mixture was subjected to pyrolysis at 800 °C for a duration of 2 h in a tubular furnace under inert atmosphere. The heating rate used was 10 °C/min. It was observed that the biochar was a mesoporous and had a fluffy structure, with high specific surface (1156.25 m^2/g). The average diameter of pores was about 1.5 μm, which was confirmed by TEM images of biochar.

Choudhary et al. [25] used biochar and its activated form produced from the cladodes of *Opuntia ficus-indica* for the efficient adsorption of malachite green. Biochar was produced by pyrolysis at 400 °C for a duration of 1 h under inert nitrogen atmosphere. Sodium hydroxide was used for the activation of the biochar with two different dosages (1:1 and 1:2). The mass yield of activated biochar reduced to 43% and 37% upon activation using 1:1 and 1:2 ratios, respectively. The biochar structure was graphitic before activation. The activation resulted in two peaks in XRD pattern (at 25° and 48°), which were related to planes (0 0 2) and (1 0 0) diffractions. In addition, the peak at $2\theta = 43°$ was related (1 0 1) plane showed that the activated biochars had amorphous nature. The highest uptake of malachite green was 341 mg/g.

Hou et al. [27] used bamboo shoot shells as the raw material for producing biochar and used for removal of Rhodamine. Around 3 g of bamboo powders were transferred to 60 ml water and pyrolyzed at 200 °C for a duration of 5 h to produce biochar. Then, it was filtered, dried at 100 °C for a day, and pyrolyzed at 300, 600 and 800 °C for 2.5 h under inert atmosphere. The biochar produced at 800 °C had highest surface area of 513 m^2/g.

3.1.5 Acid Orange G

Wang et al. [28] used corncob to synthesize green recyclable magnetic biochar for acid orange dye removal. 10.0 g corncob powder in 60 ml of $FeCl_3$ solution was subjected to calcination in 710 °C with a heating rate of 10 °C/min for 3.5 h to produce biochar MC_1. Similar experiment was done using biochar of corncob instead of corncob powder to produce another biochar MC2. To produce magnetic biochar, MC_1 was added to solution containing 0.58 g of nickel nitrate hexahydrate, 0.38 g of aluminium nitrate nonhydrate and 0.41 g of urea and stirred for 2 h. After filtration, the black solid was dried to obtain biochar MC1/NiAl-LDH composite material. Similar experiment was done with MC_2 to obtain MC1/NiAl-LDH. Further, modified magnetic biochar was synthesized by pyrolysis of Ni–Al layered double hydroxides at 700 °C for duration of 3.5 h. The surface area of MC1/NiAl-LDO was greater than that of MC_1 alone. MC1/NiAl-LDO was used for acridine orange (AO) and found to exhibit 90% greater adsorption capacity than MC_1.

3.1.6 Reactive Red

Xu et al. [29] studied the effect of inert atmosphere during pyrolysis on quality of biochar and used for reactive red removal. Biochars were produced in nitrogen atmosphere, vacuum atmosphere and in presence of air. O_2-limited atmospheres were more favourable for decreasing biochar size as compared to air atmosphere. Surface area of biochar produced under nitrogen atmosphere and vacuum was higher than that produced in air. Ball milling enhanced the adsorption potential of biochars from 1.70–3.60 to 9.20–34.8 mg/g.

3.1.7 Methyl Violet and Basic Red Dye

Zazycki et al. [30] produced biochar using chitin as a feedstock in a quartz reactor and used for methyl violet removal. The conditions during pyrolysis were: nitrogen flow rate −0.25 L/min, heating rate −10 °C/min, temperature: 800 °C. The produced biochar had high surface area of 275 m^2/g. The adsorption of methyl violet on the chitin biochar was favoured under alkaline conditions and ambient temperature. The optimal conditions for adsorption were: pH 6.8, dosage −0.5 g/L, with an adsorption capacity of 95.64 mg/g and removal percentage of 96.8%. Côrtes et al. [31] produced biochar from animal wastes (bovine bones and fish scales) and evaluated their efficiency in removing basic red dye. The quartz reactor was used had length of 98.1 cm, outside diameter 49 mm and internal diameter of 43 mm. The reactor was heated electrically by two resistors each with capacity of 1900 W. The ash content of fish scale was higher than bovine bone which was due to presence of minerals. Due to presence of proteins, the fish scales had higher volatile matter. The mesoporous biochars had a surface area of 90 m^2/g. Both the biochars had good adsorption capacity for basic red dye (49.4 mg/g for bovine bones biochar and 52.2 mg/g for fish scale biochar).

4 Mechanism of Adsorption of Various Dyes on Different Biochars

Park et al. [14] studied the underlying mechanisms of methylene blue, congo red and orange G adsorptions on biochars obtained from steam activated spent mushroom substrates. The characteristics of SB 600 and SB 900 obtained by pyrolysis at 600 and 900 °C have been discussed in Sect. 3.1.1. For methylene blue, the adsorption capacity of SB600 and SB900 increased when pH was increased from 3 to 5 and further increased rapidly from 5 to 10. In general, the factors influencing dye adsorption on biochars are: (i) charge of dye, (ii) molecular weight, (iii) the charge density of biochar, (iv) the cation exchange capacity and (v) the π-π interactions. Basically, methylene blue is positively charged at pH < 3.8 and neutral at pH > 3.8. On the other hand, biochar is positively charged at pH < 5.0 and negatively charged at pH > 5. SB900 was positively charged in the pH range of 2.5–4.5, while a negatively charged at pH 5.5. In spite of non-charge at pH 3, a high adsorption capacity was observed at this pH, which was attributed to by π-π interactions.

The anionic orange G adsorption by SB600 and SB900 exhibited differing pH dependency. With increasing from pH from 3 to 5, the quantity of orange G adsorbed by SB900 reduced faster than for SB-600 and reduced to 25.1–21.7 mg/g, when pH was increased from 5 to 10. However, the total difference in adsorption quantities of two adsorbents over entire pH range was insignificant. For the congo red R adsorption, low pH favoured higher adsorption. However, SB 600 and SB900 exhibited differences in adsorption rates. Due to presence of sulfonate group on congo red, the dye was a negatively charged at pH > 4. With increasing pH, the congo red adsorption decreased. An adsorption quantity of 14.1 mg/g was observed for congo red at all values of pH and was attributed to π-π interactions.

Yek et al. [17] investigated the mechanisms for congo red and crystal violet removal using steam and carbon dioxide activated biochars. The data obtained by conducting adsorption experiments at initial dye concentration of 1000 ppm of congo red, 2000 ppm of crystal violet at pH 6 and temperature of 30 °C was tested by intraparticle diffusion models. For congo red adsorption, diffusion profiles showed multi-linearity and the intraparticle diffusion rate constant reduced at each stage from 18.2 to 0.824 mg/(g $min^{0.5}$) for steam activated biochar. The transition times of adsorption were much quicker in steam activated biochar than inactivated (199 v/s 869 min), which was due to higher surface and porosity of steam activated biochar, which provided free paths for adsorption.

For crystal violet adsorption, the breakpoint was at 171 and 268 min for inactivated biochar and at 73.4 and 325 min for steam activated biochar. Hence, congo red exhibited two stages of diffusion while crystal violet showed three stages of diffusion, suggesting that intraparticle diffusion occurs successively in small pores of biochar and continues for mesopores for congo red. While intraparticle diffusion occurs in micropores of biochar for crystal violet. These observations were described by considering dimension of both dyes. The congo red has dimensions of (0.7 × 2.5 nm) and crystal violet dimensions are (0.87 × 1.03 × 1.14 nm). This suggests that adsorption of congo red in micropores (<2 nm) was poor for both inactivated and steam activated biochars.

The methylene blue, rhodamine and malachite green adsorption on biochar material derived from wakame were explained by Yao et al. [26]. It was concluded that the adsorption of all three involved complex physical and chemical interaction of dye with biochar. The BET surface area of biochar was very high (1155.24 m^2/g) and furthermore had highly mesoporous structure. Thus, a significant number of active sites were available for adsorption of dyes. The porous nature also helps in reduction of resistance for diffusion and assists in the adsorption of dyes. Hence, the biochar was able to uptake the dye by phenomenon like surface adsorption and pore diffusion. Analysis of FTIR of biochar showed abundant active groups (–CO, –OH, and –CH) which contributed to adsorption of dyes by means of π-π interactions, Van der Waals force and hydrogen bond. The influence of pH on adsorption of dyes showed

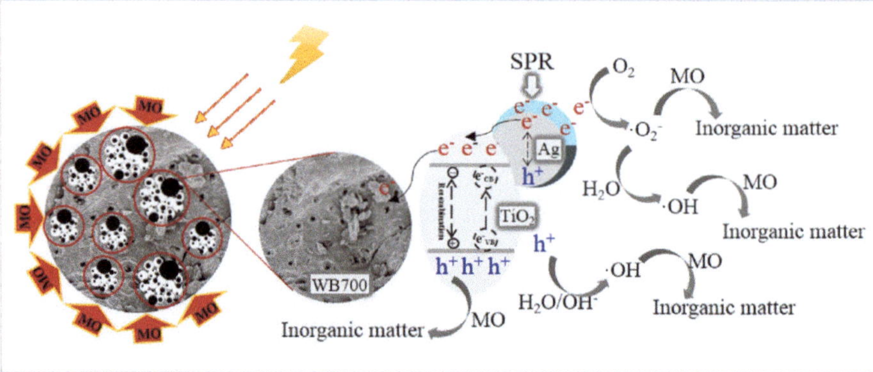

Fig. 4 Catalytic degradation of MO by CTAg1 proposed by Shan et al. [32] (Reprinted with permission from Elsevier)

the role of electrostatic attraction was responsible for the adsorption of the all dyes on biochar.

Wang et al. [28] studied the mechanism of adsorption of acridine orange on double oxide (NiAl layered) magnetic corncob biochar. The highest adsorption was reported at pH 10 due to high electrostatic interactions between biochar MC1/NiAl-LDO and dye solution. This was due to the protonation of acridine orange. When the pH of dye solution was lower than pKa, protonated AOH+ formed dimeric species. On the other hand, due to large number of micropores on the surface of MC1/NiAl-LDO, the molecules of dye could quickly adsorb on biochar surface. Finally, π-π interactions existed between magnetic biochars and acridine orange due to presence of C = C and in the benzene ring of dye molecule.

Shan et al. [32] demonstrated the adsorption mechanism of methyl orange by Ag/TiO_2/biochar composite catalysts in aqueous solutions (Fig. 4). Initially, methyl orange was adsorbed on biochar (due to oxygen containing groups). When the biochar was exposed to UV light, TiO_2 was excited to produce electrons from valence band to conduction band, while h+ were left behind in valance band. Meanwhile, the silver particles got photoexcited, and then finally, h+ , $O_2.^-$ and OH acted as active sites for interaction with dye.

Choudhary et al. [25] explained the mechanism of adsorption of malachite green dye onto activated biochar produced from *Opuntia ficus-indica.* It was suggested that the adsorption of dye was due to pore diffusion, π-π EDA interaction and cation-π interaction along with hydrogen bonding. Among these, the pore filling mechanism dominated due to wide pore size distribution. Also, the physical adsorption was high due to highly mesoporous nature of adsorbent and reduced steric hindrance. Chemisorption was due to π-π interactions, π-cation interactions and hydrogen bonding. Further, since the biochar had graphitic nature and electron-rich functional groups and the dye processes electron-deficient groups, the interactions were stronger (Fig. 5).

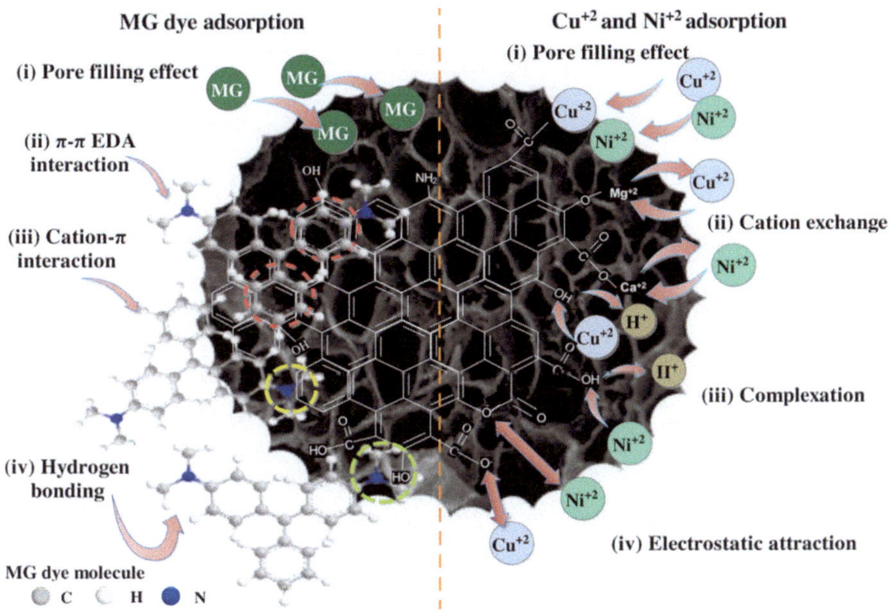

Fig. 5 Adsorption mechanism of malachite green dye by biochar produced from *Opuntia ficus-indica* (Choudhary et al. [25], reprinted with permission from Elsevier)

Zhu et al. [33] demonstrated the mechanism of adsorption of acid orange 7 using wood biochar. Wood char was used activator for reducing peroxydisulphate. The dye removal was effective in pH range of 3–10. Nearly 99.3% of dye was removed after 14 min of reaction. The sulphate and hydroxyl radicals were observed on biochar surface. Analysis of biochar indicated that density of π-electron functional groups containing oxygen (C–OH) acted as active centres for catalytic reaction. Further, it was shown that the surface reaction was controlling step during adsorption process (Fig. 6).

Xu et al. [29] suggested a possible mechanism for degradation of reactive red using biochar of Hickory chips (HC). The biochars were ball milled under nitrogen and vacuum atmospheres and used for removal of reactive red. Due to ball milling under nitrogen and vacuum, the particle size was reduced significantly which promoted external diffusion process. Also, surface O-containing functional group was formed on biochar surface. The presence of inert atmosphere led to lower oxygen containing moieties on biochar surface. On the one hand, ball milling facilitated the formation NH_2 groups on biochar which facilitated electrostatic adsorption (Fig. 7).

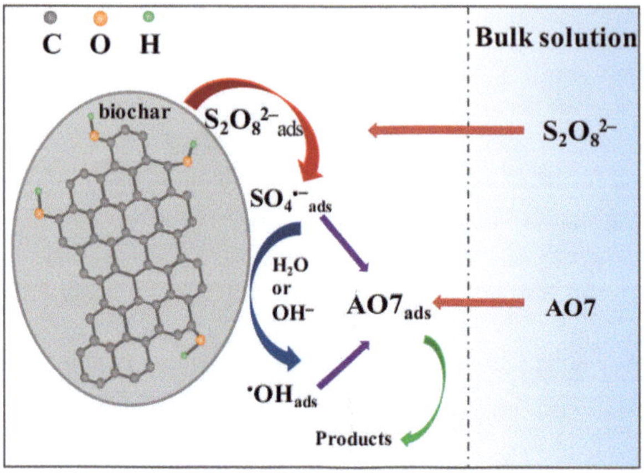

Fig. 6 Adsorption mechanism for acid orange 7 using biochar from wood (Zhu et al. [33], reprinted with permission from Elsevier)

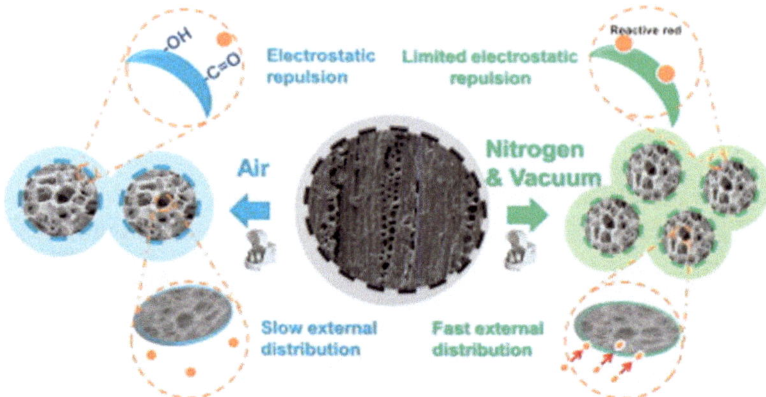

Fig. 7 Mechanisms of adsorption of reactive red on Hickory chips derived biochar (Xu et al. [29], reprinted with permission from Elsevier)

5 Adsorption Isotherms for Dyes on Biochars

Most of the adsorption isotherms of dyes on biochar surfaces have been successfully fitted to Langmuir and Freundlich isotherms.

The Langmuir equation is given by:

$$\frac{C_e}{q_e} = \frac{1}{q_e K_L} + \frac{C_e}{q_m} \tag{1}$$

The term q_e in above equation is the quantity of dye adsorbed at equilibrium. The term q_m is the quantity of dye adsorbed at saturation and is in mg/g. The concentration of dye at equilibrium is C_e (mg/L) and K_L is Langmuir constant, related to binding potential of dye on adsorbent. On the other hand, the Freundlich model of isotherm is given by:

$$q_e = K_f C_e^n \tag{2}$$

where n and K_f are the Freundlich isotherm constants, which indicate the intensity of dye adsorption and adsorption uptake capability, respectively. If $n = 1$, adsorption is a linear, while if $n < 1$, physical absorption, and $n > 1$ indicates that adsorption is chemisorption. Khan et al. [13] studied the adsorption isotherms of congo red on rice husk and cow dung biochars. Both the isotherms were satisfactory for the biochars. For Freundlich isotherm model ($R^2 = 0.883$ and 0.842) and for Langmuir model ($R^2 = 0.902$ and 0.977), the ads orption isotherm constants are reported in Table 2. The value of n was greater than 1 indicating chemisorption. The highest adsorption uptake was reported as 933.6 and 580.4 mg/g for methyl orange and methylene blue dye, respectively.

Park et al. [14] shown that the adsorption data indicated that SB900 had higher adsorption capacity than SB600 for all three dyes (methylene blue, congo red, methyl orange). The adsorption isotherms of all three dyes were fitted better to Langmuir model ($R^2 = 0.9637$–0.9995) for both SB900 and SB 600. Freundlich isotherm did not show a good fit (0.7103). Thus, adsorption behaviour of all dyes was dominated by physical adsorption. The highest uptakes of methylene blue, orange G and congo red were reported as 37.6, 8.1 and 8.2 mg/g for SB600, and 196, 38.1 and 22.6 mg/g for SB900, respectively.

Chahinez et al. [24] applied Langmuir, Freundlich, and Redlich–Peterson models for understanding the equilibrium behaviour of crystal violet adsorption on biochar obtained from date palm petioles. Based on R^2 (standard deviation), it was suggested that experimental data was well described by the Redlich–Peterson as well as Langmuir models, while the fit was poor for Freundlich model. As per Langmuir model, the highest uptake capacity of the biochar was calculated as 208 mg/g.

Yao et al. [26] used Langmuir, Freundlich isotherms for fitting the adsorption data for biochar material derived from wakame. For Freundlich model, the R^2 value of methylene blue, rhodamine B, malachite green were 0.94, 0.95 and 0.98, respectively. R^2 for Langmuir model were 0.98 0.99 and 0.99, respectively. Thus, the Langmuir model was more appropriate model to explain the adsorption of the three dyes using wakame biochar. The highest adsorption uptakes were found to be 841.34 mg/g, 574.71 mg/g and 4303.71 mg/g at 20 °C for three dyes, which were in good agreement with experimental data (841.62 mg/g, 531.78 mg/g, and 4066.96 mg/g). Hence, it was shown that adsorption of the three dyes using wakame biochar was monolayer chemical adsorption process.

Hou et al. [27] investigated rhodamine B removal using bamboo shoot shell biochar. It was demonstrated that Freundlich model was more appropriate than

Langmuir isotherm for BHC-600 and BHC-800 ($R^2 > 0.94$), while for BHC-300, Langmuir model was found to give a good fit ($R^2 > 0.97$). A high value of K_F value (6.31 for BHC-800) was attributed to its higher surface area, pore volume and low diffusional resistance for dye in inner pores of biochar. Also, the adsorption process was energetically favourable.

Wang et al. [28] used the Langmuir, Freundlich, Temkin and Dubinin–Radushkevich adsorption isotherm models to fit the experimental data of MC1/NiAl-LDO at 303, 313 and 323 K. An analysis of parameters of adsorption models suggested that experimental data was better fitted by Langmuir isotherm equation than other models.

Xu et al. [29] showed that the adsorption isotherm of reactive red onto biochars was fitted by Langmuir and Freundlich models ($R^2 = 0.84$–0.97). The pristine biochars (HC450 and HC600) exhibited lower adsorption uptake of 1.72–3.58 mg g^{-1} for the dye reactive red (Table 2), due to lower surface area and negative surface charge. However, ball milling enhanced the uptake capacities to 9.21 and 16.8 mg g^{-1} for BMHC450-A and BMHC600-A, respectively, which was attributed to the greater surface area in spite of higher negative charge.

Côrtes et al. [31] investigated the adsorption isotherms of biochars from bovine bones (BB) and fish scales (FS) derived biochar for the basic red removal. The isotherm for bovine bones followed Langmuir-type curve. However, the shape of BET adsorption isotherm for fish scale biochar was found to be of "type II". The initial linear portion of the curve refereed to a formation of a monolayer of dye onto biochar surface and the second step corresponds to the multilayer adsorption.

6 Kinetics of Adsorption of Dyes on Biochars

Sewu et al. [16] studied kinetics of congo red and crystal violet adsorption onto steam activated spent mushroom substrate. It was observed that pseudo-first and pseudo-second-order models fitted well to congo red adsorption, but the more favourable for former. For adsorption of crystal violet, both pseudo-first- and pseudo-second-order models fitted on both biochars. However, pseudo-second-order kinetics exhibited a better fit based on the higher R^2 value (0.906). The rate constants are reported in Table 2. However, Yek et al. [17] showed that shows that kinetics of adsorption of both biochars was better explained by pseudo-second order ($R^2 > 0.95$). Iqbal et al. [34] demonstrated that adsorption of congo red on composite of nano-zerovalent manganese (nZVMn) and phoenix dactylifera leaves biochar fitted better to first-order kinetics.

Rashid et al. [22] investigated the kinetics of methylene blue on pumpkin peel waste derived biochar. It was shown that adsorption kinetics followed pseudo-second order and the biochar obtained at 250 °C highest reaction rate constant value of 0.0095 min^{-1} and correlation coefficient value of 0.999.

Sun et al. [20] showed that kinetics of methylene blue on biochar from eucalyptus followed Pseudo-second-order kinetics better described the adsorption than pseudo-first-order, Elovich, and intraparticle diffusion models. The q_e values for the pseudo-first-order kinetic model were less than the corresponding experimental values. Hence, pseudo-first-order model could not accurately predict the adsorption kinetics of methylene blue activated biochars. Also, the R^2 values for Elovich kinetic model were in the range of 0.8836–0.9943, which were lower than values for pseudo-second-order kinetic model. However, the calculated and experimental values of q_e were closer for second-order model. Furthermore, it was suggested that rate-controlling step involved both intraparticle diffusion and boundary layer diffusion to some extent.

Chahinez et al. [24] investigated the kinetics of palm petiole-derived biochar for adsorption of biochar. The kinetics was fast and approximately 93% of initial dye was removed within 5 min of process. Within 15 min, the adsorption process attained equilibrium. The experimental data was explained well by four different models (pseudo-first-order, second-order, Elovich, and Avrami model), as R^2_{adj} of all models were greater than 0.997. But, Avrami model was found to give the best fit based on the least value of the error functions (chi-square and standard deviation).

Yao et al. [26] studied the kinetics of three dyes (methylene blue, rhodamine B and malachite green) on wakame biochar. It was noticed that pseudo-second order fitted the experimental data of all dyes. In addition, it was reported that the adsorption uptake capacity (q_e) increased with increasing initial dye concentration, while the rate constant (k_2) decreased, due to the severe competition among cations for surface active sites. Dai et al. [15] studied the kinetics of malachite green and congo red using biochar from crab shell. It was shown that malachite green adsorption was described by pseudo-second-order model, while congo red adsorption was fitted by both first order and second order models. Hence, for malachite green, chemical adsorption was rate limiting step and for congo red, physical process was rate limiting. Also, a much lower value of k_1 for malachite green indicated the higher affinity of biochar to negatively charged congo red dye.

Gokulan et al. [35] studied the kinetics of remazol dyes by biochar derived from *caulerpa scalpelliformis.* Experiments were conducted at pH 2.0 and 30 °C with various initial dye concentrations (0.1–1 mmol/L). 90% adsorption was observed within 90 min of process at initial dye concentration of 0.1 mmol/L. Equilibrium was attained at 300 min at various initial concentrations for all four dyes. Pseudo-first-order model fitted the kinetics for all dyes with high R^2 (>0.983) values. As the initial dye concentration increased, adsorption capacity of the biochar increased pseudo-second-order model was better fitted to experimental data as compared to the pseudo-first-order model on the basics of (high R^2).

Mubarak et al. [36] studied the kinetics of orange G on magnetic chicken biochar. Second-order kinetics was fitted to experimental better than first order. The % removal of the orange G increased with decreasing pH.

7 Thermodynamics of Adsorption of Dye on Biochars

Gokulan et al. [35] reported thermodynamics of adsorption of four different remazol dyes (remazol brilliant blue R, remazol brilliant orange 3R, remazol brilliant violet 5R and remazol black B) onto a biochar produced from *Caulerpa scalpelliformis*. The standard enthalpy ($\Delta H°$) and standard entropy ($\Delta S°$) were calculated at temperatures of 293 K, 303 K and 313 K for all dyes. Negative values of changes in $\Delta G°$ were obtained for all four dyes which suggested a favourable adsorption. Further, the positive values of $\Delta H°$ indicated that the process was endothermic. Hence, the adsorption capacity increased with increasing temperature. The magnitude of $\Delta H°$ also indicates the type of sorption. The heats of chemisorption usually are in the range of 80–200 kJ mol^{-1}. However, lower values of $\Delta H°$ (<80 kJ mol^{-1}) indicated that adsorption was physical. A positive value of $\Delta S°$ indicated that significant interactions existed between dyes and biochars. Basically, the positive value of $\Delta S°$ indicated that randomness increased at biochar and dye interface with little changes in structures of dye and biochar. The solvent molecules which are adsorbed are dislodged by adsorbed dye molecules and gain higher translational entropy than the dyes molecules loose. Hence, disorder of system increases. A positive value of $\Delta S°$ indicated that the adsorption process involves a dissociative mechanism.

Zazycki et al. [30] studied thermodynamics of methyl violet dye on chitin derived biochar. The values of standard free energy were in the range of 3.08–0.12 kJ mol^{-1}. The negative values indicated that the methyl violet adsorption was spontaneous. The standard entropy changes were negative (-0.032 kJ mol^{-1} K^{-1}). The negative sign indicated that the disorder at dye-biochar interface declined upon adsorption after the adsorption. A negative standard heat of reaction (-2.61 kJ mol^{-1}) suggested that adsorption was exothermic in nature. The magnitude of standard heat of reaction showed that physical interactions between dye and biochar were presumably hydrogen bonds.

Choudhary et al. [25] investigated the thermodynamic feasibility studies of malachite green on *Opuntia ficus-indica.* The standard Gibbs free energy changes remained almost constant with increasing temperature for adsorption of malachite green dye on biochar suggesting that spontaneity of malachite green G dye adsorption is independent of temperature. The negative values of standard enthalpy of reaction and standard entropy change explained that the adsorption was spontaneous.

Chahinez et al. [24] showed that the thermodynamics of crystal violet on palm petiole-derived biochar was favoured at a higher temperature, thus Van der Waals forces were less significant. Negative values of $\Delta G°$ indicated spontaneity of adsorption. However, ΔH° was positive and low, implying an endothermic process and removal mechanism of crystal violet by palm petiole derived biochar was basically due to pore filling (physical adsorption). Furthermore, a positive value of $\Delta S°$ indicated enhanced random motion crystal violet molecules at biochar.

Meili et al. [19] demonstrated that standard heat of reaction for methylene blue adsorption onto MgAl-LDH/biochar composite was positive which suggested that ion exchange process which occurred was physical process and endothermic. The positive value of ΔS indicated that degrees of freedom increased during adsorption.

8 Recyclability of Biochars Adsorbents

Choudhary et al. [25] studied the recyclability of biochars derived from *Opuntia ficus-indica* in removal of malachite green dye. 0.1 M sodium hydroxide solution was used as an eluant for regeneration. The regeneration of used activated biochar was confirmed as malachite green dye desorbed in NaOH solution. It was observed that % removal of dye nearly constant for first three cycles and only 15% reduction was observed after third cycle. Hence, it was confirmed that physical adsorption was dominant during adsorption. The efficiency of biochar reduced significantly after recycling for six cycles which could be due to the loss of active sites and blockage of pores biochar.

Shan et al. [32] studied the potential of recycling Ag/TiO_2/biochar in removing methyl orange dye. It was shown that the removal of dye decreased from 97.48 to 96.10% after five cycles, indicating an excellent capacity of adsorbent. The reduction in number of adsorption active sites was due to difficulty in removing by-products of dye degradation. In order to structural stability of catalyst, the biochar was tested by XRD. The recycled biochar after 5 cycles showed anatase TiO_2 peak and elemental Ag peak in the XRD pattern, thus confirming a good retainment of activity of catalyst.

Gokulan et al. [35] investigated the possibility of recycling *Caulerpa scalpelliformis* biochar in removing four remazol dyes. Various regenerating solvents were used such as 0.01 M NaOH, Na_2CO_3, NH_4OH, HCl, CH_3OH and EDTA were used for desorption studies. The elution efficiencies were reported as 99.8, 99.1 and 91.9%, respectively, for the three eluants. However, 0.01 M HCl was inefficient with an removal capacity of less than 2.6%. The performance of methanol and EDTA were also poor. Thus, 0.01 M NaOH was best eluting agent for desorption of remazol dyes. Sewu et al. [16] reported the steam activated spent mushroom derived biochar's reusability. The q_e values decreased by 50% for all biochars after first recycling. However, thereafter, q_e remained nearly for both inactivated and activated biochars. It was observed that, except 50% ethanol, the other regenerating agents showed poor efficiency in desorption of congo red and crystal violets (<5%). The % desorption did not alter significantly for congo red for all cycle of inactivated biochar (49.3–58.8%). But surprisingly, the efficiency increased to 92.3% at the end of third cycle for steam activated biochar. For crystal violet desorption, higher desorption efficiencies were reported for activated biochar than inactivated biochar for biochars for all three cycles. For first two cycles, the adsorbents exhibited nearly same efficiency (56.0 and 55.0%). Hence, it was concluded that for first cycle, the chemisorption dominated and hence the chemisorbed congo red and crystal violet retained their structure.

Park et al. [14] evaluated the recyclability of SB600 and SB900 for (methylene blue, orange G and crystal violet). SB900 was better biochar as evident from its high adsorption rate for methylene blue (78.5 mg/g after fourth cycle). For orange green and congo red, the desorption efficiency decreased significantly with increasing number of cycles. Iqbal et al. [34] explored the potential of nZVMn/Phoenix dactylifera leaves biochar. A removal efficiency of 64% was observed after seven cycles of adsorption. Thus, a drop of 78% and 33% drop in congo red removal efficiency was noted for Phoenix dactylifera leaves biochar and nZVMn/Phoenix dactylifera leaves biochars, respectively. Also, due to its magnetic nature, the nZVMn significantly increased the separation of biochar from dye. It was concluded that the excellent desorption capacity was due composite nature of biochar. Further, the use of ethanol improved the stability of biochar prepared till seventh cycle.

Dai et al. [15] studied the recyclability calcium-rich biochar for removal of malachite green and anionic congo red. Ethanol was used as eluant. The adsorption capacities were significant (>3500 and >6000 mg/g for malachite green and congo red) after 5 adsorption–desorption cycles. These values were superior to other reagents.

9 Conclusions and Future Perspectives

As illustrated in preceding sections, biochar derived from various biomass feedstocks can act as an excellent adsorbent for the removal of dyes. The properties of biochar are dependent on pyrolytic conditions (temperature, pressure, heating rate) and also on feedstock composition. Accurate optimization of these parameters can lead to enhanced yield of biochar. Further, pre- or post-treatment methods can exert significant effect on qualities of biochar. Various pre- and post-modification of biochars have been proposed in literature for improving the quality of biochar. Pre-modification steps involve treatments given to biomass feedstock before pyrolysis, while post-modification steps involve treatments applied to biochar. Among pre-modification of biochar, pre-treatment of biomass to partially remove hemicellulose and lignin can be a future perspective in this research area. The partial removal of lignin and hemicellulose can allow for better heat penetration to cellulose layer, thus allowing it to undergo pyrolytic reactions more efficiently. Although lignin is major pre-cursor to formation of biochar, the qualities of biochar from a mixture of cellulose and lignin can also result in better quality of biochar. Furthermore, a gap exists in this area. Further, the quality of biochar obtained from copyrolysis of biomass with plastics is not explored significantly so far. Plastics are hydrogen rich and addition of these during pyrolysis can improve the quality of biochar.

The third research area which can be explored is pre-treatment of biomass followed by copyrolysis with waste plastics (combination of pre-treatment and copyrolysis approach). Since the biomass structure is disrupted well during pre-treatment, it can be expected that its interactions with pyrolytic intermediates of waste plastics can be more efficient as compared to untreated biomass.

References

1. Crini G (2006) Non-conventional low-cost adsorbents for dye removal: a review. Bioresour Technol
2. De Gisi S, Lofrano G, Grassi M, Notarnicola M (2016) Characteristics and adsorption capacities of low-cost sorbents for wastewater treatment: a review. Sustain Mater Technol [Internet] 9:10–40. Available from: https://doi.org/10.1016/j.susmat.2016.06.002
3. Katheresan V, Kansedo J, Lau SY (2018) Efficiency of various recent wastewater dye removal methods: a review. J Environ Chem Eng 6(4):4676–4697
4. Rodríguez-Couto S, Osma JF, Toca-Herrera JL (2009) Removal of synthetic dyes by an eco-friendly strategy. Eng Life Sci 9(2):116–123
5. Morais da Silva PM, Camparotto NG, Grego Lira KT, Franco Picone CS, Prediger P (2020) Adsorptive removal of basic dye onto sustainable chitosan beads: equilibrium, kinetics, stability, continuous-mode adsorption and mechanism. Sustain Chem Pharm
6. Rafatullah M, Sulaiman O, Hashim R, Ahmad A (2010) Adsorption of methylene blue on low-cost adsorbents: a review. J Hazard Mater
7. Yagub MT, Sen TK, Afroze S, Ang HM (2014) Dye and its removal from aqueous solution by adsorption: a review. Adv Colloid Interface Sci
8. Ding L, Zeng HB, Wang W, Yu F (2014) Improved stability criteria of static recurrent neural networks with a time-varying delay. Sci World J
9. Li Y, Xing B, Ding Y, Han X, Wang S (2020) A critical review of the production and advanced utilization of biochar via selective pyrolysis of lignocellulosic biomass. Bioresour Technol [Internet] 312(May):123614. Available from: https://doi.org/10.1016/j.biortech.2020.123614
10. Shen Y (2020) A review on hydrothermal carbonization of biomass and plastic wastes to energy products. Biomass Bioenergy [Internet] 134(August 2019):105479. Available from: https://doi.org/10.1016/j.biombioe.2020.105479
11. Liu H, Ma M, Xie X (2017) New materials from solid residues for investigation the mechanism of biomass hydrothermal liquefaction. Ind Crops Prod
12. You S, Ok YS, Chen SS et al (2017) A critical review on sustainable biochar system through gasification: Energy and environmental applications. Bioresour Technol
13. Khan N, Chowdhary P, Ahmad A, Shekher Giri B, Chaturvedi P (2020) Hydrothermal liquefaction of rice husk and cow dung in Mixed-Bed-Rotating Pyrolyzer and application of biochar for dye removal. Bioresour Technol [Internet] 309(March):123294. Available from: https://doi.org/10.1016/j.biortech.2020.123294
14. Park JH, Wang JJ, Meng Y, Wei Z, DeLaune RD, Seo DC (2019) Adsorption/desorption behavior of cationic and anionic dyes by biochars prepared at normal and high pyrolysis temperatures. Colloids Surfaces A Physicochem Eng Asp [Internet] 572(April):274–282. Available from: https://doi.org/10.1016/j.colsurfa.2019.04.029
15. Dai L, Zhu W, He L et al (Nov 2018) Calcium-rich biochar from crab shell: an unexpected super adsorbent for dye removal. 267:510–516
16. Sewu DD, Jung H, Kim SS, Lee DS, Woo SH (2019) Decolorization of cationic and anionic dye-laden wastewater by steam-activated biochar produced at an industrial-scale from spent mushroom substrate. Bioresour Technol [Internet] 277(November 2018):77–86. Available from. https://doi.org/10.1016/j.biortech.2019.01.034
17. Yek PNY, Peng W, Wong CC et al (2020) Engineered biochar via microwave CO2 and steam pyrolysis to treat carcinogenic Congo red dye. J Hazard Mater 395(March)
18. Liu Y, Zhao X, Li J, Ma D, Han R (2012) Characterization of bio-char from pyrolysis of wheat straw and its evaluation on methylene blue adsorption. Desalin Water Treat 46(1–3):115–123
19. Meili L, Lins PV, Zanta CLPS et al (2019) MgAl-LDH/Biochar composites for methylene blue removal by adsorption. Appl Clay Sci [Internet]. 168(May 2018):11–20. Available from. https://doi.org/10.1016/j.clay.2018.10.012

20. Sun L, Chen D, Wan S, Yu Z (2015) Performance, kinetics, and equilibrium of methylene blue adsorption on biochar derived from eucalyptus saw dust modified with citric, tartaric, and acetic acids. Bioresour Technol [Internet] 198:300–308. Available from: https://doi.org/10.1016/j.biortech.2015.09.026
21. Hoslett J, Ghazal H, Ahmad D, Jouhara H (2019) Removal of copper ions from aqueous solution using low temperature biochar derived from the pyrolysis of municipal solid waste. Sci Total Environ
22. Rashid J, Tehreem F, Rehman A, Kumar R (2019) Synthesis using natural functionalization of activated carbon from pumpkin peels for decolourization of aqueous methylene blue. Sci Total Environ [Internet] 671:369–376 (2019). Available from: https://doi.org/10.1016/j.scitotenv.2019.03.363
23. Ozer C, Imamoglu M, Turhan Y, Boysan F (2012) Removal of methylene blue from aqueous solutions using phosphoric acid activated carbon produced from hazelnut husks. Toxicol Environ Chem 94(7):1283–1293
24. Chahinez HO, Abdelkader O, Leila Y, Tran HN (2020) One-stage preparation of palm petiole-derived biochar: characterization and application for adsorption of crystal violet dye in water. Environ Technol Innov [Internet] 19:100872. Available from: https://doi.org/10.1016/j.eti.2020.100872
25. Choudhary M, Kumar R, Neogi S (2020) Activated biochar derived from *Opuntia ficus-indica* for the efficient adsorption of malachite green dye, Cu+2 and Ni+2 from water. J Hazard Mater [Internet] 392(February):122441. Available from: https://doi.org/10.1016/j.jhazmat.2020.122441
26. Yao X, Ji L, Guo J et al (2020) An abundant porous biochar material derived from wakame (Undaria pinnatifida) with high adsorption performance for three organic dyes. Bioresour Technol [Internet] 318(August):124082 (2020). Available from: https://doi.org/10.1016/j.biortech.2020.124082
27. Hou Y, Huang G, Li J, Yang Q, Huang S, Cai J (2019) Hydrothermal conversion of bamboo shoot shell to biochar: preliminary studies of adsorption equilibrium and kinetics for rhodamine B removal. J Anal Appl Pyrolysis [Internet] 143(September), 104694. Available from: https://doi.org/10.1016/j.jaap.2019.104694
28. Wang H, Zhao W, Chen Y, Li Y (2020) Nickel aluminum layered double oxides modified magnetic biochar from waste corncob for efficient removal of acridine orange. Bioresour Technol [Internet] 315(May):123834. Available from: https://doi.org/10.1016/j.biortech.2020.123834
29. Xu X, Xu Z, Huang J et al (2021) Sorption of reactive red by biochars ball milled in different atmospheres: co-effect of surface morphology and functional groups. Chem Eng J [Internet] 413:127468. Available from: https://doi.org/10.1016/j.cej.2020.127468
30. Zazycki MA, Borba PA, Silva RNF et al (2019) Chitin derived biochar as an alternative adsorbent to treat colored effluents containing methyl violet dye. Adv Powder Technol [Internet] 30(8):1494–1503. Available from: https://doi.org/10.1016/j.apt.2019.04.026
31. Côrtes LN, Druzian SP, Streit AFM et al (2019) Biochars from animal wastes as alternative materials to treat colored effluents containing basic red 9. J Environ Chem Eng 7(6)
32. Shan R, Lu L, Gu J et al (2020) Photocatalytic degradation of methyl orange by Ag/TiO2/biochar composite catalysts in aqueous solutions. Mater Sci Semicond Process [Internet] 114(March):105088 (2020). Available from: https://doi.org/10.1016/j.mssp.2020.105088
33. Zhu K, Wang X, Chen D, Ren W, Lin H, Zhang H (2019) Wood-based biochar as an excellent activator of peroxydisulfate for Acid Orange 7 decolorization. Chemosphere [Internet] 231:32–40 (2019). Available from: https://doi.org/10.1016/j.chemosphere.2019.05.087
34. Iqbal J, Shah NS, Sayed M et al (2021) Nano-zerovalent manganese/biochar composite for the adsorptive and oxidative removal of Congo-red dye from aqueous solutions. J Hazard Mater [Internet] 403(May 2020):123854. Available from: https://doi.org/10.1016/j.jhazmat.2020.123854

35. Gokulan R, Avinash A, Prabhu GG, Jegan J (2019) Remediation of remazol dyes by biochar derived from *Caulerpa scalpelliformis*—an eco-friendly approach. J Environ Chem Eng [Internet] 7(5):103297. Available from: https://doi.org/10.1016/j.jece.2019.103297
36. Mubarak NM, Fo YT, Al-Salim HS et al (2015) Removal of methylene blue and orange-G from waste water using magnetic biochar. Int J Nanosci 14(4)

Application of Rice Husk-Based Biomaterial in Textile Wastewater Treatment

Chijioke Elijah Onu, Paschal Enyinnaya Ohale, Ifeoma Amaoge Obiora-Okafo, Christian Oluchukwu Asadu, Chukwunonso Chukwuzuloke Okoye, Emeka Victor Ojukwu, and Ebuka Emmanuel Ezennajiego

Abstract Rice husk-based biomaterial (RHBB) is a waste from the rice milling industry which can be effectively used as a low-cost material in treating textile wastewater. Textile industries are known for their significant water demand, as much as 200 L used per kg of dyestuff. This leads to the generation of a large amount of water which is contaminated majorly by dyestuff as well as heavy metals, complex salts, acids, etc. As expected, this will create pollution-related problems for the receiving water bodies. Several works reviewed showed that this challenge is been addressed through different techniques such as; ozonation, adsorption, coagulation, chemical oxidation, catalysis, membrane separation, advanced oxidation, etc. with inherent difficulty, ineffectiveness, and eco-unfriendliness observed in some of them. RHBB can be modified for better performance in the treatment of textile wastewater. Among the three approaches, chemical modification was recommended for

C. E. Onu (✉) · P. E. Ohale · I. A. Obiora-Okafo · C. C. Okoye · E. V. Ojukwu
Department of Chemical Engineering, Nnamdi Azikiwe University, Anambra State, P.M.B. 5025, Awka, Nigeria
e-mail: ce.onu@unizik.edu.ng

P. E. Ohale
e-mail: pe.ohale@unizik.edu.ng

I. A. Obiora-Okafo
e-mail: ia.obiora-okafo@unizik.edu.ng

C. C. Okoye
e-mail: cc.okoye@unizik.edu.ng

E. V. Ojukwu
e-mail: ve.ojukwu@unizik.edu.ng

C. O. Asadu
Department of Chemical Engineering, Gregory University Uturu, Abia State, P.M.B. 1012, Uturu, Nigeria
e-mail: a.christian@gregoryuniversityuturu.edu.ng

E. E. Ezennajiego
Department of Chemical Engineering, Chukwuemeka Ojukwu Odumegwu University, Anambra State, P.M.B 01, Uli, Nigeria

S. S. Muthu and A. Khadir (eds.), *Textile Wastewater Treatment*, Sustainable Textiles: Production, Processing, Manufacturing & Chemistry,
https://doi.org/10.1007/978-981-19-2852-9_12

better result in the treatment of textile wastewater. RHBB can be used in conjunction with other processes to achieve concerted degradation of the target pollutants. The thermodynamics, kinetics, isotherms, optimization studies, and artificial intelligence approach were dwelled upon in this review. Therefore, RHBB is proposed as a granular, chemically stable, efficient, and cost-effective material for the treatment of textile wastewater.

Keywords Rice husk-based biomaterial · Textile effluent · Treatment · Dyes · Wastewater · Pollution · Modification · Artificial intelligence · Regeneration · Optimization

1 Introduction

Though industrialization has benefitted man in so many ways, however, it is not without its attendant unpleasant consequences. In the wake of intense industrial activities, humans have started to battle environmental pollution which is having adverse effect on the ecosystem, affecting both plants and animals and even man [37, 47]. Most of the environmental pollution problems are due to the release of pollutants into the water bodies, air, etc. These include dyes, phenolic compounds, and heavy metals are part of the effluent from textile industries. Dyes are most prominent in textile effluent because of their high solubility [40]. The decolourations of the water bodies by these dyes often destroy or slow down the interaction of the ecosystem. Furthermore, it has strong mitigating effect on man and the environment while greatly reducing the rate of photosynthesis [34, 39]. They constitute pollution and destroy the esthetic nature of the receiving water bodies when such effluents are not adequately treated before discharging into water bodies [2, 36]. Large quantity of textile materials that are produced annually results to the discharge of large amount of textile effluent.

About 100,000 different types of dyes are known and about 1 million tons are produced annually worldwide [31, 47]. The major source of dye contamination is associated with the textile industry where water from printing and dyeing processes containing residual dyestuffs, heavy metals, complex mixture of salts, acids, organ-chlorine-based pesticides, and pigments are released into water bodies with little or no little treatment [25, 46]. Textile industries use large amount of water, up to 200 L/kg of dyestuff used. They produce unsafe effluents on account of high chemical oxygen demand (COD) concentration, strong color, high pH and temperature, and low biodegradability [8, 54]. Therefore, there is a need to treat the wastewater before it is released into water bodies. Techniques used in dye control and treatment of textile effluent include: ion exchange, photo-catalytic process, flocculation, electrochemical process, reverse osmosis, chemical-coagulation, membrane based separation method, ozonation, activated carbon adsorption, sonochemical process, biological treatment and catalytic reduction [1, 6, 9–11, 13, 18, 25, 29, 44, 45]. In this review, various ways of achieving the removal of pollutants from textile wastewater using rice husk and in conjunction with other approaches will be investigated. Also, thermodynamics, kinetics, optimization and the application of artificial intelligence in wastewater treatment process were equally covered.

2 Dyes

Dyes are pigments which absorb some levels of light waves that impart color to them. They are broadly classified into natural and synthetic dyes [51]. Natural dyes are those that are obtained from natural sources such as animals, plants, flowers, roots, minerals, etc. while synthetic dyes are unsaturated organic molecules. There are inorganic and organic dyes. These dyes are utilized greatly in the textile, paper, paint, printing, leather, food, cosmetic industries, etc. These industries use them to color their product but unfortunately a good amount of it escape into their wastewater and constitute environmental pollution. Dyes are particularly problematic because they have low degree of attachment to the fiber, hence about 30% are hydrolyzed and released into waterways [47]. These dyes are not potentially harmful in themselves but they can be converted to harmful amines and thus interfere with the process of oxygen replenishment in water bodies [8, 25, 43, 47]. They are generally non-biodegradable [2]. Hence efforts are now being made to remove dyes from synthetic water [33, 38]. The removal of these dye materials from solutions has been given great consideration in literature.

2.1 Classification of Dyes

Generally, dyes are classified as can be seen in Table 1.

Table 1 General classification of dyes with examples

S/N	Dye classification	Examples	References
1	Acidic (cationic) dyes	Crystal violet, rosaniline hydrochloride	[28]
2	Basic (anionic) dyes	Congo red dye, methyl blue	[25, 33]
3	Disperse dyes	Azo, anthraquinone	[51]
4	Direct dyes	Stilbene, phthalocyanine, oxazine, thiazole, Congo red (mainly azo group)	[51]
5	Reactive dyes	Azo group (blue and green)	[51]
6	Vat dyes	Indigo	[14]

3 RiceHusk Base Biomaterials (RHBB)

Rice husk is produced in very large quantities globally. Rice husk is about 20 wt% of the rice crop itself and about 500 million tons of rice are produced in developing countries hence 100 million tons of rice husk are available in these countries [25]. There are different ways rice husk is being used namely; in block molding factories, as fuels for boilers and furnaces among others [25]. Even with these usages, rice husk is still causing grave environmental pollution. It is important that the waste be channeled into removal of dye from water bodies [25]. Rice husk is a good candidate for dye removal from aqueous media for these reasons; its granular structure, chemical stability, efficiency, and cost effectiveness [25, 26]. Rice husk is composed of silica, cellulose, hemi cellulose, and lignin [43]. The wide application of rice husk is due to its high availability, toughness, resistance to weathering, and abrasion [17].

3.1 Modification of RHBB

Modification of rice husk can lead to increased capacity of adsorption, that is, the amount of the adsorbate molecule adsorbed on the surface [17, 43]. For rice husk to perform better it has to be modified thermally or chemically to activate it for the removal of dye from aqueous media using adsorption approach [25]. Thermal modification of rice husk leads to the denaturation of the lignin on the surface of the biomass thus increasing the adsorption capacity of the material [43].

Rice husk can be thermally modified in a furnace by heating at a temperature of 350 °C for 1 h to produce rice husk char (activated rice husk) [25]. Rice husk char (RHC) can be further modified by dispersing 50 g of RHC in 100 mL of KOH solution (1 M) and stirring continuously at 80 °C for 6 h [25]. After drying and grinding, firing will take place in the furnace at 450 °C for one hour. The prepared product (chemically modified rice husk char) will be grinded to powder form and used as adsorbent for the removal of dye from aqueous media [25].

Chemical modification of rice husk char can raise the adsorption efficiency of the material to a maximum of 96.6% in 20 min [25]. Even at a very high initial dye concentration, chemically modified rice husk (CMRH) char exhibits high adsorption capacity up to 90% [25]. CMRH works well in a near neutral medium and can remove dye with up 98.6% efficiency [25]. Effect of temperature on the adsorbent material is dependent on the chemical nature of the substance, that is, if the process is exothermic, increase in temperature will reduce the rate of adsorption and vice versa [25]. Although rice husk, RHC, and CMRH have close chemical behavior, yet, the difference in their surface structure is responsible for their varied adsorption capacity [25]. Since the adsorbent is negatively charged alkaline treatment will lower the adsorption capacity of the adsorbent while acid treatment will increase it [43]. Also, cationic surfactant can increase the protonation of the biomass surface thereby increasing the amount of dye molecules adsorbed due to electrostatic force of attraction. On the other hand, organic treatment with ethanol or methanol will lower the adsorption capacity due to the production of esters on the surface of the biomass [43]. Alkaline and Maleic anhydride treatment imparts roughness, interfacial bonding capacity, and high mechanical properties to rice husk for water absorption application [17]. However, this is yet to be considered for application in dye removal from textile wastewater. Modification with saturated (stearic) and unsaturated (oleic) fatty acid can lead to the formation of ester linkage on the rice husk biomaterial and hence increase the capacity for adsorption [42]. This led to removal efficiency of 81% and 77% on oleic and stearic fatty acid, respectively.

Rice husk can be transformed into ash and then converted to mesoporous silica materials with large surface area and nanosized pore sizes ranging from 20 to 100 Å. This type of rice husk ash can easily be regenerated with either acid or alkaline solution [33]. Mesoporous silica materials can be synthesized to create adsorbents using a variety of silica precursors such as n-alkylamines, sodium water glass, and aerosol. Mesoporous materials can be prepared in various morphologies including powders and films where pore alignment and orientations are controlled. This type of material can be produced by mixing silicate solution from rice husk ash and surfactant solution [33].

3.2 BET Analysis of RHBB

Brunauer-Emmet-Teller (BET) theory aims to explain the physical adsorption of gas molecules on a solid surface and serves as the basis for an important analysis technique for the measurement of the specific surface area of materials. The BET theory applies to systems of multilayer adsorption and usually utilizes probing gases (called the adsorbent) that do not chemically react with material surfaces as adsorbates to quantify specific surface area.

BET analysis is used when mass transfer is surface driven, hence the need to evaluate the gas accessible surface area [19]. BET surface area (SBET) can be used to determine micropore surface areas of microporous materials and of total surface area of microporous/mesoporous materials [12]. This is the routine analysis for microporous materials and usually accessed from adsorption experiments before capillary condensation. In BET analysis, it is assumed that the total adsorption energy of all molecules adsorbed is constant and equal to the cohesion energy in the bulk liquid [12]. The BET surface area of RHBB was reported by Iheanacho et al. [20] to be 471.67 m^2/s which is very good for adsorption.

The pore size distribution, micropore volume, and surface area can also be estimated according to the quenched solid density functional theory (QSDFT) equilibrium model for slit pores using the Autosorb 1.56 software from Quantachrome [47]. The surface area and pore size of adsorbents can be determined with a nitrogen adsorption/desorption isotherm analyzer, nitrogen adsorption/desorption isotherms of adsorbent samples are normally carried out at 77 K according to Eqs. (1) and (2) [33]. Specific surface area is a scalable property, it is undefinable and hence will depend on the gas molecule used and the adsorption cross-section. Also it is unreliable when applied to ultra microporous molecules, which contains pores of molecular dimensions (e.g., molecular sieve carbon or zeolite) [49]. BET analysis is now being replaced by an anion adsorbate approach which can be applied directly to aqueous medium at ambient temperature and ease of modification of analysis scale by variation of adsorption cross-section [19].

$$\theta = C_p/[(P_o - P)(1 + (C - 1)(P/P_o))] \tag{1}$$

where c_p is the specific heat, C is the BET constant, P and P_o are the pressure and vapor pressure respectively of the adsorptive bulk liquid phase which would be at the temperature of the adsorbate and θ is the "surface coverage"; defined as;

$$\theta = \frac{n_{\text{ads}}}{n_m} \tag{2}$$

where n_{ads} is the amount of adsorbate and n_m is called the monolayer equivalent. The n_m is the amount that if it were all in a monolayer (which is theoretically impossible for physical adsorption) would cover the surface with exactly one layer of adsorbate.

4 Methods of Treating Textile Wastewater Using RHBB

The treatment technology for dye removal be broadly classified into three categories; biological, chemical, and physical [33]. There are several methods of removal of dyes from wastewater they include coagulation, adsorption, UV/H_2O_2, photocatalysis among others.

4.1 Coagulation

Coagulation is an essential process in the treatment of both surface water and industrial wastewater [7]. Its application includes removal of dissolved chemical species and turbidity from water via addition of conventional chemical-based coagulants, namely, alum ($AlCl_3$), ferric chloride ($FeCl_3$), and poly aluminum chloride (PAC). The traditional use of alum (aluminum-based) coagulants has been linked to diseases like Alzheimer's disease hence the need to replace it with natural ones like rice husk and its ash. The use of natural coagulants will also help in overcoming ineffectiveness in low-temperature water, and reduction of procurement costs associated with conventional method [7]. The turbidity and pH parameters will be greatly reduced when rice husk and its ash are used as coagulants compared with conventional ones. Also, increase in coagulation time and dosage of rice husk and rice husk ash can also help in reducing the turbidity and pH of the effluent water [7].

4.2 Adsorption

Among all the treatment methods, adsorption using activated carbon remains the best for its versatility and effectiveness in the decontamination of industrial effluents especially, the removal of both organic and inorganic materials [5, 21, 33]. It is usually based on the preference of adsorbents to remove specific substances from solutions unto their surfaces [35]. Modified rice husk was used as a low-cost adsorbent in the removal of dyes from textile wastewater treatment sludge (WTS) produced during the pre-treatment of Nile water before demineralization as a feed-water for boilers at power stations and at the domestic water treatment plants [47]. The cost and the efficiency of adsorbent material are important factors that determine the effectiveness of the adsorption process [25].

The sludge for adsorption contains a large concentration of salt, organic, and inorganic materials. Adsorption of large molecules like drug delivery will require a new type of adsorbent with extended pore diameter in the mesoporous range (pore dimensions above 2 nm) which improves the adsorption kinetics. Another cheap, promising, and green approach is the use of rice husk which are produced annually worldwide but grossly underutilized. The use of rice husk can be used to improve water treatment sludge for wastewater treatment in line with the "waste-waste self-cleaning" [47]. Raw sludge (5 wt%) combined with rice husk can be reused with a result up to 91% of the fresh adsorbent material [47].

Some factors affecting adsorption include the ionic radius of the component to be removed time, adsorbent dose, and pH. The higher the radius, the lesser the adsorption capacity and vice versa. Another important factor is the hydration energy, the lower the hydration energy the faster the hydrolysis and hence the adsorption rate [47]. Initially, the adsorption process is controlled by ion exchange and monolayer if the influent is at low concentration. Other factors affecting adsorption of dyes are the type of adsorbents used, experimental variables such as time, adsorbent dose, dye concentration, pH, and temperature [25].

The pH plays an important role in the adsorption of dyes on the adsorbents as it influences the surface polarity of the adsorbents, ionic mobility and degree of ionization of the pollutants. When pH of solution is greater than pHpzc (pH at point of zero charge), the adsorbents surface will acquire negative charge and will take up cationic dye while at pH of solution less than pHpzc, the adsorbent surface acquires positive charge and adsorbs the anionic dye [25, 33]. Some workers reported that phenol removal rises for increase in pH (3–5) [26]. However, if the pH is increased further (5–9), the removal decreases substantially due to the amphoteric behavior of the adsorbent. If the increase in the temperature of adsorption process increases the rate of adsorption, then the process is said to be endothermic but if increase in temperature shows little or no change in the rate of adsorption, then the process is an exothermic one [43].

4.3 UV/H_2O_2

Electromagnetic waves like microwaves can be used to irradiate activated carbon from rice husk. This method was used by Ahmad et al. [2] in heating activated carbon from gasifying with CO_2 for the removal of Methylene Blue (MB). It is important to note that high microwave power increases the pores spaces on rice husk activated carbon hence reducing the yield because of the release of volatile matters [31].

The mechanism here is the production of free radicals from the interaction between photons of a proper energy level and the chemical species present in the solution [15]. The generation of free radicals through UV irradiation of oxidizing species like ozone or hydrogen peroxide has been reported to be more effective than using only UV or such oxides. If high UV absorbing components are present, it is necessary to pretreat the wastewater with an oxidant component e.g., ozone to eliminate the high UV absorbing components thereby increasing UV penetration. This approach, however, is costly and hence the need to utilize gamma rays or transition metals [15].

4.4 Photocatalysis

Rice husk can be used in conjunction with photocatalysis to achieve dye mineralization in textile wastewater [22]. This method has received attention in recent times because it is relatively easier than the conventional chemical, biological and physical methods. Most effective photocatalyst used in this operation is ZnO and TiO_2 [22]. The non-hazardous photocatalyst is usually sensitized by UV irradiation so as to initiate the degradation [3]. Composite of rice husk and ZnO and TiO_2 will provide for dual degradation of textile dyes by photocatalysis and adsorption, respectively [22]. In addition, silica extracted from rice husk ash can be used in improving the extent of photodegradation of the wastewater in the presence of transition metals under UV [53]. The extent degradation can be evaluated using the Eq. (3),

$$\%\textbf{percentage of photodegradation} = (A_o - A_t)/A_o * 100 \tag{3}$$

where A_o is the initial absorbance, and A_t is the absorbance at time t.

The degradation process usually comprises of anodic and cathodic reaction pair according to equation

$$\text{Anodic (Oxidation) reaction}: \boldsymbol{h} + \mathbf{H_2O} \rightarrow \mathbf{H^+} + \mathbf{OH^{\cdot}} \rightarrow \text{Degradation of MB} \tag{4}$$

$$\text{Cathodic(Reduction)reaction}: \boldsymbol{e^-} + \mathbf{O_2} \rightarrow \mathbf{H_2O_2} \rightarrow \mathbf{H_2O} \tag{5}$$

4.5 Biosorption

The conventional methods for removal of dyes and other pollutants from aqueous solution such as ozonation, adsorption, coagulation, chemical oxidation, catalysis, membrane separation, and advanced oxidation are difficult, ineffective, and not eco-friendly [43]. This approach can be classified based on oxygen requirement as aerobic and anaerobic meaning biodegradation in the presence and absence of oxygen, respectively [46]. Biosorption is considered as an alternative and promising technique for the removal of pollutants due to its cost effectiveness and efficiency [46]. In this approach, microorganisms such as bacteria, algae, and fungi have the capacity of degrading dyes from industrial effluents and aqueous media (with immobilized adsorbent) [28]. Advantages of the immobilization of biosorbent include excellent reuse of the biosorbent, controlling of biosorbent particle, increasing mechanical strength, and less flocculation in the continuous system [43]. Immobilization of dead biomass is more effective than living biomass because complication in biosorbent kinetics due to cell growth can be eliminated. The amount of biosorbed dye can be calculated using Eq. 6.

$$q = \frac{(C_o - C_e)V}{W} \tag{6}$$

where, q is the amount of dye biosorbed on the biosorbent (mg/g), C_o and C_e are the initial and equilibrium concentration of dye solution, respectively. V is the volume of dye solution (ml) and W is the amount of the rice husk (g).

5 Thermodynamics, Kinetics, and Isotherm Models in Treating Textile Effluent

Change in temperature is an important factor that helps determine thermodynamic parameters like ΔH (enthalpy change), ΔS (entropy change), ΔG (Gibb's free energy change) [25]. Gibb's free energy change can be used to evaluate thermodynamic properties of the treatment process according to Eq. (7)

$$\Delta G = -RT \ln K \tag{7}$$

where R is the universal gas constant (8.314 J mol^{-1} K^{-1}), T is the temperature in Kelvin and K is the equilibrium constant which is given by

$$K = \frac{C_{\text{ads}}}{C_e} \tag{8}$$

In Eq. 8, C_{ads} is the concentration of dye adsorbed on the adsorbent and C_e is the equilibrium dye concentration. The relationship between free energy and temperature is given by the following equation [25]

$$\Delta G = \Delta H - T\Delta S \tag{9}$$

To determine the enthalpy (ΔH) and entropy (ΔS) which are the intercept and the gradient respectively, ΔG will be plotted against ΔS. Negative ΔH implies an exothermic process while positive means endothermic process. The ΔS depicts the degree of randomness of adsorbent material at the adsorbent/adsorbate interface. Negative ΔG value indicates a spontaneous process while a positive value shows a non-spontaneous process.

5.1 Kinetics

The kinetics of dye removal can be effectively studied using the fixed-bed desorption column under dynamic conditions (Fig. 1). The process is carried out continuously until the effluent (C_t) concentration equals that of the influent (C_o) at time, t, that it $C_t/C_o \approx 1$ [47]. For a constant flow rate and at a given dye influent concentration, the total quantity of dye retained in the column can be found graphically by numerically integrating the area under the curve of $(C_o - C_t)/C_o$ versus the service time. The dye concentration of the influent acts as a driving force that will help overcome the mass transfer resistance between the solid and the aqueous medium [47]. It is noteworthy that the high rate of dye removal witnessed at the start of adsorption process is due to the availability of binding sites on the adsorbent but subsequently it decreases and the rate of removal falls [2, 25, 26]. This fall is due to the movement of the adsorbed dye molecule into the interior of the bulk adsorbent [43]. Mandal et al. [26] stated that variation in adsorbent concentration will lead to initial increase in adsorption rate because available of pores. The adsorption efficiency of rice husk for 500 ml of effluent has been reported to be 86.66% when only 15 gm of rice husk was used [5].

Adsorption capacity according to Khan et al. [25] can be calculated by,

$$q_e = \frac{C_o - C_t}{C_o} \tag{10}$$

where, q_e = adsorption capacity (mg/g), V = volume of metal ion solution in liters, C_o and C_f is the initial and final/equilibrium dye concentration (mg/L), respectively and m = adsorbent dose (g).

The kinetics of absorption process depends on the physical and chemical characteristics of the adsorbent materials. Safa and Bhatti [43] used both pseudo first and second order kinetics to model the treatment of textile effluent.

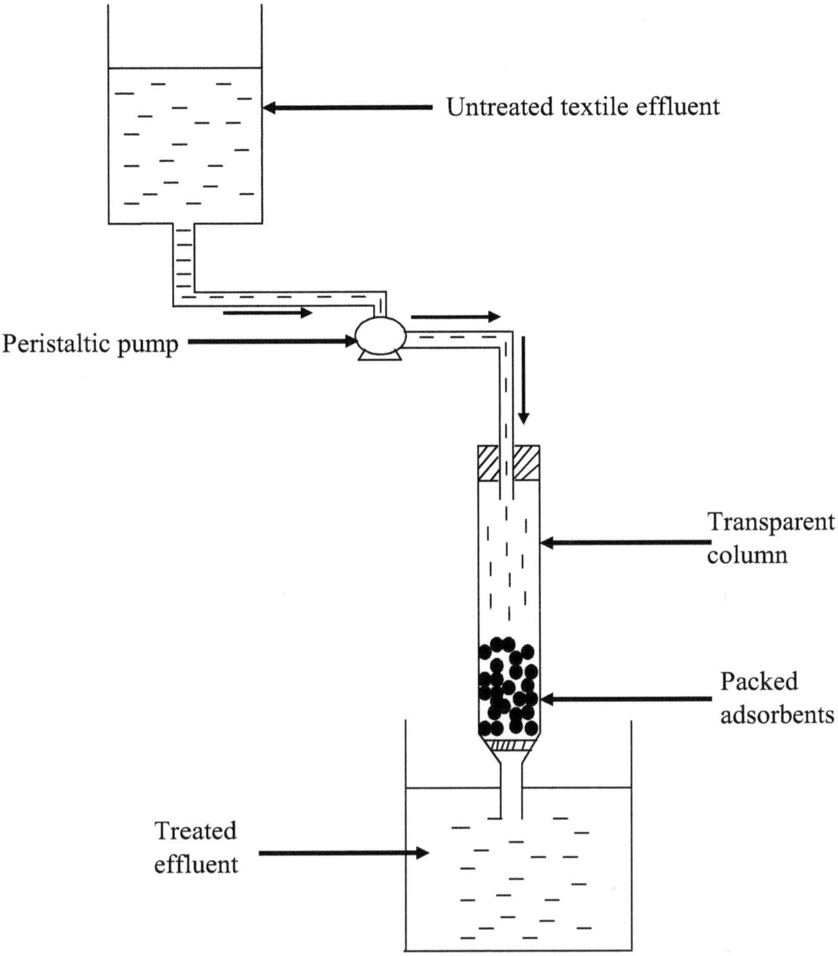

Fig. 1 Schematic diagram of fixed-bed adsorption column

5.1.1 Pseudo First Order Kinetics

This kinetics assumes the rate of adsorption capacity is directly proportional to the change in the dye concentration as given below;

$$\frac{dq_t}{d_t} = k_1(q_e - q_t) \quad (11)$$

where, q_e and q_t are the biosorption capacity (mg/g) at equilibrium and time t, respectively, k_1 is the rate constant (L min^{-1}) of pseudo first order kinetic model. After integrating the just stated equation and applying boundary conditions $t_0 = 0$ with $t = t$ and $q_0 = 0$ with $q_e = q_t$ and rearranging, the equation becomes:

$$\log(q_e - q_t) = \log(q_e) - k_1/2.303.t \tag{12}$$

5.1.2 Pseudo Second Order Kinetics

Pseudo second order kinetic model is also based upon the biosorption capacity of the biosorbent material [43]. The biosorption mechanism over a complete range of the contact time is explained by the pseudo second order kinetic model:

$$\mathrm{d}qt/\mathrm{d}t = k_2(qe - qt)^2 \tag{13}$$

where, k_2 (g/mg min) is the second order rate constant of biosorption process. By integrating and applying boundary conditions $t_0 = 0$ with $t = t$ and $q_0 = 0$ with $q_e = q_t$, the stated equation can be written in linear form as given in Eq. 14:

$$t/q_t = 1/k_2 q_e^2 + 1/q_e.t \tag{14}$$

Safa and Bhatti [43] found that pseudo second order kinetics fitted well to the kinetic data due to its high correlation coefficient. Also, in their work, Khan et al. [25] confirmed that the pseudo second order kinetics best fitted the removal of Congo red dye. Halim et al. [16] investigated the removal of Fe^{2+} and Pb^{2+} ions from wastewater using rice husk. It was found that the pseudo second order kinetic gave a better R^2 (0.99 and 0.976, respectively) value than the pseudo first order kinetic model.

5.2 *Isotherms*

Various isotherm models exist for the study of the interaction between the dye and the adsorbent material in view of optimizing the adsorption process, determining the adsorption capacity of the molecules and the nature of the process [25]. They are Langmuir adsorption isotherm, Freundlich adsorption isotherm, Temkin adsorption isotherm, and D-R isotherm.

5.2.1 Langmuir Model

In this type of isotherm, the active sites of the absorbent materials have the same affinity for the adsorbate molecule and hence constant activation energy and enthalpies. The linear form of adsorption isotherm is given by Eq. 15.

$$\frac{C_e}{q_e} = \frac{C_e}{q_m} + \frac{1}{K_{q_m}} \tag{15}$$

where, C_e is the equilibrium concentration of adsorbate in the solution (mg L^{-1}), q_e is the amount of adsorbate adsorbed on the adsorbent at equilibrium (mg g^{-1}), q_m is the adsorption capacity (mg g^{-1}) and K is the Langmuir adsorption equilibrium constant which is related to the energy of adsorption. The Langmuir constants q_m can be determined from the intercept and the slope of the straight line obtained by plotting C_e/q_e versus C_e [25].

5.2.2 Freudlich Model

This isotherm is used to show the non-ideal, reversible, and multilayer adsorption [25]. The model pays particular attention to the heterogeneous surface of the adsorbent [25]. The linear form is as shown in Eq. 16

$$\text{Log}q_e = \text{Log}K_F + 1/n\text{Log}C_e \tag{16}$$

where q_e is the quantity of dye adsorbed at equilibrium (mg/g). C_e adsorbate at equilibrium (mg/L). K_f and n are Freundlich constants related to the capacity and adsorption intensity of the adsorbent, respectively. The Freundlich constants are calculated from the plot of $\log q_e$ versus $\text{Log}C_e$ [25].

5.2.3 Temkin Model

This model examines the effect of adsorbent-adsorbent interaction in the adsorption process. This model depicts that the heat of absorption of all molecules in the layer decreases linearly with coverage [25].

The linear form of Temkin equation is given as in Eq. 17

$$q_e = B \ln K_{KT} + B \ln C_e \tag{17}$$

where $B = RT/b$, is the Temkin constant related to the heat of adsorption and K_{KT} isotherm binding constant (L/mg) related to maximum binding energy [25]. The values of these constants can be calculated from the slope and intercept obtained from plot of q_e versus $\ln C_e$.

The data generated from the adsorption process will be fitted into any of the above isotherms and the R^2 value evaluated to check the degree of fitness.

6 Optimization of Textile Wastewater Treatment

Different methods of optimization and modeling have been used in textile wastewater treatment. One of them is the Response Surface Method (RSM) approach is usually used to study the interaction between two or more variables [23]. It is a combination of mathematical and statistical techniques for designing experiments, building numerical models, evaluating the effects of variables, and searching for the optimum conditions. The advantage of this is reduced the number of experimental trials that are needed to analyze multiple variables and their interactions. [23], used CCD to study the effect of three numerical activated carbon preparation variables, Pyrolyzed temperature, carbonization time and chemical impregnation ratio. The optimum conditions for the process were; pyrolyzed temperature 700 °C, carbonization time of 45 min and chemical impregnation ratio of 0.5 while the optimum BET surface area for the process was 336 m^2/g with Synozol Blue Reactive (RSB) removal of 8.5383 mg/g. Murat et al. [31] investigated the optimal conditions for preparation of rice husk-based activated carbon (RHAC) using carbon dioxide gasification of rice husk char under microwave irradiation. The central composite design (CCD) was used to determine the effects of the two preparation variables, microwave power and activation time on MB removal and RHAC yield. Based on the CCD, a quadratic model and linear models were developed for MB removal and RHAC yield, respectively. The optimum conditions for RHAC preparation were obtained by using activation time of 4.28 min and microwave power of 440 W, which resulted to MB removal of 83.05% and RHAC yield of 31.98%. Karam et al. [24] carried out a comparative study of different adsorbent materials for the removal of color from textile wastewater. RSM was used in which there were five variables (pH, dose, time, stirring rate, and initial color concentration) and one response (percentage of color removal). It was observed that the RSM approach was efficient in predicting and simulating the process based on the observed R^2 of over 90%.

7 Application of Artificial Intelligence in Textile Wastewater Treatment

Intelligence systems are generally applied to complex processes like the treatment of textile wastewater [52]. One of the methods here is the Artificial Neural Network (ANN). A typical ANN architecture is given in Fig. 2. This has been in use in the last two decades for analyzing and developing models for complex systems. It is an information processing system derived from the biological nervous system of the brain [41]. They can be used for better prediction of system performance due to their high accuracy, adequacy, and promising applications. There are two stages of application in ANN, they are the training and the testing stages [52]. Also, three layers exit with many neurons, namely, the input, output, and hidden layers as shown in Fig. 2. The neurons consist of five main elements including; inputs, weights, summing function, activation (transfer) function, and output. After the calculation in any neuron, the output is sent to the neurons in the next layer.

The summation is given as follows;

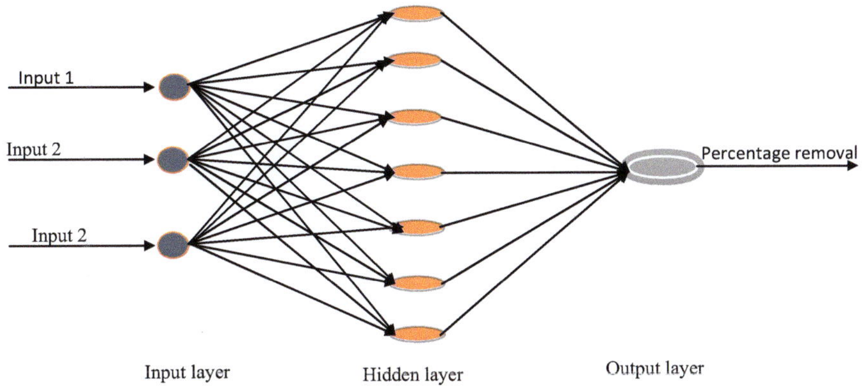

Fig. 2 Typical ANN architecture of textile effluent treatment process

$$f(x) = \sum_{i=1}^{n} (w_{ix} + a_i) \tag{18}$$

where w is the weight factor and a is the input value.

In the same vein, the Mean Squared Error (MSE) can be estimated using the formula below;

$$\text{MSE} = \frac{1}{N} + \sum_{n=1}^{n} (q_{(i,\text{pre})} - q_{(i,\text{exp})}) \tag{19}$$

where N is the data points, $q_{i,\text{pre}}$ is the network prediction, $q_{i,\text{exp}}$ is the experimental response while i is the index data.

The models are designed to have maximum correlation coefficient (R-value) and minimum mean squared error (MSE) values in the neurons. Tumer and Edebali [52] used three input, three hidden layers, and one output model (3-3-1) ANN-structure and obtained a maximum correlation value between measured and predicted output was up to 0.99 indicating neural network can be used to effectively predict the performance of the konya wastewater treatment plant.

Another statistical approach in the treatment of waster waste is the support vector machine (SVM). SVM is mostly used for multivariate function estimation and non-linear regression problem because of its ability to escape the local minima, improved generalization capability, and spare representation of the solution [27]. It is based on the structural risk minimization principle thus addressing the problem of over fitting by balancing model complexity [27].

Adaptive neuro-fuzzy inference system (ANFIS) is a hybrid fuzzy logic-based technique integrated with the learning power of artificial neural network that improves the performance of any kind of intelligent system by utilizing knowledge acquired after learning [27]. The ANFIS architecture is shown in Fig. 3. It uses backpropagation gradient descent and least square method to frame a fuzzy inference system

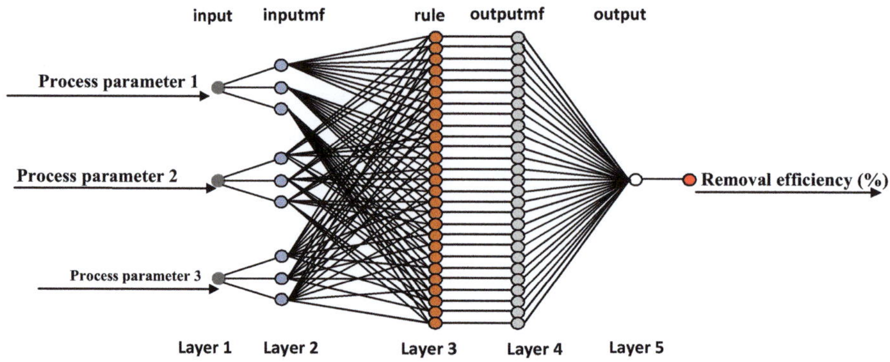

Fig. 3 Typical ANFIS architecture for textile effluent treatment

whose membership functions (MFs) parameters are iteratively tuned and adjusted [27]. ANFIS consists of mainly five layers—rule base, database, fuzzification interface, defuzzification interface, and decision making unit [27]. Manu and Thalla [27] used SVM and ANFIS models with Gbell and trapezoidal MFs for the treatment of wastewater. The performances of the proposed models were evaluated using Correlation coefficient (CC), root mean square error (RMSE) and Nash–Sutcliffe error (NSE). It was found that SVM was a better model than ANFIS while among the ANFIS models, the Gbell was a better than the MF model in modeling non-linear time series. Karam et al. [24] used the backpropagation algorithm of ANN architectures of input, hidden environment, and output layers to study the effect of nano zero valent Iron, activated carbon, and green nano zerovalent Iron on the removal of color from textile wastewater. It was observed that the ANN architecture was able to predict and simulate the process.

8 Regeneration

Regeneration is the instant recovery of the spent adsorbent through practically and economically viable techniques [30, 48]. Regenerated rice husk has been reported to show desirable adsorptive stability after several cycles of reuse [4, 30, 32, 48, 50]. Regeneration process is important for two major reasons namely: the process saves the cost incurred from the production of new absorbent and it also presents a means for recycling or safe disposal of adsorbed pollutants [4, 32, 48]. Several regeneration techniques such as steam regeneration, thermal regeneration, chemical regeneration, and ultrasound regeneration have been adopted to retain the re-usability of adsorbent [30]. For rice husk adsorbent, the documented regeneration techniques are explained below.

Chemical regeneration necessitated the use of chemical solvents in the desorption of adsorbed species from the surface of the adsorbent [32, 50]. The regeneration efficiency of an adsorbent is largely dependent on the solution pH of the chemical species and the rate of degradation [30]. The regeneration efficiency can be preserved by using a suitable chemical specie in which the solution pH enhances the exchange of ions (cation and anion) with the adsorbed pollutant. Some of the commonly used reagents for the regeneration of rice husk adsorbent include buffer solution (pH 4.0), NaOH, H_2O_2, O_3, and HCl [4, 30, 32, 48, 50].

According to Nabieh et al. [32], a complex mixture of buffer 4.0 and Pb^{2+} has been employed to desorb palladium ions from rice husk activated carbon. The authors reported stability of the adsorbent after ten successive cycles of adsorption. Sivalingam and Sen [50] successfully used ultrasound technique for the regeneration of spent rice husk-based nZSM—5 adsorbent using deionized water with 0.1 M NaOH. The authors tested the re-usability of the regenerated adsorbent using crystal violet (CV) dye. They observed that the adsorption capacity was reduced by only 9.7% after five cycles [50].

For easy desorption of dye from adsorbent, hydrochloric acid is mostly used because it can easily elute more than 90% of dye from the adsorbent [4]. Malachite green dye adsorbed on rice husk activated with tartaric was reportedly desorbed using 0.1 M HCl. According to the result, the regenerated rice husk was efficiently stable (~95%) up to five repeated cycles of adsorption [48].

9 Conclusion

This work reviewed the use of rice husk-based biomaterials (RHBB) in the treatment of textile wastewater. The works done in this area touched on the modification of the rice husk biomaterial for better performance especially in adsorption and then other approaches like coagulation, use of UV/H_2O_2, photocatalysis, and biosorption. Physical, chemical, and thermal processes are ways through which rice husk biomaterials can be modified. Among these processes, chemical process seems to be popular and effective in modification of rice husk for treatment of textile wastewater. Optimization approaches can be used to optimize the process as it is a complex multivariate system. RSM, ANN, ANFIS, and SVM were highlighted in this work and have shown good results for modeling and predicting the process behavior. Particularly, SVM deals with the problem of over fitting which is common with the use of ANFIS by balancing the model complexity. The spent RHBB can be recovered using chemical regeneration.

Acknowledgements The authors will like to thank the Department of Chemical Engineering, Nnamdi Azikiwe University Awka for providing enabling environment to carry out the work.

References

1. Abid MF, Abdulla ON, Kadhim AF (2019) Study on removal of phenol from synthetic wastewater using solar photo catalytic reactor. J King Saud Univ Eng Sci 31(2):131–139. https://doi.org/10.1016/j.jksues.2017.03.002
2. Ahmad MA, Basir NI, Yahaya NKE (2018) Microwave assisted rice husk based activated carbon for adsorption of methylene blue dye. Int J Petrochem Res 2(2):162–164. https://doi.org/10.18689/ijpr-1000128
3. Ahmed S, Jahan SA, Zaman N, Ahsan M (2011) Photodegradation of methylene blue mediated by rice husk ash (RHA)-TiO_2 composite. Bangladesh J Sci Ind Res 46(1):17–26
4. Alam MM, Hossain AM, Delowar Hossain MAH, Johir JH, Saifur R, John LZ, Kamrul H, Aneek KK, Mohammad BA (2020) The potentiality of rice husk-derived activated carbon: from synthesis to application. Processes 8:203–210
5. Anaswara NS, Arun KP, Deva PB, Ranjith KM, Subitha T (2017) Decolourisation of textile effluent by adsorption using carbonated rice husk. Malaya J Biosci 4(1):01–06. ISSN 2348-6236 print/2348-3075 online
6. Anisuzzaman S, Bono A, Krishnaiah D, Tan YZ (2016) A study on dynamic simulation of phenol adsorption in activated carbon packed bed column. J King Saud Univ Eng Sci 28(1):47–55. https://doi.org/10.1016/j.jksues.2014.01.001
7. Anjitha A, George D (2016) Comparative study using rice husk and its ash as natural coagulants for waste water treatment. Int J Sci Eng Res 7(4):232–237. ISSN 2229-5518
8. Arslan S, Eyvaz M, Gürbulak E, Yüksel E (2016) A Review of state-of-the-art technologies in dye-containing wastewater treatment. In: The textile industry case. https://doi.org/10.5772/64140
9. Bi X, Wang P, Jiao C, Cao H (2009) Degradation of remazol golden yellow dye wastewater in microwave enhanced ClO_2 catalytic oxidation process. J Hazard Mater 168(2–3):895–900. https://doi.org/10.1016/j.jhazmat.2009.02.108
10. Billal K, Belkacem M, Aissa C, Rabah Z, Jean PL, Francois L (2018) Removal of a textile dye using photovoltaic electrocoagulation. Sustain Chem Pharm 7(1):27–35. https://doi.org/10.1016/j.scp.2017.11.004
11. Feng J, Hu X, Yue PL (2004) Novel bentonite clay-based Fe nanocomposite as a heterogeneous catalyst for photo-fenton discoloration and mineralization of orange (II). Environ Sci Technol 38(1):269–275. https://doi.org/10.1021/es034515c
12. Galarneau A, Mehlhorn D, Guenneau F, Coasne B, Villemot F, Minoux D, Aquino C, Dath J (2019) Specific surface area determination for microporous/mesoporous materials: the case of mesoporous FAU-Y zeolites. Langmuir Am Chem Soc 34(47):14134–14142. https://doi.org/10.1021/acs.langmuir.8b02144.hal-01938089
13. Guenfoud F, Mokhtari M, Akrout H (2014) Electrochemical degradation of malachite green with BDD electrodes: effect of electrochemical parameters. Diam Relat Mater 46:8–14. https://doi.org/10.1016/j.diamond.2014.04.003
14. Gupta VK (2019) Chemical and technology of natural and synthetic dye and pigments. Fundamentals of natural dyes and its application on textile substrate. IntechOpen. https://doi.org/10.5772/intechopen.89964
15. Hai FI, Yamamoto K, Fukushi K (2007) Hybrid treatment systems for dye wastewater. Crit Rev Environ Sci Technol 37(4):315–377. https://ro.uow.edu.au/scipapers/4887
16. Halim AA, El-zaby KH, El-Gamal MI, Saber HM (2019) Removal of Fe^{2+} and Pb^{2+} ions from wastewater using rice husks-based adsorbents. J Egypt Acad Soc Environ Develop 20(1):47–60. https://doi.org/10.21608/jades.2019.67694
17. Halip JA, Lee SH, Tahi PM, Chuan LT, Selimin LA, Saffian HA (2021) A review: chemical treatment of rice husk. Biointerface Res Appl Chem 11(4):12425–12433. https://doi.org/10.33263/BRIAC114.1242512433
18. Hamad HT (2020) Removal of phenol and inorganic metals from wastewater using activated ceramic. J King Saud Univ Eng Sci. https://doi.org/10.1016/j.jksues.2020.04.006

19. Hanaor DAH, Ghadiri M, Chrzanowski W, Gan Y (2014) Scalable surface area characterization by electrokinetic analysis of complex anion adsorption. Am Chem Soc 30:15143–15152. https://doi.org/10.1021/la503581e. (Langmuir 2014,30,15143−15152)
20. Iheanacho OC, Nwabanne JT, Onu CE (2019) Optimum process parameters for activated carbon production from rice husk for phenol adsorption. Curr J Appl Sci Technol 36(6):1–11. https://doi.org/10.9734/CJAST/2019/v36i630264
21. Iheanacho OC, Nwabanne JT, Obi CC, Onu CE (2021) Packed bed column adsorption of phenol onto corn cob activated carbon: linear and nonlinear kinetics modeling. S Afr J Chem Eng 36:80–93. https://doi.org/10.1016/j.sajce.2021.02.003
22. Kabir MH, Kabir MF, Nigar F, Ahmed S, Mustafa AI, Ahsan M (2013) Preparation and characterization of rice husk ash (RHA)-TiO_2/ZnO compositesand its application in treating effluents from textile industries. Bangladesh J Sci Ind Res 47(4):445–448
23. Kacan E (2016) Optimum BET surface areas for activated carbon produced from textile sewage sludges and its application as dye removal. J Environ Manage 166:116–123. https://doi.org/10.1016/j.jenvman.2015.09.044
24. Karam A, Zaher K, Mahmoud AS (2020) Comparative studies of using nano zerovalent iron, activated carbon, and green synthesized nano zerovalent iron for textile wastewater color removal using artificial intelligence, regression analysis, adsorption isotherm, and kinetic studies. Air Soil Water Res 13:1–19. https://doi.org/10.1177/117862212090827
25. Khan AMA, Anwar N, Naeem M (2020) Comparative study of Congo red dye on rice husk, rice husk char and chemically modified rice hush char from aqueous solution. Bull Chem Soc Ethiop 34(1):41–54. https://doi.org/10.4314/bcse.v34i1.4
26. Mandal M, Mukhopadhyay P, Das SK (2019) Efficiency analysis of rice husk as adsorbent for removal of phenol from waste water. J Environ Anal Toxicol 9(605):1–8
27. Manu DS, Thalla AK (2017) Artificial intelligence models for predicting the performance of biological wastewater treatment plant in the removal of Kjeldahl Nitrogen from wastewater. Appl Water Sci 7:3783–3791. https://doi.org/10.1007/s13201-017-0526-4
28. Mary JI, Gowrisanka L, Renugadevi N (2021) Adsorptive removal of dye unto a cost effective biomaterial: a review. J Environ Treat Tech 9(1):218–223. https://doi.org/10.47277/JETT/9(1)223
29. Ming C, Kai SW, Chin LH, Chih WC, Tsung JC, Shiuan SL, Shih HC (2012) Improvement of textile dye removal by electrocoagulation with low-cost steel wool cathode reactor. Chem Eng J 192(1):37–44. https://doi.org/10.1016/j.cej.2012.03.086
30. Momina I, Mohammad S, Suzylawati I (2018) Regeneration performance of clay-based adsorbents for the removal of industrial dyes: a review. RSC Adv 8:24571
31. Murat MN, Ahmad MA, Idris MN (2018) Optimization of preparation conditions for rice husk based activatedcarbons for the removal of methylene blue dye. Int J Petrochem Res 2(2):186–188. https://doi.org/10.18689/ijpr-1000133
32. Nabieh KA, Wael IM, Tamer E, Helmy IMM, Kenawy YG, Abou ER (2021) Chemically modified rice husk as an effective adsorbent for removal of palladium ions. Heliyon 7:e06062
33. Nguyen NT, Chen SS, Nguyen NC, Nguyen HT, Tsai HH, Chang CT (2016) Adsorption of methyl blue on mesoporous material using rice husk ash as a silica source. J Nanosci Nanotchnol 16(4):4108–4114. https://doi.org/10.1166/jnn.2016.10704
34. Nwabanne JT, Okpe EC, Igbokwe PK, Asadu CC, Onu CE (2016) Isotherm and kinetic modelling of adsorption of dyestuffs onto kola nut (cola acuminate) shell activated carbon. J Chem Technol Metall 51(2):188–201
35. Nwabanne JT, Onu CE, Nwankwoukwu OC (2018) Equilibrium, kinetics and thermodynamics of the bleaching of palm oil using activated nando clay. J Eng Res Rep 1(3):1–13. https://doi.org/10.9734/JERR/2018/42699
36. Nwobasi VN, Igbokwe PK, Onu CE (2020) Removal of methylene blue dye from aqueous solution using modified ngbo clay. J Mater Sci Res Rev 5(2):33–46
37. Ohale PE, Onu EC, Ohale NJ, Oba SN (2020) Adsorptive kinetics, isotherm and thermodynamic analysis of fishpond effluent coagulation using chitin derived coagulant from waste Brachyura shell. Chem Eng J Adv 4:1–13. https://doi.org/10.1016/j.ceja.2020.100036

38. Okoye CC, Onukwuli OD, Okey-Onyesolu CF (2019) Utilization of salt activated raphia hookeri seeds as biosorbent for Erythrosine B dye removal: kinetics and thermodynamics studies. J King Saud Univ Sci 31:849–858
39. Onu CE, Oguanobi NC, Okonkwo CO, Nnamdi-Bejie J (2020) Application of modified agricultural waste in the adsorption of bromocresol green dye. Asian J Chem Sci 7(1):15–24. https://doi.org/10.9734/AJOCS/2020/v7i119011
40. Onu CE, Nwabanne JT (2014) Adsorption kinetics for Malachite green removal from aqueous solution using Nteje clay. J Environ Hum 1(2):133–150. https://doi.org/10.15764/EH.2014.02015
41. Onu CE, Nwabanne JT, Ohale PE, Asadu CO (2021) Comparative analysis of RSM, ANN and ANFIS and the mechanistic modeling in eriochrome black-T dye adsorption using modified clay. S Afr J Chem Eng 36:24–42. https://doi.org/10.1016/j.sajce.2020.12.003
42. Rafique U, Zulfiqar S (2014) Synthesis and technological application of agro-waste composite for the treatment of textile waste water. Int J Agric Innovation Res 3(3): 2319–1473. ISSN (online)
43. Safa Y, Bhatti HN (2011) Adsorptive removal of direct dyes by low cost rice husk: effects and treatments. Afr J Biotechnol 10(16):3128–3142. https://doi.org/10.5897/ajb10.1302
44. Saleh TA, Gupta VK (2012) Photo-catalyzed degradation of hazardous dye methyl orange by use of a composite catalyst consisting of multi-walled carbon nanotubes and titanium dioxide. J Colloid Interface Sci 371:101–106. https://doi.org/10.1016/j.jcis.2011.12.038
45. Saravanan R, Gupta VK, Narayanan V, Stephen A (2014) Visible light degradation of textile effluent using novel catalyst $ZnO/\gamma\text{-}Mn_2O_3$. J Taiwan Inst Chem Eng 45(4):1910–1917. https://doi.org/10.1016/j.jtice.2013.12.021
46. Sghaier I, Guembri M, Chouchan H, Mosba A, Hada O, Jaouani A, Cherif A, Neifar M (2019) Recent advances in textile wastewater treatment using microbial consortia. J Textile Eng Fashion Technol 5(3):134–146
47. Shalaby NH, Ewais EMM, Elsaadany MH, Ahmed A (2017) Rice husk templated water treatment sludge as low cost dye and metal adsorbent. Egypt J Pet 26:661–668. https://doi.org/10.1016/j.ejpe.2016.10.006
48. Shamsollahi Z, Partovinia A (2019) Recent advances on pollutants removal by rice husk as a bio-based adsorbent: a critical review. J Environ Manage 246(2019):314–323
49. Sing KSW (1998) Adsorption methods for the characterization of porous materials. Adv Coll Interface Sci 76–7:3–11
50. Sivalingam S, Sen S (2020) Rice husk ash derived nanocrystalline ZSM-5 for highly efficient removal of a toxic textile dye. J Market Res 9(6):14853–14864
51. Srivatsav P, Bhargav SP, Shanmugasundaran V, Arun J, Gopinath KP, Bhatnagar A (2020) Biochar as an eco-friendly and economical adsorbent for the removal of colorants (dyes) from aqueous environment: a review. Water 12(3561). https://doi.org/10.3390/w12123561
52. Tumer AE, Edebali S (2015) An artificial neural network model for wastewater treatment plant of Konya. Int J Intell Syst Appl Eng 3(4):131–134. https://doi.org/10.18201/ijisae.65358
53. Vinoda BM, Vinuth M, Yadav DB, Manjanna J (2015) Photocatalytic degradation of toxic methyl red dye using silica nanoparticles synthesized from rice husk ash. Environ Anal Toxicol 5:6. https://doi.org/10.4172/2161-0525.1000336
54. Yeseen DA, Scholz M (2019) Textile dye wastewater characteristics and constituents of synthetic effluents: a critical review. Int J Environ Sci Technol 16:1193–1226. https://doi.org/10.1007/s13762-018-2130-z

Application of Hydroxyapatite in Textile Wastewater

Madeeha Batool and Hafiza Sana

Abstract Textile industry is emerging as a primary source of wastewater that causes permanent damage to the environment and human health. Textile effluents have high pollution potential and require adequate pretreatment before their discharge into the environment. Several methods, including adsorption, have been explored for the remediation of pollutants of textile wastewater. Adsorption serves as one of the most preferred option because of its design simplicity, low operational cost, high efficiency, and negligible sludge production. So far, various adsorbents, including natural and artificial materials have been explored. Among the available adsorptive materials, the immobilization of priority pollutants using phosphate minerals such as hydroxyapatite ($Ca_{10}(PO_4)_6(OH)_2$) is a promising, recent, and effective strategy for wastewater remediation. Hydroxyapatite (HAP) is a vital component of vertebrate bone minerals and can easily be extracted from natural sources. However, its properties can further be enhanced using appropriate synthetic procedures and suitable modifications. The primary advantage of using HAP is its functionality, porosity, and the crystal structure. Its promising adsorption behavior helps in effectively removing the priority pollutants of textile wastewater e.g., sulfates, oils, phenols, dyes, and toxic heavy metals. The mechanism of adsorption process of HAP is well explained by isothermal and thermodynamic studies and Equilibrium Kinetic Modeling. Thus, its efficiency, cost-effectiveness, and reusability provide viable option to target textile waste pollutants.

Keywords Hydroxyapatite · Textile wastewater · Wastewater remediation · Adsorption · Inorganic adsorbent · Heavy metals · Dyes · Oils · Phenols · Sulfates

M. Batool (✉)
School of Chemistry, University of the Punjab, New Campus, Lahore 54590, Pakistan
e-mail: madeeha.chem@pu.edu.pk

H. Sana
School of Chemical Engineering, Zhengzhou University, Zhengzhou 450001, Henan, China

Institute of Energy and Environmental Engineering, University of the Punjab, New Campus, Lahore 54590, Pakistan

S. S. Muthu and A. Khadir (eds.), *Textile Wastewater Treatment*, Sustainable Textiles: Production, Processing, Manufacturing & Chemistry,
https://doi.org/10.1007/978-981-19-2852-9_13

1 Introduction

Textile industry, ranked as one of the main industries worldwide, contributes massively to global pollution. Textile industries discharge an excessive amount of wastewater that comprises of noxious contaminants due to wet fiber processing (size reduction, scrubbing, and decolorizing), which utterly damage the atmosphere [1]. Primary pollutants in textile wastewater consist of various organic (dyes, formaldehyde, phenols, mineral oils, surfactants, perfluorooctanoic acid, etc.) and inorganic substances (heavy toxic metals like lead, cadmium, arsenic, chromium, zinc and nickel, etc.), which make this water a threat to environment. Numerous wastewater remediation approaches namely; precipitation, solvent extraction, ion exchange, electrochemical separation, membrane filtration, oxidation–reduction system, reverse osmosis, flotation, coagulation, flocculation, electro-coagulation, ozonation, biosorption, and adsorption are in practice [21]. Among all, the adsorption process holds some attractive properties e.g., process ease, removal proficiency, and practice of low-cost adsorbents [14]. The attachment of contaminants on the adsorbent's exterior can be physical or chemical. Adsorption process not only eliminates the impurities but also de-sorbs them for recycling. However, other conventional remediation approaches are incompetent to completely degrade the recalcitrant compounds (e.g., dyes and oils). Considering this, efficient, cost-effective, and continuous adsorbents are required to target these priority pollutants [39, 42]. In this regard, various adsorbents such as biopolymers, biochar, activated carbon, hydroxyapatite (HAP), and its composites (with sodium alginate, cellulose, biochar, activated carbon, chitosan, etc.), are being employed for eradicating heavy metals, dyes, organics, and other pollutants [39]. The high affinity of HAP toward pollutants, refers to its specific surface area and its stability under oxidizing and reducing conditions, which makes HAP a highly effective adsorbent against water treatment. In this chapter, application of hydroxyapatite in Textile Wastewater Remediation is being discussed.

2 Primary Pollutants in Textile Wastewater and Environmental Issues

Textile wastewater is one of the nastiest contaminators of our valuable aquatic and land ecosystems. It spreads carcinogenic, mutagenic, genotoxic, cytotoxic, and allergenic threats to the living beings. The accurate characterization of textile wastewater relies on the nature of method practiced and the fabric produced. Generally, it covers a variety of organic (dyes, formaldehyde, phenols, mineral oils, surfactants, perfluorooctanoic acid, etc.) and inorganic components (heavy toxic metals like lead, cadmium, arsenic, chromium, zinc, and nickel, etc.) [21]. Textile wastewater is further characterized by its high dye content, high pH, high chemical and biochemical oxygen demand, extensive total solids, total organic carbon, chlorides, and sulfates [1].

Organic impurities such as textile dyes are emerging as a prime source of wastewater pollution. The initial concentration of dyes in fabric wastewater may range up till 1000 mg/L [21]. Dyes are categorized into natural and artificial dyes, as per their mode of synthesis [27]. Natural dyes are extracted from natural resources such as plants and animals, however, the artificial dyes are synthetically prepared [39]. Artificial dyes are generally easy to synthesize, available in varied colors, and further categorized by their fastness, chemical structure (e.g., azo, sulfur, phthalocyanine anthraquinone, triarylmethane, etc.,) and usage [1]. There are three types of dyes as per ionic charge they possess i.e., cationic, anionic, and non-ionic. Cationic dyes (e.g., methylene blue, Nile blue, and methyl violet) are considered as extremely poisonous, owing their aromatic structure and ability to get stored into food cycle [15]. Anionic dyes are water soluble and comprise of acidic, azo, direct, and reactive dyes. The elimination of anionic dyes is the most problematic job, as they leave very bright tint in water and are acidic in nature [43]. On the other hand, non-ionic dyes are significantly water insoluble and occur as colloidal particles in solutions. They consist of disperse and vat dyes, which normally possess anthraquinone and sulfide structures with –CaO, –NH– and aromatic moieties [32].

In the coloring unit of a textile industry, ca. 1000 L of water is consumed per 1000 kg garments treated. The presence of dyes in wastewater, even in minute quantity, is highly evident and unwanted. The existence of dyes in the aquatic streams may tend to; make chelated-metal ion compounds which cause micro toxicity to aquatic wildlife, limit light penetration, and reduce the photosynthetic activity [4]. These textile dyes have high aromaticity and low biodegradability and their use is still a major menace to the human health. Other than dyes, Phenols are also noxious and indestructible pollutants of textile wastewater, that are prone to cause cancer, mutation, and embryo-toxicity. "United States Environmental Protection Agency" has designated the phenols as an essential element to target in textile wastewater treatment [33]. Textile wastewater also contains sulfate-based compounds which raise serious health issues and cause ecological worsening. When sulfates entities are discharged into the environment, they may rise the acidity in soil and water bodies. High concentration of sulfates in drinking water may spread diarrhea and dehydration. Sulfates are designated as the high potential and the most common organic pollutants, which may enter into the environment by either domestic or industrial mean. Sodium lauryl sulfate is usually used as an anionic surfactant in various detergents and shampoos, because of its cost-effectiveness. It triggers eye or skin irritation among animals and humans [19].

Heavy metal contamination in the aqueous stream has undesirable effects on flora and fauna. The adverse properties of heavy metals on plants include; inhibition of photosynthesis by copper and mercury, growth cease and lessening in chlorophyll by lead, and reduced enzymatic activity by chromium. The heavy metals are also reported to have harmful effects on animals, e.g., ceased body growth, cancer development, organ impairment, malfunctioned nervous system, and even death under severe conditions [52]. Arsenic exists in wastewater, both in inorganic (i.e., As^{3+} and As^{5+}) and organic (i.e., methylarsonic acid, methylarsinic acid) forms though,

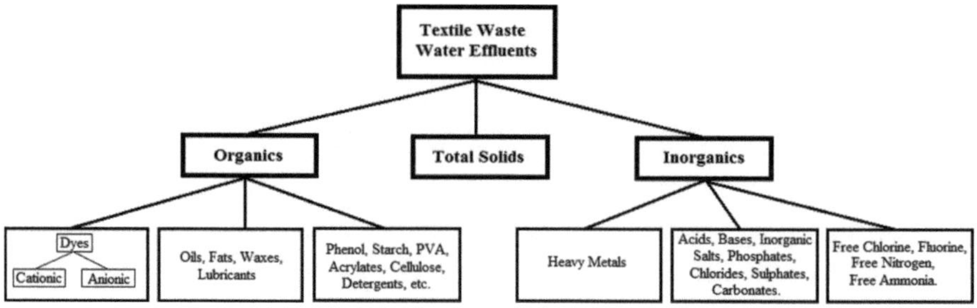

Fig. 1 General composition of textile wastewater

inorganic-As is considered as extremely lethal component and demands great consideration [22]. In contrast to As^{5+}, As^{3+} exhibits the maximum poisonousness on human health and it is about 60–80 times more deadly. The removal of As^{3+} from wastewater is more problematic than As^{5+}. That is why, World Health Organization grades arsenic as class 1 carcinogenic contaminant for humans [23]. Similarly, lead is considered as a non-biodegradable and bio-accumulative heavy metal and even its traces can damage liver, kidney, central nervous system, and reproductive system, particularly, among pregnant females and kids [55]. Apart from heavy metals, excessive ingestion of fluorine may cause bone fluorosis, which can also be dangerous in severe cases [33].

Textile wastewater is a complex of organic and inorganic compounds, where oils and lubricants also exist in the treated water. Treated water can be recycled and reclaimed for water re-injection in other applications, consequently needs to be processed [12]. The oily wastewater has serious and harmful effects on marine ecosystem and human well-being [53]. Therefore, all the organic and inorganic harmful components must need an appropriate handling, prior to their final discharge into water bodies or reuse for other intended applications [28] (Fig. 1).

3 Hydroxyapatite—An Effective Adsorbent

The HAP belongs to the apatite family and considered as the best-known calcium phosphate compound with the highest stability and the least solubility [18]. HAP has widespread applications in industry, environment, and medical. Calcium phosphates, particularly appetites, are extensively used for ceramics, chromatographic purposes (to separate protein and enzyme), methane oxidation, acting as a carrier for drug delivery, catalyst for dehydration and de-hydrogenation of alcohols, and excellent alternative for mock teeth and bone paste [36]. Moreover, HAP has been extensively studied in the pursuit of wastewater remediation owing to its low solubility at neutral and basic pH and holds a high adsorption efficiency. HAP is a significant inorganic material with wide availability, three-dimensional structure, ion exchangeability, and property to establish bonds with organic molecules of dissimilar sizes. During the last two decades, many studies have conferred this material, being equally compatible for the removal of inorganic and organic toxins [4].

Nano-hydroxyapatite (n-HAP) has more advanced activity for pollutants, because it possesses fine particle size and more active sites with distinguished surface area. These n-HAP particles are more susceptible to agglomerate, owing to have small crystals and more reactivity. Dropping the tendency of agglomeration is vital and several physical/chemical modifications like magnetic, ultrasonic, and ball milling methods are exploited to cut nano-powder agglomeration and to amplify adsorbability toward target analyses [47, 56]. However, nanoparticles need to get mounted on a carrier medium prior to be practically used in the filtering material. Doping or compositing of HAP with biopolymers or heavy metals is an entirely novel approach for attaining more efficient removal rates [39]. Coupling of HAP with biopolymers designs more versatile and advanced adsorbing species [14]. To increase the ligand binding tendency of HAP, an extensive variety of functionalization is practiced through various ions and functional moieties e.g., F^-, $Fe^{2+/3+}$, Mn^{2+}, and $CO_3{}^{2-}$, which resultantly generate a supple and exceedingly firm framework of HAP [42]. Hence, HAP may act as an idyllic adsorbent for continuous adsorption system, since it has low water solubility, high stability, high accessibility, low harmfulness, and versatile nature with high adsorption proficiency toward positive and negative ions [36] (Fig. 2).

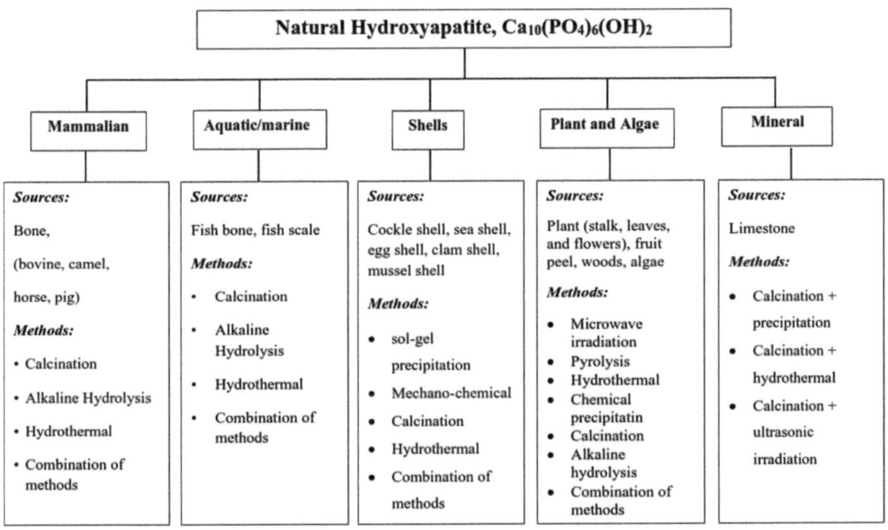

Fig. 2 Natural sources of HAP and summary of synthetic protocols. Reproduced with permission from Ref. [37]

3.1 Sources, Synthesis, and Extraction of Hydroxyapatite

HAP can either be extracted from natural resources or synthesized artificially by mixing Ca and P substrates. Natural resources of HAP could be either biological (e.g., mammalian and fish bones, shells, plants) or minerals ($CaCO_3$). Natural HAP is non-stoichiometric and consists of trace elements (i.e., Na, Mg, Ba, F, Zn, Si, K, and CO_3^{2-}), which make its composition identical to the human bone. On the contrary, artificial HAP can be fabricated through several approaches such as; dry (solid state and mechano-chemical), wet (chemical precipitation, hydrolysis, sol–gel, hydrothermal, sono-chemical, etc.), and high temperature methods (combustion and pyrolysis) [37]. Both natural and artificial HAPs have been equally compatible, though former are relatively economical but artificial HAP performs very well due to its high adsorb-ability, purity, appropriate Ca/P ratio, and even chemical structure [10]. A distinct devotion is paying by researchers to extract the apatite via natural sources and cheap processes. Among commonly used methods, the most inexpensive and commonly used methods with the most comprehensive and adsorptive properties are; hydrothermal [57], sol–gel [55], and precipitation method [14].

Hydrothermal is a simple and low temperature method with mild operating parameters where calcination and size reduction are not required. Resultantly, a relatively pure, fine, crystalline product with complete crystal growth and manageable shape and size is tailored, though, synthesized particles are prone to agglomerate [35].

Solvothermal method is similar to hydrothermal, where water is replaced by organic or non-aqueous solvent. Advanced term of solvothermal method is called sol–gel method. The fundamental principle of sol–gel method refers to hydrolysis of metal-alkoxide or inorganic salt, polymerization of the solute, drying and roasting the Gel, and ultimately receiving the inorganic products. n-HAP synthesis by sol–gel method has emerged recently, which gives a product with small size, comparatively complete lattice, and high purity. However, sol–gel method encounters various issues; complexity of chemical process, inferior crystallinity and sintering, high cost, use of toxic solvents, and tendency to agglomerate during calcination, which greatly limit the application of sol–gel method [35] (Fig. 3).

Precipitation method is carried out at specific pH, temperature, and Ca/P molar ratio with constant stirring, to generate colloidal HAP precipitates. Afterward, drying and calcination (ca. 900–1200 °C) are practiced to prepare crystalline and ultrafine particles of HAP [31]. Precipitation is extensively used owing to its simple operation, low reaction temperature, cost-effectiveness, fine product, and high purity. Low temperature precipitation results bar like crystals of HAP, where their diameter ranges from 50 to 80 nm and length 250 nm [13]. However, HAP tailored by this method holds poor consistency, agglomerating tendency, defects in lattice and less than 1.67 Ca/P ratio. Therefore, the control on operational parameters ought to be relatively high [35]. Irrespective of the type of HAP synthesis, there are few intricacies in regulating microstructure, stoichiometry, phase impurities, and agglomeration. To overcome these difficulties, some novel protocols are needed to be practiced [39]. For example, n-HAP prepared via precipitation and ultrasonic technique helps to

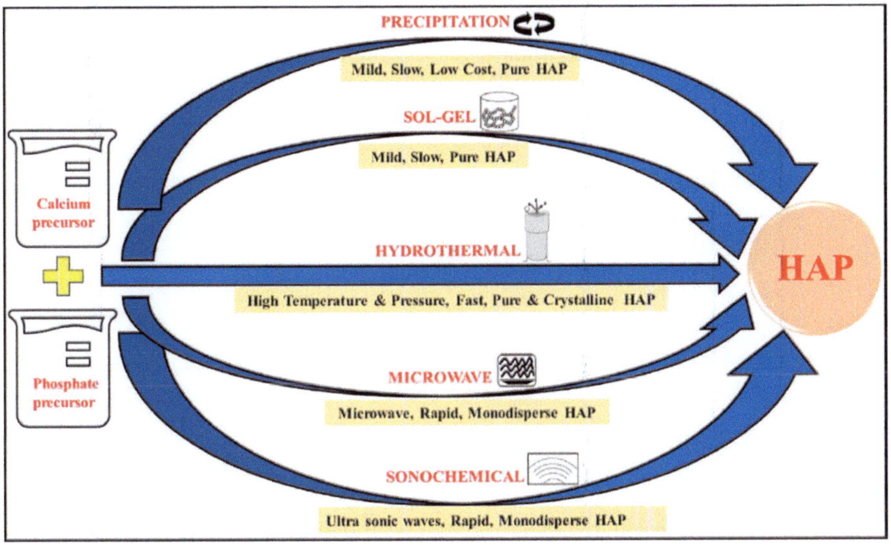

Fig. 3 HAP fabrication via various wet methods for environmental application. Reproduced with permission from Ref. [49]

decrease the agglomerating tendency of n-HAP. The ultrasound activation enhances the porosity, surface area, pore volume, and active sites of HAP [56].

3.2 Chemistry of Hydroxyapatite

The physical and chemical features of any adsorbent are significant because the structure, purity, and flowability rely on those characteristics. Hydroxyapatite ($Ca_{10}(PO_4)_6(OH)_2$) is thermo-dynamically stable, three-dimensional crystal-like solid. It contains 39.7% Ca and 18.5% P, with a mass and molar ratio of Ca/P up to 2.15 and 1.667, respectively [18]. HAP has various crystalline forms and the most common form of it is hexagonal. The space group of HAP has following crystallographic parameters: $a = 9.418$ Å; $c = 6.881$ Å and $= 120$ Å [18]. In a study, HAP composites represent rod-like crystals with developed porosity, decent mass transferability, and excellent absorbability [57]. The average hexagonal crystal size for HA is observed as 38.7 nm. n-HAP with relatively low crystallinity is expected to show good adsorption properties. The degree of crystallinity in n-HAP sample is observed to be ca. 59.8% that shows good adsorption results [52].

The advantageous property of HAP is its ability to accept a large number of anionic and cationic exchangers, which confirms its widespread utilization in the separation of ionic species. In addition, HAP owns 2.6 nm^{-2} of P–OH groups as active sites for the adsorption [36]. The $PO_4{}^{3-}$ group of HAP is tetrahedral and has four vibrational modes namely; symmetric P-O stretching (v_1), O–P–O bending (v_2 and v_4), and P–O stretching (v_3), and all are possibly involved in adsorption in a diverse manner [56]. Cation with ionic radius lesser than Ca^{2+} (0.099 nm) has fewer probability to get attached on HAP's structure (e.g., Ni^{2+}, Cd^{2+}), comparing the attachment of cations with more ionic radii (e.g., Pb^{2+}, Cu^{2+}) [36]. Ca^{2+} and $PO_4{}^{3-}$ groups of HAP act as hard Lewis's basis and thus, make HAP more prone to capture hard acids like bivalent metal ions [29]. This shows that the effectiveness of HAP as an adsorbent depends on microstructure (grain size and distribution), functionality, shape, porosity, pH, and crystallinity [39].

4 Priority Pollutants Adsorbed by Hydroxyapatite

Variety of biopolymeric materials and composites is evaluated as an adsorbent now a day, owing to their cost-effectiveness, wide availability, higher absorbability, and reusability [22]. HAP is one of those biomaterials which displays high removal efficacy and has been exploited in a variety of pollutants as explained in the following sections:

4.1 Heavy Metals

Heavy metals are considered as a group of elements that are metallic in nature such as copper, lead, and zinc. They are a source of ecological contamination even in traces [45]. Any toxic metal can be named as heavy metal (HM), regardless of their weight or density. Therefore, the term "heavy" is more interrelated to their poisonous nature rather than specific weight [3]. Heavy metals are produced by natural and anthropogenetic activities and industrial surplus is their main cause. They cannot be degraded and poses a serious hazard to the living creatures. Suitable waste management and environment friendly remediation (e.g., adsorption) are essential to get rid of these hazardous metals in textile wastewater [48].

Multi-heavy metals are prevailing in the atmosphere (either in water or in soil) simultaneously, and it is noteworthy to comprehend their adsorptive features and the removal efficiency, by applying HAP. The adsorb-ability in multi-metal adsorption system is contingent on the charge density and hydrated ion diameter of each individual metal. According to the Pearson's theory of hard and soft acids and bases, the oxygen atoms of triphosphate groups attached to the HAP's surface are classified as hard bases [9]. The functional moieties existing on HAP are OH^- and $PO_3{}^{4-}$ ions which scale as hard Lewis bases, and their oxygen atoms coordinate with heavy metal ions (i.e., hard acids) [29]. Pb^{2+} scales as borderline hard Lewis's acid, whereas Cd and Ni ions are considered as soft Lewis's acids. The Pb^{2+} demonstrates superior adsorption affinity because it has higher electronegativity for inner sphere surface complexation and electrostatic reactions. Moreover, Pb ions have highest atomic radius (0.118 nm) than Cd (0.097 nm) and Ni (0.072 nm) ions; their respective adsorption efficiencies are estimated up to 1000, 143, and 40 mg g^{-1}, which exhibits a clear correlation of adsorption efficiency of HAP with atomic radii of the adsorbed metal [36]. In another research, the higher adsorption capacity of Pb^{2+}, Cu^{2+}, Zn^{2+}, and Cd^{2+} on HAP, is reported up to 1352, 272, 285, and 304 mg g^{-1} respectively. This illustrates the highest adsorption potential of HAP on Pb^{2+}, Cu^{2+}, Zn^{2+}, and Cd^{2+} than observed by any other materials [56].

In order to further enhance the appropriateness of HAP against wastewater remediation, numerous biopolymers have been coupled with it and reported in different ways [11]. Recently, HAP composites have gained more publicity as novel adsorbents for the exclusion of toxic heavy metals, owing to their amended crystal structure and functionality [49]. A fluor-HAP composites have been examined for Cd^{2+} removal, where fluorine-containing silica and hydrated lime have transformed into composites by hydrothermal method. The adsorption efficiency becomes high and the maximum value has been reported up to 236 mg/g [57]. Chitosan-HAP composites designed by wet precipitation technique have been evaluated for Pb^{2+}, Cd^{2+}, F^-, and As(V) removal and have been proved superlative even under different pHs, adsorbate concentrations, and presence of co-existing ions [14]. Microwave-augmented cellulose-HAP nanocomposites also create an impact on As(V) remediation with ca. 13 mg/g adsorption capacity [24]. Composite of HAP with silica (prepared via ultrasound-augmented sol–gel method) has also been reported for its efficient binding with Pb^{2+}. It has been observed that the composition of HAP with silica improves the adsorption efficacy for Pb^{2+} ions from 543 to 946 mg/g compared to its adsorption by pure HAP [55]. Similarly, the HAP-calcium-silicate-hydrated-biochar composites have also been addressed for competent removal of Cd ions from contaminated soils [7]. Likewise, the viability of Cu^{2+} removal from soil and groundwater has been explored on HAP-calcium silicate hydrate which show higher adsorption efficiency of 138 mg/g than untreated HAP (90.3 mg/g) [34]. HAP-sludge-biochar composites have also been exploited to enhance Cu^{2+} and Cd^{2+} adsorption from aqueous solutions. Their adsorption fallouts have revealed higher adsorption efficacy (114.68 mg/g) than raw biochar (89.98). Zeta potential of HAP-modified sludge-biochar is 3.44, which is advantageous for its practical application across a broad pH range [8]. Similarly, HAP-activated carbon nanocomposites have been practiced as good adsorbent [13]. Clam shell biomass derived n-HAP-cellulose-composite hydrogels have been used to eliminate heavy metal ions (Cu^{2+}, Pb^{2+}, Fe^{2+}, Cd^{2+}, and Zn^{2+}) from palm oil mill effluents. This composite hydrogel has managed to get rid of approximately 70, 58, 49, 27 and 26% of Cu^{2+}, Pb^{2+}, Fe^{2+}, Cd^{2+}, and Zn^{2+} ions, respectively. The $PO_4{}^{3-}$, $CO_3{}^{2-}$, and OH^- groups of the composite hydrogel have been predominantly involved in their removal. Fine particles of HAP composites have agglomerated to a size greater than 100–250 nm. This agglomeration has been risen because of water molecules that have not been removed completely from HAP during freeze drying. For the cellulose-n-HAP composite hydrogels, the adsorption results followed the sequence of $Cu^{2+} > Pb^{2+} > Fe^{2+} > Cd^{2+} > Zn^{2+}$, which is correlated with metal's electronegativity [52].

Literature shows that isolation of HAP from post-adsorption solutions can be difficult and it may lead to secondary pollution in water [40]. This problem can be overcome by using magnetite as a support in HAP composites, which enhance the recovery of adsorbent by applying some external magnetic field. Fe_3O_4 Magnetic-HAP composites can eliminate more than 99% of Pb^{2+}, with a higher adsorption capacity of 599 mg/g, which is higher than previously published data by other magnetic composites. Herein, the presence of Ca^{2+} in HAP has not affected the removability of Pb^{2+}, but phosphate ions have been responsible for the stable Pb-phosphates formation [11]. Very few magnetic-HAP composites have been documented against heavy metals removal and still there is room to work on these HAP composites. The HAP has also been combined with activated carbon, magnetic-zerovalent copper nanoparticles, and sodium alginate for the removal of As^{3+}. The magnetic-zerovalent copper nanoparticles advance the stability and adsorption efficiency of HAP, activated carbon, and alginate beads. Furthermore, it has great catalytic activity and non-toxic nature. The removal of As^{3+} onto HAP-alginate beads has reached to 95%, after coupling with activated carbon and n-zerovalent copper [22]. The magnetic n-HAP-alginate beads (synthesized thru in-situ precipitation followed by hydrothermal method) have also been employed for efficient Cr(VI) ions removal, than individual HAP or alginate components [40].

Usually, HAP has negative zeta potential under neutral conditions which poses a great challenge in removing chromate ions. Therefore, numerous surface modifications are accomplished on HAP so far. Chitin functionalized iron-enriched-HAP nanocomposites have positive Zeta potential and therefore capture chromate ions well by electrostatic interaction. Zeta potential of chitin nanocomposite observes to be greater than +13 mV that lowers to −9 mV confirming its strong electrostatic attraction with chromate ions [41].

As mentioned earlier, artificial HAPs give better adsorption efficacy than natural apatite, though high manufacturing cost of former may limit their consumption [15, 39]. Concerning this, n-HAP has extracted thermo-mechanically by bovine bone for Pb removal and a higher adsorption efficacy ca. 200 mg g^{-1} has been observed. The removability of the Pb^{2+} has prejudiced mildly through milling time (14.79%), but mainly by dosage (38.2%) and initial metal concentration (23.64%). Ball milling has reduced the crystallinity up to certain time period but thermo-mechanical treatments have no impact on the chemistry of HAP, although, milling may originate size reduction and agglomeration of adsorbent [16]. HAP-derived from calcined chicken waste has also resulted more than 99% Pb^{2+} removal and about 185 mg/g removal efficiency at optimal conditions. Furthermore, presence of co-existing bivalent ions, such as Ca^{2+}, Mg^{2+}, and Mn^{2+}, have no notable effect on Pb^{2+} removal [50]. As per literature, HAP extracted from fish scales can also be used for heavy metals removal [25]. HAP derived from eggshell has been employed against Ni^{2+} adsorption. The maximum Ni^{2+} adsorption has reported ca. 109 mg/g, which is greater than any other adsorbents derived from waste. The presence of arupite [$Ni_3(PO_4)_2 \cdot 8H_2O$] has been observed onto post-sorbed HAP. A substantial reduction in Ca ions and involvement of more than one characteristic sites of $PO_4{}^{3-}$ groups have been confirmed after spectroscopic results of post-adsorption HAP [10]. The removal of As^{3+} using HAP-derived from bone char (pyrolysis at 900 °C) has also been investigated for the first time as per author's facts [2]. In short, HAP and its composites are efficiently used to remove heavy metals from textile wastewater.

4.2 Dyes

Dyes are used to color the fabric materials in textile industry and are not easily changed by washing or simple treatment. About 70% of the dyes are utilized in the fabric industry, making it the major consumer. The remediation of dyes has gained interest by employing various adsorbents; amid hydroxyapatite and its derived products is a recent approach. Chitosan-Fe-HAP hybrid beads have been gauged for removing alkaline and cationic dye i.e., Methyl Blue [42]. Where, n-HAP has been fabricated by micro-emulsion method and Fe-substitution has been done by ion exchange system. The supreme adsorption capacity of methyl blue is found to be 1324 mg g^{-1}, which is notably higher than other reported results. The removal of Methyl Blue has also been assessed for HAP-sodium alginate composite, which is made through co-precipitation process followed by loading varied content of biopolymer up to 20% [17]. The highest adsorption efficiencies of HAP and HAP-alginate are ca. 78 and 143 mg g^{-1}, respectively, which shows that the altered HAP is more efficient than virgin form and can be affordably employed as an adsorbent against dyes from wastewater. Similarly, a partially crystalline HAP, manufactured by co-precipitation method, has been used to study the adsorption of non-ionic dye (i.e., Disperse Blue SBL) [4]. It has been observed that the removal rate of dye increases with the dosage of adsorbent, until the complete discoloration appears. The extreme adsorption capacity of Disperse Blue SBL is varied from 244 to 435 mg/g, by rising the solution's temperature from 30 to 60 °C. Textile dyes i.e., Blue Hydron, Blue Solophenyl, and Turquoise Solophenyl have been removed by adsorption onto HAP under neutral conditions at room temperature. The results demonstrate that Blue Hydron and Solophenyl Blue give more adsorb-ability than Solophenyl Turquoise [30]. HAP-based-ceramic membranes have also been employed to remove ca. 99.9% of coloring. Cow bone waste derived HAP-based-ceramic membrane has been synthesized through combined phase inversion and sintering technique. No HAP-based ceramic membrane has been made-up from animal bone, prior to this study. This HAP membrane has fine pore size and shows optimal dye separation efficiency and therefore, acts as an outstanding alternative to expensive commercial ceramic membrane [21].

Surface modification of HAP by various functional groups acts as an effective approach to enhance dye removal proficiency. In a study, fish scale HAP has been functionalized by 3-mercaptopropyl-trimethoxysilane to incorporate a thiol group (–SH). Surface doping may result the coverage of HAP's crystal structure by amorphous groups and the predominant thiol, phosphate, and carbonate groups situated on the surface of designed HAP discloses an excellent application potential as an adsorbent material. This functionalized HAP has been utilized to adsorb various anionic (e.g., Congo Red and Coomassie Brilliant Blue) and cationic (i.e., Malachite Green) dyes.

The detected supreme adsorption efficiencies are ca. 500, 235, and 625 mg g^{-1}, for Congo Red, Coomassie Brilliant Blue, and Malachite Green, respectively [47]. Foroutan et al. have synthesized biogenic-HAP from poultry bone and modified its surface by magnesium oxide nanoparticles. For Methyl Violet adsorption, this surface modification of biogenic-HAP moderately improves the adsorption potential from 38 to 43 mg/g, which is interrelated with the high surface area and mesoporous nature of Hap-MgO composites [15]. Graphene oxide can also be used to modify HAP with suitable biopolymeric adsorbents (e.g., chitosan) and augment the porosity, surface area, thermal stability, and absorbency against dyes i.e., Congo Red, Acid Red 1, and Reactive Red 2. Chitosan-graphene oxide-HAP composite has efficiently been prepared under inert atmosphere via co-precipitation technique. Here, chitosan acts as a host surface for composite generation. Batch adsorption process of HAP composite exhibits an adsorption efficacy of ca. 43, 41, and 40 mg/g, for Congo Red, Acid Red 1, and Reactive Red 2, respectively. The impact of co-existing anions such as Cl^-, SO_4^{2-}, and HCO_3^- ions, onto HAP composite, shows that the Cl^-, SO_4^{2-} ions have a significant negative effect on dye removal, since the latter ions have highest ionic radius and former have greater electron affinity [46].

4.3 *Sulfates*

Hydroxyapatite and its composites such as HAP-modified micro fibrillated cellulose are also utilized in adsorbing sodium dodecyl sulfates ($NaC_{12}H_{25}SO_4$) and sodium sulfates (Na_2SO_4). The highest adsorption proficiencies are reported to be ca. 35 and 7.4 mg/g for sodium dodecyl sulfate and Na_2SO_4, respectively. HAP-modified micro fibrillated cellulose acts as an easy going, natural and eco-friendly adsorbent with reasonable removal competency. The adsorption efficacy for Na_2SO_4 has been reduced to 13% in the existence of phosphates and nitrates, but remained unchanged for $NaC_{12}H_{25}SO_4$. The positively charged surface of HAP-modified micro fibrillated cellulose attracts the negatively charged sulfate ions, especially under acidic conditions. However, the final pH has been reported to be basic, due to the protonation of PO_4^{3-} and –OH groups and therefore, HAP-modified micro fibrillated cellulose has demonstrated the buffering properties [19, 44].

4.4 *Phenols*

Phenols can also be effectively removed through HAP. Up to 94% of Bisphenol-A has effectively been adsorbed on ceramic-HAP, particularly under acidic environment. The adsorption of phenols is mainly carried by their contact with functional moieties (i.e., –OH, Ca^{2+}, and PO_4^{3-}) situated onto HAP's surface. The oxygen atoms of phenols connect with Ca^{2+} ions of HAP through Lewis acid–base concept, while hydrogen groups of Bisphenol-A may interact with PO_4^{3-} and/or –OH groups via H–H bonding [26]. HAP-cellulose nanofibrils can also remove phenols from aqueous medium. Cellulose nanofibrils are super-hydrophilic in nature that facilitates the diffusion and transport of those water molecules which have been bound to phenol molecules [38]. Magnetic-HAP composites can also remove up to 90% of phenol from wastewater. The surface area of magnetic adsorbent facilitates in phenol adsorption. However, magnetic materials are not practiced widely as an organic adsorbent because of the feeble forces that exist between organic molecule and Fe_3O_4. In contrast, HAP's surface is –OH enriched, which possibly designates as a decent "bridge" for organic groups. Phenols can easily be removed through magnetic-HAP, both under acidic and neutral conditions, but the adsorb-ability of HAP reduces sharply under basic condition. Phenol adsorption by HAP can be explicated in three possible ways. Firstly, the negative charged phenols, under alkaline conditions, do not support adsorption. Next, the inter-molecular H-bonding occurs between the phenolic–OH and hydroxide radical of HAP. The phenols act as a weak acid having pKa value ~9.89 and prone to dissociate at pH > pKa, which leads to the wanning of H-bonds. Lastly, phenolic acids are less soluble in acidic solutions [51].

4.5 *Oils*

Oily contaminants and solvents are highly inflammable and extremely hazardous if exploded, therefore, it is compulsory to design thermally stable and fire-resistant adsorbents. In comparison to organic materials, inorganic materials are perfect applicants, as they possess exclusive characteristics such as good thermal stability, fire resistivity, and biodegradability. In this aspect, an ultra-long HAP-nanowires have been efficiently bound with magnetic nanoparticles and polydimethylsiloxane, to form a distinctive paper composite. A mini-boat tailored from this composite paper shows good porosity, excellent removal selectivity (>99%), and permeation flux ($2924.3\ L\ m^{-2}\ h^{-1}$) and therefore, performs as continuous oil-removing device [53]. Furthermore, $ZnFe_2O_4$-HAP core-shell nanocomposites (manufactured by precipitation process) can also eliminate spent oil from wastewater. The maximum equilibrium adsorption capacity of oil is reported to be 3500 mg/l. The contact time of oil and HAP can be reduced with the increase of bed height and reduction in oil flow rate. Moreover, the removal efficacy of fixed-bed column is directly linked to the inflowing oil concentration. The concentration gradient develops between solid and liquid adsorbate provides a driving force for the adsorption [12].

5 Adsorption Process

5.1 Equilibrium Kinetic Modeling Studies

A suitable kinetic model is frequently used for enumerating the variations in adsorption with respect to time [24]. Precisely, kinetic model provides an imperative tool to explore the adsorption rate of pollutants [2, 24]. The equilibrium contact of pollutants in an aqueous or a solid phase describes that the adsorbent's active sites are saturated with the adsorbates at fixed temperature [23]. Adsorption kinetics significantly estimates the characteristics of adsorption material and studies the possible mechanism accordingly [6]. As the kinetic study of an adsorbent depends on the number of active sites available that is why, the assessment of pollutant ions onto those active sites without any structural interruption is also possible [19].

Various kinetic models e.g., pseudo-first-order, pseudo-second-order, Elovich and diffusion models, have been cited to appropriately optimize the best possible fit, which shows complete alliance with experimental adsorption data. Pseudo-first-order kinetic model represents a physical adsorption mechanism and pseudo-second-order model refers to chemical adsorption as a rate-limiting stage [20]. Elovich model identifies the rate of chemisorption reaction and it does not account for desorption stage, by assuming that the equilibrium achieves after a long time period. Furthermore, the adsorption of targeted pollutants in aqueous medium can be occurred by intra-particle diffusion and pore diffusion, or the grouping of these two mechanisms [2].

The adsorption of bivalent metal ions e.g., Cu^{2+}, Cd^{2+}, Pb^{2+}, and Zn^{2+} on HAP is observed to follow pseudo-second-order model, which manifests their adsorption via chemisorption mechanism [11, 34, 56, 57]. Additionally, the uptake of Cd^{2+} and Ni^{2+} ions onto natural HAPs (derived from egg shells and fish scales) show high correlation coefficient and good alliance with pseudo-second-order model [10, 25]. Other than pseudo-second-order model, the adsorption of Cu^{2+} and Cd^{2+} ions has also been controlled by liquid film diffusion as the rate-limiting stage [8]. The removal rate of Pb^{2+} ions is heavily dependent on pH and has been explained well by Elovich and pseudo-second-order models. The Elovich model defines the adsorption via heterogeneous diffusion [55, 56]. The removal of As^{+3} by HAP composites has also been controlled by pseudo-second-order kinetic model [22]. Cr(VI) uptake on HAP-alginate beads goes well with pseudo-second-order and intra-particle diffusion models [40]. Similarly, removal of As^{3+} and As(V) has been led to intra-particle and pore diffusion at low concentrations while, pseudo-second-order and Elovich models have been recorded for concentrated solutions of As^{3+} and As(V) [2]. A quick equilibrium has been achieved by HAP and kinetic data of As(V) has also been observed through pseudo-first-order model [24, 41]. Concisely, bivalents metals mostly obey pseudo-second-order and Elovich models, whereas multivalent metal ions can either represent their kinetic studies via pseudo-first-order (at low adsorbate concentration), pseudo-second-order (at high adsorbate concentration), intra-particle and pore diffusion model. However, few anomalies are still expected on case-to-case basis.

The equilibrium kinetics of numerous dyes, e.g., Methyl Blue, Congo Red, Coomassie Brilliant Blue, Malachite Green, Acid Red 1, Reactive Red 2, and Methyl Violet, with all differently synthesized HAP adsorbents e.g., chitosan-Fe-HAP nano beads [42], HAP-Alginate composites [17], 3-mercaptopropyl trimethoxysilane-modified-HAP [47], Chitosan-graphene oxide-HAP composites [46] and Hap-MgO composites [15], have strictly followed the pseudo-second-order kinetic model and it is perhaps correlated to a chemical adsorption phenomenon. Both organic and inorganic sulfates indicate a quick adsorption which is well described by the pseudo-second-order model [19, 44]. However, the phenols removal usually relies on pseudo-first-order, pseudo-second order, and intra-particle diffusion models, which support chemical adsorption [26, 38].

5.2 Isothermal Studies

The adsorption isotherms designate the distribution of adsorbate species between aqueous and solid phase at equilibrium. The Langmuir isotherm represents a homogenous distribution of active sites and mono-layer chemical adsorption onto the sorbent surface. It is utilized to forecast the favorability of adsorption process under fixed conditions. Whereas, Freundlich model refers to a multi-layer and heterogenous chemisorption phenomenon. The D–R isotherm model doesn't depict a homogeneous or constant adsorption. It is normally useful to showcase the adsorption phenomenon occurring on both homogeneous and heterogeneous surfaces. The D–R isotherm model describes mechanism and mean energy of adsorption [20]. Sips model is a combination of both Langmuir and Freundlich models [2]. To improve adsorption efficiency of a system, it is vital to explore isothermal studies to create the suitable correlation for the equilibrium curves.

Adsorption of bivalent metal (Pb^{2+}, Cd^{2+}, Cu^{2+}, and Ni^{2+}) through nano and natural HAP, is mostly defined by the Langmuir isotherm [8, 10, 16, 25, 36]. In contrast, a fairly good fitting for Freundlich and Temkin model has also been noticed, expressing a hetero-layer adsorption phenomenon and strong chemical interaction between n-HAP and Pb^{2+} ions [56]. For HAP-activated carbon and HAP-Fe_3O_4 nanocomposite, the Pb^{2+} adsorption data meets Langmuir and Freundlich isotherms equally well [11, 13].

The adsorption of dyes generally follows two isotherms namely; Langmuir and Freundlich. The adsorption of Methylene Blue from aqueous solution agrees well with the Langmuir adsorption model using chitosan-Fe-HAP and HAP-Alginate composites [17, 42]. The Congo Red, Coomassie Brilliant Blue, Malachite Green and Disperse Blue SBL dyes have also followed Langmuir isotherm onto the surface of functionalized and partially crystalline HAP [4, 47]. However, the adsorption for Blue Hydron, Blue Solophenyl, Turquoise Solophenyl, Methyl Violet Congo Red, Acid Red, and Reactive Red is described well by Freundlich isotherm, using synthetic-HAP, Bio-Hap, Bio-Hap-MgO composites and CS-GO-HAP composite [15, 30, 46].

The Sip's isotherm goes well with the sulfate's removal on HAP-modified micro fibrillated cellulose [19]. Additionally, the adsorption experiments confirm Freundlich isotherm at lower phenol concentrations (i.e., –50 mg/L), and Langmuir isotherm among concentrated solutions (i.e., 50–500 mg/L) [51]. It has been reported that Thomas and Yoon-Nelson models are equally appropriate to envisage the uptake efficiency of column bed adsorption for oil removal [12].

5.3 Thermodynamic Studies

Thermodynamic studies are used to assess the practical viability of adsorption process and to authorize the nature of the adsorption [2, 24]. The thermodynamic parameters for Pb^{2+}, Cd^{2+}, Cu^{2+}, and Ni^{2+} designate the impulsive and endothermal adsorption process [34, 55, 57]. Similar to bivalent metal ions, the thermodynamical studies direct the spontaneous and heat-absorbing fashion for Cr(VI) adsorption [40]. Sips model has described As^{3+} adsorption, while As(V) has been explained by Langmuir model [2, 24]. On the contrary, Freundlich isotherm has been found as the best fit for the removal of As^{+3}, by HAP-composite beads [22]. Generally, inorganic metals follow Langmuir or Freundlich isotherm, however, some anomalies do exist. A favorable, spontaneous, and endothermic adsorption has been reviewed against heavy metals remediation.

Thermodynamic studies indicate that rising temperature leads to an upsurge in the adsorption of dyes [17, 42]. Thus, dyes adsorption is generally pH and temperature dependent [4]. Based on thermodynamics model, the adsorption behaviors of Congo Red, Coomassie Brilliant Blue, Malachite Green and Disperse Blue SBL dyes have been observed as endothermic in nature [4, 47]. Nevertheless, the adsorption of Methyl Violet, Methylene Blue, Acid Red 1, and Reactive Red 2, using Bio-HAP, Bio-HAP-MgO, HAP-Alginate, and CS-GO-HAP composites, represent exothermic, spontaneous, and desirable processes [15, 17, 46].

Furthermore, thermodynamic studies of phenols remediation via HAP designate the spontaneous and endothermic adsorption [26]. The oil removal improves with rising temperature, in fact, the rise in temperature instigates a substantial decline in oil viscidness particularly for the heavier oils. Though, it is hard to regulate the temperature of oil spill zone but oil tends to diffuse into inside the pores of HAP adsorbent at high temperature, ensuing to improve the adsorption capacity (e.g., 7000 ppm at 77 °C and 3500 ppm at 25 °C) [12].

5.4 Mechanistic Studies

The adsorption mechanism generally depends on the physiochemical properties and mass transfer of adsorbent [19]. Diverse types of adsorption mechanisms have been reported on HAP namely; physical adsorption, chemical adsorption, ion exchange, dissolution-precipitation, etc., which have been described as follows;

The main mechanism of cation adsorption on HAP from aqueous solutions is stated by ion exchange with Ca^{2+} [57], whereas anions are remediated through binding of Ca^{2+} sites and by the exchange with replaceable OH^- groups [14]. There are two mechanisms namely; dissolution-precipitation and complexation which are equally significant in lead ions adsorption by magnetic-HAP composites [11]. HAP composites also cause the uptake of Pb ions by means of dissolution-precipitation and ion exchange [13, 55]. It has been seen that the dissolution-precipitation mechanism is mainly responsible for Pb sequestration under acidic conditions (pH $\leq$ 3.0) [50]. The acidic environment significantly enhances the dissolution of HAP and tends to form stable metal-phosphates i.e., $M_{10}(PO_4)_6(OH)_2$ [16]. The acidic pH promotes the release of PO_4^{3-} and OH^- ions of HAP into the solution which in turn rise solution's pH. The basic pH favors the metal ions adsorption via complexation and deprotonation reactions, which increases attraction between bivalent metal ion and HAP. The involvement of PO_4^{3-} (v_4 bands), CO_3^{2-}, and HPO_4^{2-} groups, in bivalent metal ions adsorption specially Pb^{2+}, has been correlated to the distinguished adsorption results [56]. The removal of Cu and Ni ions include surface adsorption, dissolution-precipitation, and ion exchange [10, 34].

The complexation (via –COOH) and exchange reactions have also been observed against As^{+3} removal [2, 22], though, As(V) adsorption can be hindered by the co-existing ions e.g., SiO_3^{2-}, PO_4^{3-}, and NO_3^- [24]. The Fe^{3+} and Ca^{2+} ions of Fe_3O_4- HAP beads act as Lewis's acid, and interact with the negatively charged Lewis's base (i.e., chromate ions) via electrostatic attraction [40]. Zeta potential of HAP composites has been transformed from $+13$ mV to less than -9 mV after adsorption, which also refer to strong electrostatic attraction of chromate ions [41]. Hence, metal ions adsorption by HAP predominantly carry out via ionic exchange, surface complexation with PO_4^{3-}, Ca, and –OH groups, and/or co-precipitation of partly soluble phases, as the main removal mechanisms [8, 16, 42].

Electrostatic interaction and physical adsorption have been stated as the key adsorption mechanisms for dyes removal. However, hydrogen bonding and chemical adsorption between dye and adsorbent have also been reported as a protuberant adsorption mechanism. Some of the case studies with possible mechanism are reported here; the mechanism of dyes adsorption on Chitosan-graphene-oxide-Hap involve the electrostatic interaction, complexation, H–H bonding, and $\pi-\pi$ interactions [46]. Electrostatic interactions have been reported between Disperse Blue SBL molecules and HAP particles [4]. The oxygen functional moieties of graphene oxide bind with dye molecules through electrostatic interactions, π–π bonding, and H–H bonding. Also, the Ca^{2+} ions present in the HAP composite are inclined to form bonds with –NH_2, –OH, and –SO_4^{-2} groups of dyes through surface complexation and electrostatic force of attraction [46].

A strong interaction between phenol, sulfonate, and amine groups of the dye's molecule and Ca^{2+} ions of HAP has also been observed in literature. The low adsorption for the Solophenyl Turquois might be explained by the exchange of Cu^{2+} with Ca^{2+} [30]. Methylene Blue adsorption on HAP-Alginate may be explained via electrostatic attraction between the carboxylate groups of alginates and positively charged groups of dye [17]. Based on the isothermal parameters, the Methyl Violet adsorption using Bio-HAP and Bio-HAP-MgO composites are physical in nature [15].

Hydroxyapatite adsorbs sodium dodecyl sulfates mainly via ion exchange between dodecyl sulfate ion and phosphate ion [44]. However, the electrostatic attraction has been reported for anionic sodium dodecyl sulfates though, negatively charged exterior of HAP-modified micro fibrillated cellulose may experience a repulsive electrostatic interaction with anionic dodecyl sulfates [19]. The adsorption process of phenols on the HAP has been represented by mono-layer or multi-layer chemisorption [26, 51]. The meso-porosity of HAP composites leads the interparticle pore diffusion and an indefinite multilayered adsorption phenomenon [12].

5.5 Regeneration Studies

The reusability/regeneration of adsorbed material is a significant factor to quantify its efficiency [6]. HAP holds a great potential with remarkable adsorption efficiency and recyclability. It has been reviewed that HAP and its derived composites represent immense reusability varying from three [41] to seven cycles [22], with no significant reduction in their absorbability. For phenols, a high reusability of HAP has been reported up to 86% removal rate till third consecutive regeneration cycle for phenols [38]. Moreover, the oil can infuse speedily onto magnetic-HAP paper due to gravity, though water is excluded to go through and stays on the surface of paper. More significantly, the HAP-Fe_3O_4-polydimethylsiloxane paper can easily be washed and recycled using ethanol and the uptake efficacies are all greater than 99% up to 10 cycles [53].

6 Conclusion

Immobilization of priority pollutants using hydroxyapatite is a promising, recent, and effective strategy for textile wastewater remediation. Natural HAP is a cost-effective, equally competent, and eco-friendly biogenic material with easy availability, however, chemically synthesis modifies the functionality, porosity, and crystal structure of HAP according to the requirement. Among the synthetic procedures of HAP, the wet chemical precipitation can be considered as a simple low-cost method to produce Nano-hydroxyapatite (n-HAP). Nano-hydroxyapatite has more advanced adsorb-ability, because of fine particle size and distinguished surface area. However, n-HAP particles are more susceptible to agglomerate, owing to have small crystals and more reactivity. Several physical/chemical modifications like magnetic, ultrasonic, and ball milling methods are exploited to cut nano-powder agglomeration. Remediation of wastewater from noxious dyes and heavy metals has met with considerable success when mediated by modification and compositing. Extraction of composites from the aqueous system after the interaction is a critical challenge and requires energy; therefore, composites with magnetic properties are of interest in the scientific community. It removes toxic ions from water via ion exchange, dissolution precipitation, electrostatic interaction, and physiochemical adsorption system. Literature adsorption data shows that HAP can be used in a point-of-use water handling device at ambient temperature. Langmuir and Freundlich models are reported predominantly to elucidate the adsorption behaviors with high correlation coefficient, referring to chemical adsorption. The adsorption mostly follows pseudo-second-order kinetic model an endothermic adsorption thermodynamics study. Thus, HAP shows a great potential to be used as a simple, cost-effective, naturally accruing, biocompatible, environmentally friendly, and safe alternative adsorbent to remove both organic and inorganic pollutants from contaminated water.

References

1. Abdel-Karim A, El-Naggar ME, Radwan EK, Mohamed IM, Azaam M, Kenawy E-R (2021) High-performance mixed-matrix membranes enabled by organically/inorganic modified montmorillonite for the treatment of hazardous textile wastewater. Chem Eng J 405:126964. https://doi.org/10.1016/j.cej.2020.126964
2. Alkurdi SSA, Al-Juboori RA, Bundschuh J, Bowtell L, Marchuk A (2021) Inorganic arsenic species removal from water using bone char: a detailed study on adsorption kinetic and isotherm models using error functions analysis. J Hazard Mater 405:124112. https://doi.org/10.1016/j.jhazmat.2020.124112
3. Appenroth K-J (2010) Definition of "Heavy Metals" and their role in biological systems. In: Soil heavy metals. Springer, Berlin, Heidelberg, pp 19–29
4. Barka N, Qourzal S, Assabbane A, Nounah A, AÎT-Ichou Y (2008) Adsorption of disperse blue SBL dye by synthesized poorly crystalline hydroxyapatite. J Environ Sci 20(10):1268–1272. https://doi.org/10.1016/S1001-0742(08)62220-2
5. Bisschops I, Spanjers H (2003) Literature review on textile wastewater characterisation. Environ Technol 24:1399–1411. https://doi.org/10.1080/09593330309385684
6. Chao L, Wang Y, Chen S, Li Y (2021) Preparation and adsorption properties of chitosan-modified magnetic nanoparticles for removal of Mo (VI) ions. Pol J Environ Stud 30(3):2489–2498. https://doi.org/10.15244/pjoes/130039
7. Chen H, Yuan X, Xiong T, Jiang L, Wang H, Wu Z (2020) Biochar facilitated hydroxyapatite/calcium silicate hydrate for remediation of heavy metals contaminated soils. Water Air Soil Pollut 231(2):66. https://doi.org/10.1007/s11270-020-4425-1
8. Chen Y, Li M, Li Y, Liu Y, Chen Y, Li H, … Chen L (2021) Hydroxyapatite modified sludge-based biochar for the adsorption of Cu^{2+} and Cd^{2+}: adsorption behavior and mechanisms. Bioresour Technol 321:124413. https://doi.org/10.1016/j.biortech.2020.124413
9. Datta D (1992) On Pearson's HSAB principle. Inorg Chem 31(13):2797–2800. https://doi.org/10.1021/ic00039a025

10. De Angelis G, Medeghini L, Conte AM, Mignardi S (2017) Recycling of eggshell waste into low-cost adsorbent for Ni removal from wastewater. J Clean Prod 164:1497–1506. https://doi.org/10.1016/j.jclepro.2017.07.085
11. Dong L, Zhu Z, Qiu Y, Zhao J (2010) Removal of lead from aqueous solution by hydroxyapatite/magnetite composite adsorbent. Chem Eng J 165(3):827–834. https://doi.org/10.1016/j.cej.2010.10.027
12. El-Maghrabi HH, Hosny R, Ramzi M, Zayed MA, Fathy M (2019) Preparation and characterization of novel magnetic ZnFe2O4–hydroxyapatite core-shell nanocomposite and its use as fixed bed column system for removal of oil residue in oily wastewater samples. Egypt J Pet 28(2):137–144. https://doi.org/10.1016/j.ejpe.2018.12.005
13. Fernando MS, de Silva RM, de Silva KMN (2015) Synthesis, characterization, and application of nano hydroxyapatite and nanocomposite of hydroxyapatite with granular activated carbon for the removal of Pb^{2+} from aqueous solutions. Appl Surf Sci 351:95–103. https://doi.org/10.1016/j.apsusc.2015.05.092
14. Fernando MS, Wimalasiri AKDVK, Dziemidowicz K, Williams GR, Koswattage KR, Dissanayake DP, … de Silva RM (2021) Biopolymer-based nanohydroxyapatite composites for the removal of fluoride, lead, cadmium, and arsenic from water. ACS Omega 6(12):8517–8530. https://doi.org/10.1021/acsomega.1c00316
15. Foroutan R, Peighambardoust SJ, Aghdasinia H, Mohammadi R, Ramavandi B (2020) Modification of bio-hydroxyapatite generated from waste poultry bone with MgO for purifying methyl violet-laden liquids. Environ Sci Pollut Res 27(35):44218–44229. https://doi.org/10.1007/s11356-020-10330-0
16. Googerdchian F, Moheb A, Emadi R, Asgari M (2018) Optimization of Pb(II) ions adsorption on nanohydroxyapatite adsorbents by applying Taguchi method. J Hazard Mater 349:186–194. https://doi.org/10.1016/j.jhazmat.2018.01.056
17. Guesmi Y, Agougui H, Lafi R, Jabli M, Hafiane A (2018) Synthesis of hydroxyapatite-sodium alginate via a co-precipitation technique for efficient adsorption of methylene blue dye. J Mol Liq 249:912–920. https://doi.org/10.1016/j.molliq.2017.11.113
18. Hammouti B (2021) Toxic heavy metals removal using a hydroxyapatite and hydroxyethyl cellulose modified with a new Gum Arabic
19. Hokkanen S, Bhatnagar A, Koistinen A, Kangas T, Lassi U, Sillanpää M (2018) Comparison of adsorption equilibrium models and error functions for the study of sulfate removal by calcium hydroxyapatite microfibrillated cellulose composite. Environ Technol 39(8):952–966. https://doi.org/10.1080/09593330.2017.1317839
20. Hokkanen S, Doshi B, Srivastava V, Puro L, Koivula R (2019) Arsenic (III) removal from water by hydroxyapatite-bentonite clay-nanocrystalline cellulose. Environ Prog Sustain Energy 38.https://doi.org/10.1002/ep.13147
21. Hubadillah SK, Othman MHD, Tai ZS, Jamalludin MR, Yusuf NK, Ahmad A, …Harun Z (2020) Novel hydroxyapatite-based bio-ceramic hollow fiber membrane derived from waste cow bone for textile wastewater treatment. Chem Eng J 379:122396. https://doi.org/10.1016/j.cej.2019.122396
22. Iqbal J, Shah NS, Sayed M, Imran M, Muhammad N, Howari FM, …Haija MA (2019) Synergistic effects of activated carbon and nano-zerovalent copper on the performance of hydroxyapatite-alginate beads for the removal of As.sup.3+ from aqueous solution. J Cleaner Prod 235:875
23. Iqbal MM, Imran M, Ali B, Nawaz M, Siddique MH, Al-Kahtani AA, …Ali S (2021) Nanocomposites of sedimentary material with ZnO and magnetite for the effective sequestration of arsenic from aqueous systems: reusability, modeling and kinetics. Environ Technol Innovation 21:101298. https://doi.org/10.1016/j.eti.2020.101298
24. Islam M, Mishra PC, Patel R (2011) Arsenate removal from aqueous solution by cellulose-carbonated hydroxyapatite nanocomposites. J Hazard Mater 189(3):755–763. https://doi.org/10.1016/j.jhazmat.2011.03.051
25. Jaafar A, Darchen A, Hamzi SE, Lakbaibi Z, Driouich A, Boussaoud A, …Hachkar M (2021) Optimization of cadmium ions biosorption by fish scale from aqueous solutions using factorial

design analysis and Monte Carlo simulation studies. J Environ Chem Eng 9(1):104727. https://doi.org/10.1016/j.jece.2020.104727

26. Khallok H, Zbair M, Ojala S, Ainassaari K, Brahmi R, Keiski RL, Hatim Z (2021) Ceramic hydroxyapatite foam as a new material for Bisphenol A removal from contaminated water. Environ Sci Pollut Res 28(14):17739–17751. https://doi.org/10.1007/s11356-020-12076-1
27. Khehra MS, Saini HS, Sharma DK, Chadha BS, Chimni SS (2006) Biodegradation of azo dye C.I. Acid Red 88 by an anoxic–aerobic sequential bioreactor. Dyes Pigm 70(1):1–7. https://doi.org/10.1016/j.dyepig.2004.12.021
28. Kishor R, Purchase D, Saratale GD, Saratale RG, Ferreira LFR, Bilal M, …Bharagava RN (2021) Ecotoxicological and health concerns of persistent coloring pollutants of textile industry wastewater and treatment approaches for environmental safety. J Environ Chem Eng 9(2):105012. https://doi.org/10.1016/j.jece.2020.105012
29. Laus R, de Fávere VT (2011) Competitive adsorption of Cu(II) and Cd(II) ions by chitosan crosslinked with epichlorohydrin–triphosphate. Biores Technol 102(19):8769–8776. https://doi.org/10.1016/j.biortech.2011.07.057
30. Lemlikchi W, Drouiche N, Belaicha N, Oubagha N, Baaziz B, Mecherri MO (2015) Kinetic study of the adsorption of textile dyes on synthetic hydroxyapatite in aqueous solution. J Ind Eng Chem 32:233–237. https://doi.org/10.1016/j.jiec.2015.08.023
31. Lemlikchi W, Sharrock P, Mecherri MO, Fiallo M, Nzihou A (2012) Treatment of textile waste waters by hydroxyapatite co-precipitation with adsorbent regeneration and reuse. Waste Biomass Valor 3(1):75–79. https://doi.org/10.1007/s12649-011-9096-0
32. Li Q, Yue Q-Y, Sun H-J, Su Y, Gao B-Y (2010) A comparative study on the properties, mechanisms and process designs for the adsorption of non-ionic or anionic dyes onto cationic-polymer/bentonite. J Environ Manage 91(7):1601–1611. https://doi.org/10.1016/j.jenvman.2010.03.001
33. Li X, Yu X, Liu L, Yang J, Liu S, Zhang T (2021) Preparation, characterization serpentine-loaded hydroxyapatite and its simultaneous removal performance for fluoride, iron and manganese. RSC Adv 11(27):16201–16215. https://doi.org/10.1039/D1RA02028E
34. Liu Y, Zhang R, Sun Z, Shen Q, Li Y, Wang Y, … Wang X (2021) Remediation of artificially contaminated soil and groundwater with copper using hydroxyapatite/calcium silicate hydrate recovered from phosphorus-rich wastewater. Environ Pollut 272:115978. https://doi.org/10.1016/j.envpol.2020.115978
35. Ma G (2019) Three common preparation methods of hydroxyapatite. IOP Conf Ser Mater Sci Eng 688:033057. https://doi.org/10.1088/1757-899x/688/3/033057
36. Mobasherpour I, Salahi E, Pazouki M (2012) Comparative of the removal of Pb^{2+}, Cd^{2+} and Ni^{2+} by nano crystallite hydroxyapatite from aqueous solutions: adsorption isotherm study. Arab J Chem 5(4):439–446. https://doi.org/10.1016/j.arabjc.2010.12.022
37. Mohd Pu'ad NAS, Koshy P, Abdullah HZ, Idris MI, Lee TC (2019) Syntheses of hydroxyapatite from natural sources. Heliyon 5(5):e01588. https://doi.org/10.1016/j.heliyon.2019.e01588
38. Narwade VN, Khairnar RS, Kokol V (2017) In-situ synthesised hydroxyapatite-loaded films based on cellulose nanofibrils for phenol removal from wastewater. Cellulose 24(11):4911–4925. https://doi.org/10.1007/s10570-017-1435-2
39. Pai S, Kini SM, Selvaraj R, Pugazhendhi A (2020) A review on the synthesis of hydroxyapatite, its composites and adsorptive removal of pollutants from wastewater. J Water Process Eng 38:101574. https://doi.org/10.1016/j.jwpe.2020.101574
40. Periyasamy S, Gopalakannan V, Viswanathan N (2018) Hydrothermal assisted magnetic nano-hydroxyapatite encapsulated alginate beads for efficient Cr(VI) uptake from water. J Environ Chem Eng 6(1):1443–1454. https://doi.org/10.1016/j.jece.2018.01.007
41. Rajak JK, Khandelwal N, Behera MP, Tiwari E, Singh N, Ganie ZA, …Schäfer T (2021) Removal of chromate ions from leachate-contaminated groundwater samples of Khan Chandpur, India, using chitin modified iron-enriched hydroxyapatite nanocomposite. Environ Sci Pollut Res Inthttps://doi.org/10.1007/s11356-021-13549-7
42. Saber-Samandari S, Saber-Samandari S, Nezafati N, Yahya K (2014) Efficient removal of lead (II) ions and methylene blue from aqueous solution using chitosan/Fe-hydroxyapatite

nanocomposite beads. J Environ Manage 146:481–490. https://doi.org/10.1016/j.jenvman.2014.08.010
43. Sajda SA (2021) Classifications, advantages, disadvantages, toxicity effects of natural and synthetic dyes: a review. Univ Thi-Qar J Sci 8(1). Retrieved from https://www.jsci.utq.edu.iq/index.php/main/article/view/790
44. Shimabayashi S, Tanaka H, Nakagaki M (1986) Adsorption of dodecyl sulfate ion on hydroxyapatite and concurrent release of phosphate and calcium ions from the surface of hydroxyapatite. Chem Pharm Bull 34(11):4474–4478. https://doi.org/10.1248/cpb.34.4474
45. Singh R, Gautam N, Mishra A, Gupta R (2011) Heavy metals and living systems: an overview. Indian J Pharmacol 43(3):246–253. https://doi.org/10.4103/0253-7613.81505
46. Sirajudheen P, Karthikeyan P, Ramkumar K, Meenakshi S (2020) Effective removal of organic pollutants by adsorption onto chitosan supported graphene oxide-hydroxyapatite composite: a novel reusable adsorbent. J Mol Liq 318:114200. https://doi.org/10.1016/j.molliq.2020.114200
47. Sricharoen P, Kongsri S, Kukusamude C, Areerob Y, Nuengmatcha P, Chanthai S, Limchoowong N (2021) Ultrasound-irradiated synthesis of 3-mercaptopropyl trimethoxysilane-modified hydroxyapatite derived from fish-scale residues followed by ultrasound-assisted organic dyes removal. Sci Rep 11(1):5560. https://doi.org/10.1038/s41598-021-85206-5
48. Thakare M, Sarma H, Datar S, Roy A, Pawar P, Gupta K, … Prasad R (2021) Understanding the holistic approach to plant-microbe remediation technologies for removing heavy metals and radionuclides from soil. Curr Res Biotechnol 3:84–98.https://doi.org/10.1016/j.crbiot.2021.02.004
49. Varadavenkatesan T, Vinayagam R, Pai S, Kathirvel B, Pugazhendhi A, Selvaraj R (2021) Synthesis, biological and environmental applications of hydroxyapatite and its composites with organic and inorganic coatings. Prog Org Coat 151:106056. https://doi.org/10.1016/j.porgcoat.2020.106056
50. Wang H, Lv Z, Wang YN, Sun Y, Tsang YF (2021) Recycling of biogenic hydroxyapatite (HAP) for cleaning of lead from wastewater: performance and mechanism. Environ Sci Pollut Res Int. https://doi.org/10.1007/s11356-020-10855-4
51. Wang X (2011) Preparation of magnetic hydroxyapatite and their use as recyclable adsorbent for phenol in wastewater. CLEAN Soil Air Water 39:13–20. https://doi.org/10.1002/clen.201000241
52. Wong SM, Zulkifli MZA, Nordin D, Teow YH (2021) Synthesis of cellulose/nano-hydroxyapatite composite hydrogel absorbent for removal of heavy metal ions from palm oil mill effluents. J Polym Environ. https://doi.org/10.1007/s10924-021-02183-6
53. Yang R-L, Zhu Y-J, Chen F-F, Qin D-D, Xiong Z-C (2018) Recyclable, fire-resistant, superhydrophobic, and magnetic paper based on ultralong hydroxyapatite nanowires for continuous oil/water separation and oil collection. ACS Sustain Chem Eng 6(8):10140–10150. https://doi.org/10.1021/acssuschemeng.8b01463
54. Yaseen DA, Scholz M (2019) Textile dye wastewater characteristics and constituents of synthetic effluents: a critical review. Int J Environ Sci Technol 16(2):1193–1226. https://doi.org/10.1007/s13762-018-2130-z
55. Zhang Z, Wang X, Wang H, Zhao J (2018) Removal of Pb(II) from aqueous solution using hydroxyapatite/calcium silicate hydrate (HAP/C-S-H) composite adsorbent prepared by a phosphate recovery process. Chem Eng J 344:53–61. https://doi.org/10.1016/j.cej.2018.03.066
56. Zhou C, Wang X, Wang Y, Song X, Fang D, Ge S (2021) The sorption of single- and multi-heavy metals in aqueous solution using enhanced nano-hydroxyapatite assisted with ultrasonic. J Environ Chem Eng 9(3):105240. https://doi.org/10.1016/j.jece.2021.105240
57. Zhu X-H, Li J, Luo J-H, Jin Y, Zheng D (2017) Removal of cadmium (II) from aqueous solution by a new adsorbent of fluor-hydroxyapatite composites. J Taiwan Inst Chem Eng 70:200–208. https://doi.org/10.1016/j.jtice.2016.10.049

MIX
Papier aus verantwortungsvollen Quellen
Paper from responsible sources
FSC® C105338

If you have any concerns about our products,
you can contact us on
ProductSafety@springernature.com

In case Publisher is established outside the EU,
the EU authorized representative is:
Springer Nature Customer Service Center GmbH
Europaplatz 3, 69115 Heidelberg, Germany

Printed by Libri Plureos GmbH
in Hamburg, Germany